SAGE was founded in 1965 by Sara Miller McCune to support the dissemination of usable knowledge by publishing innovative and high-quality research and teaching content. Today, we publish more than 750 journals, including those of more than 300 learned societies, more than 800 new books per year, and a growing range of library products including archives, data, case studies, reports, conference highlights, and video. SAGE remains majority-owned by our founder, and after Sara's lifetime will become owned by a charitable trust that secures our continued independence.

Los Angeles | London | Washington DC | New Delhi | Singapore | Boston

Advance Praise

This volume is dedicated to the memory of late Professor Robert Evenson, who served for many years on the faculty at Yale University.... In keeping with the geographic breadth of Professor Evenson's research, ... it examines the innovation process at levels from the world's smallest farms to its largest industries. This book will undoubtedly stand as a valued reference for economists working on innovation and development.

—**Douglas Gollin**, Professor, Oxford Department of International Development (ODID), University of Oxford, Oxford

The collection of essays in this volume, based on serious research in areas in which Professor Evenson excelled, is truly a fitting tribute to his legacy. [It] will be found immensely useful by university students, active researchers and policy analysts. The three editors and contributors of the essays deserve to be profusely complimented for their imaginative and scholarly contributions.

—**K.L. Krishna**, currently Chairperson, Madras Institute of Development Studies, Chennai, and formerly Professor and Director, Delhi School of Economics, Delhi

TECHNOLOGY, INNOVATIONS & ECONOMIC
Development

Thank you for choosing a SAGE product!
If you have any comment, observation or feedback,
I would like to personally hear from you.
Please write to me at **contactceo@sagepub.in**

Vivek Mehra, Managing Director and CEO,
SAGE Publications India Pvt Ltd, New Delhi

Bulk Sales

SAGE India offers special discounts
for purchase of books in bulk.
We also make available special imprints
and excerpts from our books on demand.

For orders and enquiries, write to us at

Marketing Department
SAGE Publications India Pvt Ltd
B1/I-1, Mohan Cooperative Industrial Area
Mathura Road, Post Bag 7
New Delhi 110044, India

E-mail us at **marketing@sagepub.in**

Get to know more about SAGE

Be invited to SAGE events, get on our mailing list.
Write today to **marketing@sagepub.in**

This book is also available as an e-book.

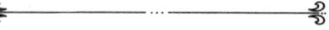

TECHNOLOGY, INNOVATIONS & ECONOMIC *Development*

Essays in Honour of Robert E. Evenson

Edited by
Lakhwinder Singh
K.J. Joseph
Daniel K.N. Johnson

www.sagepublications.com
Los Angeles • London • New Delhi • Singapore • Washington DC • Boston

Copyright © Lakhwinder Singh, K.J. Joseph, Daniel K.N. Johnson, 2015

All rights reserved. No part of this book may be reproduced or utilized in any form or by any means, electronic or mechanical, including photocopying, recording, or by any information storage or retrieval system, without permission in writing from the publisher.

First published in 2015 by

SAGE Publications India Pvt Ltd
B1/I-1 Mohan Cooperative Industrial Area
Mathura Road, New Delhi 110 044, India
www.sagepub.in

SAGE Publications Inc
2455 Teller Road
Thousand Oaks, California 91320, USA

SAGE Publications Ltd
1 Oliver's Yard, 55 City Road
London EC1Y 1SP, United Kingdom

SAGE Publications Asia-Pacific Pte Ltd
3 Church Street
#10-04 Samsung Hub
Singapore 049483

Published by Vivek Mehra for SAGE Publications India Pvt. Ltd, typeset in 10/13pt Times New Roman by Diligent Typesetter, Delhi and printed at Chaman Enterprises, New Delhi.

Library of Congress Cataloging-in-Publication Data

Technology, innovations and economic development : essays in honour of Robert E. Evenson / edited by Lakhwinder Singh, K.J. Joseph, Daniel K.N. Johnson.
  pages cm
  Includes bibliographical references and index.
1. Economic development—Technological innovations. 2. Technological innovations—Economic aspects. 3. Technology transfer. I. Evenson, Robert E. (Robert Eugene), 1934– honoree. II. Singh, Lakhwinder, editor. III. Joseph, K. J., 1961– editor. IV. Johnson, Daniel K. N., editor.
  HD82.T4193   338—dc23   2015   2015018376

ISBN: 978-93-515-0269-2 (HB)

The SAGE Team: N. Unni Nair, Alekha Chandra Jena, Nand Kumar Jha and Vinitha Nair

Contents

List of Tables	vii
List of Figures	xi
List of Abbreviations	xiii
Preface	xvii
Legacy—Robert E. Evenson	xix
Daniel K.N. Johnson	
Technology, Innovations and Economic Development:	
An Introduction	xxiii
Lakhwinder Singh, K.J. Joseph and Daniel K.N. Johnson	

SECTION I
Innovation and Economic Development

CHAPTER 1
Education Reforms, Technological Change and Economic Development 3
Leonardo A. Lanzona, Jr

CHAPTER 2
Eco-innovation: A Literature Review of the Challenges Facing the Development of Green Technologies 43
Daniel K.N. Johnson and Kristina M. Lybecker

CHAPTER 3
Social Inclusion and Institutional Innovations: Working towards a Policy-theoretical Framework 81
M.A. Oommen

SECTION II
Technological Progress and Agricultural Development

CHAPTER 4
Measuring Public Agricultural Research Capital and Its Impact on State Agricultural Productivity in the United States 107
Wallace E. Huffman

CHAPTER 5
Access to Markets and Farm Efficiency: A Study of Bicol Rice Farms over Two Decades 147
Sanjaya DeSilva

SECTION III
Technology Transfer, National Innovation Systems and Industrial Development

CHAPTER 6
Global Innovation Networks and Industry–University Interaction: A Study of India's ICT Sector 181
K.J. Joseph and Vinoj Abraham

CHAPTER 7
Globalization of Industrial R&D in Developing Countries: A Sociological Perspective 209
Binay Kumar Pattnaik

CHAPTER 8
Technological Capability, Employment Growth and Industrial Development: A Quantitative Anatomy of Indian Scenario 238
Lakhwinder Singh and Baldev Singh Shergill

CHAPTER 9
Intellectual Property Protection, Innovation and Medicines: Lessons from the Indian Pharmaceutical Industry 272
Dinesh Abrol

Appendix	309
About the Editors and Contributors	322
Index	325

List of Tables

1.1	Data Summary of International GDP Per Capita by Quartile and by Year	24
1.2	Means and Standard Deviation of Correlates of School Quality by GDP Per Capita Quartile and Year	26
1.3	Means and Standard Deviations of Explanatory Variables for Quality Assessments by GDP Per Capita Quartiles and Year	32
1.4	Random Effects Estimates of Education Quality under Two Specifications	35
2.1	Innovation Systems and Policy Agendas	46
4.1	Specific Spillin Weight by States across Top into States on Left (Geo-climate Sub-region based)	125
4.2	Average Annual Growth Rate for Farm Output, Input, TFP and Public Agricultural Research Capital by State Grouped by ERS Farm Production Regions, 1970–2004	130
4.3	Variable Names and Definitions and Summary Statistics	134
4.4	Econometric Estimates of State Agricultural Productivity Equation: Contribution of Public Agricultural Research Capital and Other Factors, 48 US States, 1970–2004	135
4.5	The Marginal Impacts of Agricultural Research and Extension on TFP: State Data	137
5.1	Simple Regression Results	155
5.2	Step-by-Step Regression Estimates of the Market Access Effect	160
5.3	Yield Equation Estimates	163
5.4	Unit Cost Equation Estimates	166
5.5	Production Frontier and Efficiency Estimates	169
5.6	Cost Frontier and Efficiency Estimates	170

6.1	Survey Design	186
6.2	Existence of R&D Activity in Firms (Per Cent)	188
6.3	Nature of R&D Activities in Firms	188
6.4	Firms That Reported Any Form of University–Industry Interaction during the Last Three Years That Was Important for an Innovation for Them	190
6.5	Occurrence of Interaction (Collaboration) with Universities/PRIs	190
6.6	Regional Variation in the Number of Firms That Reported Any Form of Interaction with University and RIs during the Last Three Years That Was Important for an Innovation	191
6.7	Firm Interaction with Local Universities	192
6.8	Number of Firms That Reported Any Form of Foreign University–Industry Interaction during the Last Three Years by Region and Country	194
6.9	Incidence of R&D Activity and University–Industry Interactions among Firms (Per Cent)	196
6.10	Performance of Various Functions of the Firm and Internationalization	198
6.11	Factors That Represent a Challenge or a Barrier to International Innovation Collaboration	199
6.12	Factors Influencing the Internationalization of Innovation Activities	200
8.1	Changing Structure of Employment across the Indian Industries: 1980–81 to 2004–05	246
8.2	Patterns of Growth in Employment, Value Added, Output and Emoluments across Indian Industries	251
8.3	Employment Elasticity across Organized Manufacturing Industry Groups	257
8.4	Decomposition of Effects on the Growth of Real Wages (1980–81 to 2004–05)	260
8.5	Estimated Fixed Effects Models (Dependent Variable Is Log of Employment)	265
8.6	Correlation Coefficients between Technology Capability Index and Growth Rate of Employment (1992–93 to 2004–05)	266

List of Tables ix

9.1	Statement on Sector-wise FDI Inflows from April 2000 to April 2009	281
9.2	Industry Analysis: No. of US FDI Projects by Activity	283
9.3	Sample Characteristics of Selected Pharmaceutical Companies	284
9.4	Intensity of R&D, Royalties and Marketing and Advertising (2006–08)	284
9.5	Directions of Innovative Activities of Foreign Firms (1999–2009)	288
9.6	Disease Type-wise Product R&D Activities of Foreign Firms Active in India (1999–2009)	289
9.7	Pharmaceutical Companies in CRAM Activities in India	292
9.8	Disease Type-wise Product-specific R&D Activities of Domestic Firms Active in India (1999–2009)	296
9.9	Clinical Phases of Compound for Various Diseases by Foreign and Domestic Pharmaceutical Industry during 2007–2009	297
9.10	Therapeutic Area-wise Estimation of Pharmaceutical Projects and Patents and the Pattern of Matches with the National Burden of disease (1992–2007)	298
9.11	Comparison with Disease Burden of Public Sector Projects from 1992 to 2007	301

List of Figures

1.1	A General Model of Household Decisions	9
1.2	Average School Assessment by Level and GDP Per Capita Quartiles	25
1.3	Enrolment Rates by Schooling Level and GDP Per Capita Quartiles	28
1.4	Assessment of School Quality by Type and by GDP Per Capita Quartile and by Year	29
1.5	Assessment of Math and Science Education and Training and Vocational Education by GDP Per Capita Quartile and by Year	30
4.1	US Agricultural Geo-climatic regions and Sub-regions	114
4.2	Public Agricultural Research Timing Weights: Trapezoidal Pattern	115
4.3	Public Agricultural Research Capital, 1970–2004: Idaho	128
4.4	Public Agricultural Research Capital, 1970–2004: Alabama	128
4.5	Public Agricultural Research Capital, 1970–2004: S. Carolina	129
4.6	Public Agricultural Research Capital, 1970–2004: Louisiana	129
4.7	Public Agricultural Research Capital, 1970–2004: Connecticut	130
5.1	Yield and Distance to Poblacion	152
5.2	Yield and Population Density	153
5.3	Unit Cost and Distance to Poblacion	154
5.4	Unit Cost and Population Density	154
5.5	Production Efficiency and Distance to Poblacion	157
5.6	Production Efficiency and Population Density	157
5.7	Cost Efficiency and Distance to Poblacion	158
5.8	Cost Efficiency and Population Density	159
8.1	Low-tech Industrial Employment Shares	247
8.2	Medium Low-tech Industrial Employment Shares	247

8.3	Medium High-tech Industrial Employment Shares	248
8.4	High-technology Industrial Employment Shares	249
8.5	Low-technology Industrial Employment Growth Rates	254
8.6	Medium Low-technology Industrial Employment Growth Rates	254
8.7	Medium Low-technology Industrial Employment Growth Rates	255
8.8	High-technology Industrial Employment Growth Rates	255

List of Abbreviations

7-ACCA	7-Amino-3-Chloro Cephalosporanic
ANDAs	abbreviated new drug applications
ARS	Agricultural Research Service
ASI	Annual Survey of Industries
BC	Beneficiary Committee
BLEC	Block Level Expert Committee
BPL	below poverty line
BPO	business process outsourcing
CBO	community-based organization
CDS	Centre for Development Studies
CDS	community development society
CEO	chief executive officer
CFCs	case of chlorofluorocarbons
CO_2	carbon dioxide
CRAMS	contract research and manufacturing services
CRIS	Current Research Information System
CSIR	Council of Scientific and Industrial Research
CSSs	centrally sponsored schemes
CTRI	Clinical Trials Registry India
DBT	Department of Biotechnology
DICE	Dynamic Integrated model of Climate and the Economy
DLEC	District Level Expert Committee
DMF	drug master filings
DPC	District Planning Committee
DPCO	Drug Prices Control Order
DPRP	Drug Promotion Research Programme
DPRP	Drugs and Pharmaceuticals Research Programme
DRDA	District Rural Development Agency
DST	Department of Science and Technology
DVDs	digital videodiscs

EIU	Economic Intelligence Unit
EMRs	exclusive marketing rights
EPA	Environmental Protection Agency
EPO	erythropoietin
ERS	Economics Research Service
EU	European Union
FAO	Food and Agriculture Organization
FDI	foreign direct investment
GATT	General Agreement on Tariffs and Trade
GCI	global competitiveness index
GDP	gross domestic product
GE	General Electric
GIN	global innovation network
GM	General Motors
GM	genetically modified
GMP	good manufacturing practices
GoI	Government of India
GoK	Government of Kerala
GP	gram panchayat
GSPOA	Global Strategy and Plan of Action
HCFCs	hydrofluorocarbons
HDR	Human Development Report
IAS	Indian Administrative Service
IBM	International Business Machines
ICICI	Industrial Credit and Investment Cooperation of India
ICMR	Indian Council of Medical Research
ICT	information and communications technology
ID	index of deprivation
IICT	Indian Institute of Chemical Technology
IISc	Indian Institute of Science
IIT	Indian Institute of Technology
IP	intellectual property
IPRs	intellectual property rights
IT	information technology
KDS	Kudumbashree
LDCs	less developed countries
LDF	Left Democratic Front

LGs	local governments
MDGs	Millennium Development Goals
MNC	multinational corporation
MNEs	multinational enterprises
NAS	National Academy of Sciences
NASSCOM	National Association of Software and Services Companies
NCEs	new chemical entities
NCL	National Chemical Laboratory
NDDS	new drug delivery system
NIC	National Industrial Classification
NMITLI	New Millennium Indian Technology Leadership Initiative
NRC	National Research Council
NRIs	non-resident Indians
NSS	National Sample Survey
OECD	Organisation for Economic Co-operation and Development
OLI	ownership–location–internalization
OLS	ordinary least squares
OSDD	open-source drug discovery
PCs	personal computers
PCSE	panel-corrected standard errors
PFC	pharmaceutical fine chemicals
PPC	People's Plan Campaign
PPPs	public–private partnerships
PRIs	Panchayati Raj Institutions
PWD	Public Works Department
R&D	research and development
RIs	research institutions
RPA	research problem area
S&T	science and technology
SAESs	State Agricultural Experiment Stations
SBIRI	Small Business Innovation Research Initiative
SCOPE	Standing Conference of Public Enterprises
SCP	special component plan
SCs	schedule castes

SFC	State Finance Commission
SMEs	small- and medium-scale enterprises
SO_2	sulpher dioxide
SPB	State Planning Board
STs	scheduled tribes
TAG	technical advisory group
TCS	Tata Consultancy Services
TDB	Technology Development Board
TFP	total factor productivity
TiE	The Indus Entrepreneurs
TRIMs	Trade-Related Investment Measures
TRIPS	Trade-Related Aspects of Intellectual Property Rights
TSP	tribal sub-plan
TVET	technical and vocational education and training
UNCTAD	United Nations Conference on Trade and Development
UNDP	United Nations Development Programme
UPA	United Progressive Alliance
URAA	Uruguay Round Agreement Act
USDA	US Department of Agriculture
USFDA	US Food and Drug Administration
UTI	Unit Trust of India
VC	venture capital
VoIP	Voice over Internet Protocol
VTC	voluntary technical corps
WEF	World Economic Forum
WTO	World Trade Organization

Preface

The global economy has been transforming at a rapid rate. Technology and innovations have played a pivotal role in this process of economic transformation. The nature of technological progress and institutional arrangements affects the direction of economic development. There is increasing interest among economists to examine the creation and dissemination of innovations across farms, firms, regions and over time. Historically, the innovations and technological progress have remained the domain of the state because of its public goods properties. The large proportion of research and development expenditure was incurred by the government, and government usually set the priority areas. Agriculture research and extension system is one such example. Robert E. Evenson was one of the pioneers who had examined this system and its impact on the agricultural productivity. Coming from a farming background, he had the passion to improve the well-being of poor farmers in both developed and developing countries. His research was filled with insights on how to do this. His work on the determinants of agriculture productivity and economic returns to agriculture research was pioneering and remained a core interest throughout his life. He was Professor of Economics at the Yale University from 1977 to 2007, and made enduring contributions in many other areas of economics such as biotechnology, intellectual property rights (IPRs), industrial innovations, development, education and genetically modified foods. Bob, as he was referred to, cared deeply about the poor and devoted much of his working life to the development of better and stronger yields of rice on smaller plots of land. The essays in this volume are a fitting tribute to Robert E. Evenson for his long and outstanding selfless service to the academic world. These essays have been written by friends, collaborators and students, each distinguished in his or her own right.

The editors express heartiest thankfulness to the contributors for their support in terms of conceding to the request on a short notice to write research papers. As and when approached for clarification and revisions, they cooperated and observed the time schedule. The anonymous referee of the volume helped the editors in organizing it as well as suggesting revisions for the improvement. His strategic help is appreciated and gratefully acknowledged. The editors acknowledge the help and direct and indirect support of several towering personalities of the economics discipline such as K.L. Krishna, D.B. Gupta, S. Mahendera Dev, Douglas Gollin, Ariel Pakes, late Jean Lanjouw, Abhijit Banerjee and Anita Gill. Special thanks go to Judy Evenson for her care and hospitality to many scholars from developing countries, including the editors. During the preparation and revision stages, unfailing secretarial assistance was provided by Mr Baltej Singh Bhathal; his services are gratefully acknowledged. The cooperation and advice of the editorial team at SAGE Publications (India) in bringing out this volume is highly acknowledged.

Lakhwinder Singh
K.J. Joseph
Deniel K.N. Johnson

Legacy—Robert E. Evenson
Daniel K.N. Johnson

As is true for any life well spent, it is impossible to do justice to the man Robert Evenson in a few short words. He has, quite simply, done too much for too many, for me to hope to offer a summary here.

Bob, as he was referred to, found his passion in agricultural policy, earning his BA from the University of Minnesota in 1961, his MSc from the same in 1964 and finishing his PhD at the University of Chicago in 1968 while already teaching at the University of Minnesota. Yale scooped him up soon thereafter, and aside from a three-year tour in the Philippines with the Agricultural Development Council (a period that would figure prominently in his research path thereafter), he called Yale University home for his entire professional life.

I first met Bob Evenson 20 years ago, in the spring of 1991, visiting Yale as part of a graduate school shopping tour. That started a close professional and personal relationship that has transcended the decades since. He struck me then as he always has since: honest and open, compassionate and dedicated, scholarly and inquisitive. Several years later, when I expressed pleasure at his agreement to serve as my primary dissertation advisor, he was characteristically humble, understating his ability to direct my work. He was always in high demand, with students waiting outside for his attention, his desk a jumble of papers, his floor-to-ceiling bookshelves crammed, his filing cabinets bursting with potential ideas to explore. A scale model of an agricultural implement, made long ago for someone's application at the US Patent and Trademark Office, sat among the scattered papers. In the one clear space on the wall opposite his desk, there was a painting of a farmer, one that I always thought suited Bob inordinately well: casual yet focused, professional but without personal ambition, intentional and without pretence.

Isaiah Berlin famously described creative thinkers as following one of the two great paradigms: the fox, who wanders broadly and reads widely, and the hedgehog, who delves deeply into one subject in order to master the nuances. In my mind, Bob is the best kind of hedgehog, one who is eternally focused on one pivotal question—how do agricultural productivity gains accrue to less developed nations—and he answers it by looking outward, by travelling the world, by collaborating with as many diverse scholars as possible, by training the next generation to think broadly about related issues.

Bob has always emphasized that good research questions should be important, interesting and answerable: *important*, because to spend your time otherwise would be inefficient; *interesting*, because to spend your time otherwise would be boring; *answerable*, because to spend your time otherwise would be fruitless. To answer his agricultural productivity questions, Bob went quite literally to the ends of the earth.

Even among those who travel broadly, Bob's travels were legendary. He always told his students that there was no substitute for on-site observation. He wanted to know what the important local problems were, what the relevant data might be, what the challenges of policy implementation would be. So Bob advised, consulted and lectured wherever he went; and he went everywhere. I was lucky enough to accompany him on long trips to Brazil and India, where he busied himself in arcane texts and data sets with the enthusiasm of an addict. His willingness to serve others, to be flexible with travel plans beyond any expectations, saw him arrive jet-lagged in the middle of the night to share my hotel room after a reservation went awry; and on another occasion, I saw him give a 60-minute presentation without preparation due to a scheduling error. Bob did not show any stress of agitation, but simply took it in his stride with a characteristic 'Ummmm…. Okay'.

Bob has over 70 refereed journal articles in print that directly address agricultural productivity, and he was a frequent contributor to the *American Journal of Agricultural Economics*. More than that though, he has edited numerous volumes (many published by the Centre for Agricultural Bioscience International), including the third volume in North-Holland Elsevier's prestigious 'Handbooks of Agricultural Economics' series. Then one must consider the books he authored, the conference sessions he organized, the collaborations he brokered, the data sets he fathered (Appendix-selected publications of Robert E. Evenson). For example, his work with Jon Putnam and Sam Kortum and me started an entire cottage industry in patent concordances.

As a master of reduced-form analysis, Bob looks to reality for inspiration, and immediately wants to know what the data say. Can it be quantified? If not, why not? His work is elegant in a shortest-distance-to-the-answer sense, eschewing highly mathematical models and keeping the analysis as close to the raw data as possible. He was a great believer in the importance of patents, both as a tool for innovation and as a tool for productivity researchers. He was forever looking for another way to count patents, to weigh their importance, to impute their value, to measure their impact. The same can be said of his devotion to extension research, to public agricultural research funding and to crop enhancements via high-yield varieties or genetic modification.

Bob was convinced of the ability of humankind to innovate our way out of problems, as long as policy surmounts the obvious market problems of public goods, externalities and uncertainty. He was also an ardent supporter of education, a technical education to be sure, but also a broader education based on newspaper articles and archives, and conversations and experience. He asserted that you could never tell from whence the next research idea might arrive (his own, or an agricultural innovator's). In his words, 'the only difference between the bleeding edge and the leading edge is the current direction of motion.'

Among those who read this text, I am surely not the one who knew him longest or best. Nor am I the most eloquent, or the most qualified to offer a worthy tribute. I am merely one of many who hold him constantly in my thoughts as an inspiration, as a mentor and as a friend.

Technology, Innovations and Economic Development: An Introduction

Lakhwinder Singh, K.J. Joseph and Daniel K.N. Johnson

The pace of knowledge generation and diffusion, a basic responsibility of the nation state, has fundamentally determined the pace of economic development across countries and over time (Kuznets, 1966; Ruttan, 2001; Singh, 2007; Solow, 1957). Therefore, there has been a close relationship between innovation and the government's innovation policies governed by the national goals and expectations. Globalization, however, has dramatically altered the innovation system in terms of not only the institutional context wherein the different actors interact but also the nature and scope of interactive learning; the rules of the game, increasingly enacted and implemented at the instance of global institutions, appears to be less suitable for developing countries trying to catch up. With the removal of trade barriers, the domestic firms are exposed to international competition and that the infant industry protection and government subsidies, widespread in most of the earlier catch-up episodes, are having limited role today. The unprecedented exposure to international competition in turn influences the innovative behaviour and competitive strategies, leading to increasing incidence of joint ventures and takeover of local firms by foreign firms. At the same time, innovation systems in general and R&D in particular is increasingly becoming a global activity with its effects on the developing country participants. Similarly, the strong intellectual property right (IPR) regime being imposed on the developing countries entails an environment significantly different from the ones that was confronted earlier (Nelson, 2007). Hence, there is little scope for learning from and then imitating, widespread in the earlier catch-up episodes. Interaction between industries (read as productive sectors, including agriculture), universities and public research laboratories that had significant role in innovation and development is apparently playing limited role as there has been significant cut in social sector expenditure in a context wherein countries are forced to adhere to fiscal prudence. In addition, the heightened concern for environmental sustainability sets further limits to the innovative options open to the developing countries. With the growing concern for development with human face and role of science and technology therein, evolving an innovation policy in the fast pace of globalization has emerged as a matter of great concern.

Issues pertaining to innovation, technology and economic development have been at the core of different heuristic paradigms like classical legacies, Schumpeterian paradigm, Arrovian legacy, endogenous growth models to the national innovation systems approach and has had their influence in policy. In such a context, this book is concerned with issues that have been at the centre of scholars' attention for many years. With the growing body of literature and changes that have taken place in the context wherein technological change and innovation takes place, a number of new issues have been discovered. As a result, some of the traditional issues have become redundant and new paradigms have emerged that influence public policy both by national governments and by international organizations. Some of these issues remained close to the heart of Professor Robert E. Evenson and were reflected in the research agenda he pursued since the 1960s and 1970s.

Robert Evenson's contribution to the understanding of the influence of innovations and technological progress on economic development has been widely recognized and has attracted the attention of scholars from Latin America, North America, Europe and Asia. He was a successful and influential economist in terms of creating a team of scholars, expanding and developing applications and extensions to his ideas, who cut across various continents and narrow boundaries of economic disciplines. His contributions to agriculture research, education, extension, IPRs, consumer acceptance of genetically modified foods, innovation exhaustion potential, biotechnology, the international exchange of germplasm, international diffusion of technology and policy are unrivalled. The scholars who have worked closely with Evenson as students, colleagues and collaborators carry his ideas as established scholars in their chosen field of study as well as informed policymakers in numerous countries throughout the world. In honour of Professor Robert Evenson's lifelong contributions in key areas of innovation, technological progress and economic development in the rapidly changing world economy, this volume brings together essays by leading experts in a memorial volume to celebrate the long (but not long enough) life of Bob Evenson, collecting articles in the fields of innovation, technology and economic development. This volume is distinct in terms of focusing on problems from the perspective of developing countries. The contributions of this volume are classified in three sections:

I. Innovation and Economic Development
II. Technological Progress and Agricultural Development
III. Technology Transfer, National Innovation Systems and Industrial Development

Section I: Innovation and Economic Development

A set of three chapters included in Section I addresses the core issues of innovation and economic development. During the recent phase of globalization and the era of economic reforms, it is pertinent to ask questions regarding how reforms have impacted technological change and economic development. Sustainability of economic development at a current pace and level has also remained a matter of global debate. To some extent, the answer lies in eco-innovations. These issues are indicative of the state of public policy and suggest a paradigm shift that has occurred in economic reforms attempting to achieve more inclusive and sustainable economic development.

In his chapter 'Education Reforms, Technological Change and Economic Development', Leonardo Lanzona highlighted the impact of educational reforms made to meet the growing needs of new technologies for economic development and welfare in developing countries. The author, while reviewing theoretical literature on education and development, has developed a conceptual framework that links skill formation to labour market outcomes and technological progress. The identification of the key elements has been done by examining successful experiences of countries that have conducted educational and structural reforms. Lanzona has developed an econometric model to estimate the relationship between education reforms and technological change on economic development, using data from the World Economic Forum for two years, 2009 and

2011. The results from the panel data clearly show a positive relationship between increased expenditure in education and better quality of higher education. Empirical estimates show that higher quartile countries devoting more resources to education possess higher quality of education and are also able to implement educational reforms. An important explanatory variable that emerges from the empirical exercise undertaken by the author is that outward-looking trade policies positively affect the demand for higher-quality education. The most significant finding that emerges from the empirical estimates is that R&D expenditure has an insignificant impact on gross domestic product (GDP) per capita. This implies that in the absence of superior R&D institutions, low private investments in R&D in the developing countries result into inferior schooling quality. Therefore, the author suggests that it is necessary to encourage a developing country to invest in innovations that increase the demand for quality schooling. Furthermore, he has asked that international organizations support the efforts of the developing countries to improve higher education, further technological innovations, reduce poverty and not merely provide aid to make adjustments towards newer technologies.

One of the most pressing problems faced by the humanity is the sustainability of the ongoing economic development process while minimizing destruction of the environment. A substantial number of positive efforts have been made under the leadership of the United Nations to tackle this problem. Economic literature on environmental innovations has been growing by leaps and bounds to shed light on the problem of environmental degradation since the last decade of the 20th century. Dan Johnson and Kristina Lybecker, in their chapter 'Eco-Innovation: A Literature Review of the Challenges Facing the Development of Green Technologies', have described and evaluated the literature surrounding the economics of environmental innovations. In the chapter, the authors focused on the constraints to successful development, diffusion and financing of eco-innovation. After presenting the wide variety of definitions of eco-innovations, the authors argued that environmental innovation faces a high degree of uncertainty related to the end product of a research process, market acceptability and ability to appropriate the returns to research. IPRs in the form of patents are the legal rights to exclude others from an activity that allows the producer of environmental innovation to reduce uncertainty, secure returns on developmental costs and encourage further innovations. The authors examined the literature on the use of IPRs, specifically patents, for encouraging innovations to solve the problem of growing environmental degradation. There exists a division among economists with regard to the allocation of IPRs. On the one hand, they increase the costs of adoption by consumers and intermediary producers, but, on the other hand, they provide a virtually unavoidable piece of protection for future revenues of inventors. Thus, it is suggested by the authors that achieving efficient scale and reduction of per unit production costs is critical for the success of the most eco-innovative products and processes. The authors argue that regulations are critical to the development and success of eco-innovation. Environmental regulations might lead to cost-saving innovation if (a) the fixed costs of innovations are lower than compliance plus production costs, or (b) spillover effects make innovation strategically a bad idea for the firm but a good idea for society, or (c) regulation helps to fix incentive problems between managers and owners, or (d) regulation helps to promote information flows. The authors cautioned policymakers that the solution requires flexibility and vigilance so that innovators can creatively fill in the frame and build out in unpredictable directions.

M.A. Oommen is one of the oldest friends of Robert Evenson and worked at Yale University with Evenson starting in 1974–75 and reflected in the theme 'Social Inclusion and Institutional Innovations: Working Towards a Policy-Theoretical Framework'. The capitalist development model based on high rates of economic growth achieved by one of the emerging economic global powers (India) has not benefited the larger section of the Indian population. Policymakers in India have accepted this fact and worked to reorient their economic development strategy, coining the term 'inclusive growth/development'. It seems that the new strategy has been drawn from the widely discussed success story of the decentralized economic development experienced by Kerala (southern state of India) since the mid-1990s. Oommen has deeply examined the concept of social inclusion to draw the boundaries and rationale for developing the policy framework. The author has followed the philosophical foundations based on the freedom–capability approach created by Amartya Sen to operationalize the guidelines for making economic growth more inclusive. In this approach, the prerequisite for social inclusion and inclusive growth is the expansion of capabilities and freedom of individuals. The full benefits of this approach, according to the author, can be realized if matching institutional innovations could be developed so that people can act as an agent of change rather than as mere beneficiaries of development. He has argued that to make economic development truly more inclusive, it is necessary to involve the third tier of government (Panchyati Raj Institutions). The claim of the author is that new institutional innovations, such as constituting beneficiary committees (KDS), Kudumbashree (women-oriented anti-poverty programme) and community-based organizations (CBO), have ensured transparent participation and accountability in public works programmes while reducing poverty along with increasing the rate of economic growth of the state. Local democratic systems can be used as an instrument for development through adaptations and innovative actions.

Section II: Issues in Technological Progress and Agriculture Development

The growing global population has put tremendous strain on the food supply, causing a dramatically changed perspective on agriculture development. The supply-side constraints of agriculture production have occurred due to the shrinking availability of arable land, increasing diversification of agriculture towards non-food grain production and climate change. Technological progress, institutional innovation, organized research and agricultural extension programmes are vital for providing food security to the global population. Historically speaking, public agriculture research has played a pivotal role in raising agriculture production and productivity for a sustained period of time while also giving rise to large number of research institutions (RIs) in advanced and developing countries. This is popularly described as the agriculture innovation system. Robert Evenson has contributed immensely, enhancing our understanding about the functioning of agricultural research, education and extension and their impacts on agricultural development across countries and over time.

In this vein, Wallace Huffman, in his chapter 'Measuring Public Agricultural Research Capital and Its Impact on State Agricultural Productivity in the US', examined the relationship between public agricultural research capital and agricultural productivity across US states.

Huffman believes that agricultural innovations and technological progress have been solely dependent on public investment in research. This happens because farm-firms do not undertake organized research due to the large fixed costs, long gestation periods and very specialized talent needed to successfully undertake research in a manner cost-effective relative to farm sales. Another reason for public investment in research underlined by the author is the 'public good' nature of the discoveries. Therefore, the author stressed the need to develop a more representative method of measuring public agriculture research capital that captures the contribution of research and development expenditure incurred by the 48 contiguous states of the United States. Huffman developed an econometric model to explain total factor productivity by the state using independent variables such as public agricultural research capital, and spill-in agriculture research capital, controlling for trend and state dummy variables. The results obtained from the model were highly significant and relevant, especially for public research capital and spill-in research capital. In further estimations, Huffman showed that the state's own public agricultural research capital and spill-in research capital are complementary. Another important finding was that the real investment in public agricultural productivity enhancing research capital peaked in the United States in 2004 and has declined since then. The author argued that the impact of this decline in public agricultural investment in research capital will reduce the potential for productivity growth in agriculture for the US economy over the next three decades.

Agricultural development in the Philippines was always close to the heart of Robert Evenson. He intensively examined the impact of innovations on productivity growth in the Philippine agricultural sector as well as other problems of underdevelopment from the mid-1970s until his death. Evenson encouraged researchers to work on various problems faced by the Philippines agricultural sector and even created a panel data set. Drawing on the Yale University Economic Growth Center and Bicol River Basin Development Program of the Philippines government data set, Sanjaya DeSilva has used an econometric model to examine the impact of the spread of market institutions on the improvements in the productivity and efficiency of rice farmers over a period of two decades. In his chapter 'Access to Markets and Farm Efficiency: A Study of Bicol Rice Farms over Two Decades', the author analyzed the causes and consequences of transaction costs and dysfunctional markets as well as the determinants of productivity and technical efficiency. DeSilva also included institutional factors that reduce or enhance agricultural productivity and efficiency of the farms in the villages of Bicol region. An important finding that emerges from the empirical analysis is the vindication of the Schultz hypothesis, that is, there is an inverse relationship between distance from the market (urban–industrial hypothesis) and productivity as well as efficiency of farms, at least in 1983, but analysis still firmly rejects this as a long-run relationship. The education level of farmers played a significant role in enhancing productivity and the returns to education: returns were highest for those farmers who were educated to the primary and secondary levels. The significant empirical evidence that emerges from the econometric analysis is that there is a decrease in transaction costs in densely populated villages. Another noteworthy finding of this study is that gains in productivity and efficiency of farmers located in remote and sparsely populated villages were higher. This result adheres to the convergence hypothesis due mainly to the availability of public goods such as market access, irrigation, road networks and research and extension systems for more populated areas. A significant policy lesson that emerges from the study is of great importance for less developed agricultural regions such as Africa: if the state provides essential public goods in the rural sector of the economy, it translates into a substantial rise in agricultural productivity and efficiency levels.

Section III: Issues in Technology Transfer, Innovation Systems and Industrial Development

Industrial development has remained the engine of economic growth and structural transformation in developed countries since the Industrial Revolution. In recent times, it has also proved beyond doubt, through the development of newly industrializing countries of East Asia, that industrialization is a possible pathway to economic prosperity and structural transformation (Szirmai, 2009). Technological progress has induced dynamism that has determined the pace of economic development since the Industrial Revolution. The industrial sector has emerged as the dominant sector and has generated linkages between other sectors while simultaneously changing the ways and means to induce innovation. There are several patterns of technological networking that have emerged from the innovation-producing and innovation-using sectors (Englander and Evenson, 1993). However, technology production and transfer remained the domain of developed countries. The developing countries by and large remained adopters of industrial technologies until the recent wave of globalization. The national innovation system in developing countries is evolving but is in its infancy. This system draws technological knowledge from different sources by creating various kinds of networks to share technological transformation. The arrival of multinational corporations (MNCs) and investments to exploit markets in developing countries has opened up new opportunities for establishing research laboratories in fast-growing economies such as China and India. Four chapters in this section strive to address the various aspects of technology transfer, innovation system and industrial development in the era of globalization.

In their chapter 'Global Innovation Networks and Industry-University Interaction: A study of India's ICT Sector', K.J. Joseph and Vinoj Abraham analyzed the information and communications technology (ICT) firm's interaction with university/RIs within India and abroad, hoping to shed light on technology transfer mechanisms while following the framework of global innovation networks (GIN). The authors have critically examined the existing literature on GIN with the aim to identify research gaps and have extended the GIN framework to make it more inclusive, introducing a national innovation system approach. The study is based on a survey conducted in 2010 covering 325 ICT firms from India. For exploring international knowledge flows across India's ICT firms, the authors have classified firms into three categories: stand alone, subsidiary of MNCs and headquarters of MNCs. An important finding that has emerged from the empirical evidence is that MNC headquarters in India have a high degree of interaction with university/RIs compared to MNC subsidiaries. While the MNC subsidiaries have more interaction than stand-alone firms with said institutions, the university/RIs interaction, when explored at the regional level, reveals that MNCs headquartered in India have intensive connections with national and local universities. The MNC headquarters in India collaborate with North American Universities and research institutes and this testifies to the fact that ICT exports to North America are the highest. A notable finding is that firms not doing in-house R&D are unable to enter into research collaborations with universities/RIs. The factors that prohibited the university/RIs interaction with Indian firms were the lack of relevant prerequisite knowledge and discouraging attitude of the scientific workforce towards internationalization efforts of the firms. Joseph and Abraham suggested that MNCs were not aimed at generating innovations. This finding is based on the analysis and data collected from two MNCs operating in India. The MNCs were however harnessing the skill base generated by the Indian universities/RIs for ready use in the industry so that the cost of in-house training could be curtailed. The authors have called for a suitable policy change to prevent GINs from becoming global innovation traps.

The focus of the chapter 'Globalization of Industrial R&D in Developing Countries,' by Binay Kumar Pattnaik, is on the impact of the ongoing process of globalization on industrial R&D in developing countries. Having distinguished between multi-nationalization and globalization of R&D at the outset, it further distinguishes the globalization of industrial R&D in developing countries from that of developed countries. The author conceptualizes the impact of globalization of industrial R&D as a new phenomena: (a) emerging international division of labour in R&D, (b) emerging flatter technological regime of the world, (c) growing multi-nationalization of R&D in firms from developing countries, (d) growing globalization of local R&D of firms from developing countries and (e) offshoring R&D services by firms in developing countries. These formulations are of course based on a wide spectrum of reported empirical and secondary data from developing countries. Placing the impact on the theoretical framework of the new dependency school, the author observes that the centre–periphery relations in the context of a few selected developing countries have undergone a qualitative change. Therefore, he observbed that some developing countries have successfully moved into the semi-periphery stage. It is worth noticing that the change in relation between the centre and the peripheral countries is a limited phenomenon confined to the experiences of a handful of developing countries like Singapore, China, India, Brazil and Indonesia. It may be the case that of these, a few peripheral countries have changed their relations qualitatively with the countries at the centre keeping the general nature of centre–periphery relations intact. Pattnaik concluded by suggesting that one may contest the semi-periphery status of India, but it is fast moving into the category. This transition of India can be justified on the grounds that there have been significant technological developments in areas of ICT, pharmaceuticals, biotechnology and subsequent export of industrial products and services.

Lakhwinder Singh and Baldev Singh Shergill analyzed, in their chapter 'Technological Capabilities, Employment Growth and Industrial Development: A Quantitative Anatomy of Indian Scenario', the impact of technology on employment generation using descriptive statistics and econometric models. The authors have argued that industrial development in developed countries has remained a dynamic process responsible for structural transformation of the economy and generated gainful employment opportunities for a growing workforce. The post-reform spurt of economic growth in India has been described as 'jobless growth'. Singh and Shergill examined the question of when industrial development provides required dynamism for generating desired employment opportunities for the labour force and when it does not. An industrial technological capability-based approach has been adopted to analyze the Indian industrial development experience by classifying industries on the basis of stages of technological development during the period of 1980–2005. The main finding that emerged from the empirical evidence is that the medium high-tech industries have shown consistency in terms of generating employment growth. The labour market regulation opinion put forward by various scholars supporting liberalization policies therefore does not hold up. The relationship between industrial technological capabilities and employment growth turns out to be ambiguous. This implies that weak technological capabilities adversely affect employment growth, and heavy dependence on imported technological know-how from developed countries is labour displacing. It is thus suggested that developing countries should invest in both institutions and industrial firms to build technological capabilities that suit resource endowment, local conditions and the stage of industrial development. There is a dire need to explore alternative paths of industrial and technological capability development to sustain economic transformation process for achieving prosperity and reducing the time for catch-up development.

It is widely acknowledged that the Indian pharmaceutical industry has shown dynamism in terms of innovations and making treatments available at a low cost. This industry has saved millions of lives in the poor section of Indian society (Lanjouw, 1998). This contribution has been possible due to state support in terms of incentives and instituting liberal IPRs in the form of process parenting system. The intellectual property regime under the World Trade Organization (WTO) has shifted from process patenting to product patenting that may have affected the innovation capability of the pharmaceutical industry in the developing countries and more so in the Indian economy. A contribution by Dinesh Abrol, 'Intellectual Property Protection, Innovation and Medicine: Lessons from the Indian Pharmaceutical Industry', examined the impact of the post-TRIPS (Trade-Related Aspects of Intellectual Property Rights) regime on the innovation of the pharmaceutical industry of India. He has reviewed existing literature on the relationship between strong IPRs and innovations. He has evaluated the contribution made by Robert Evenson and his associates on the issue, placing him in the school of thought that has advocated strong IPRs for harnessing innovation potentials. Abrol argued that several studies analyzing the relationship between IPRs and innovations generated evidence that was at best ambiguous. However, the Indian government has adopted the WTO-recommended IPRs regime while introducing various amendments to the Indian patent act. The analysis based on empirical evidence of Indian pharmaceutical industry, during the period of 15 years (1995–2010), indicates that the system of strong IPRs has not been able to favour India with the benefits supposedly associated with patent changes: increased access to good quality foreign direct investment (FDI), technology transfer, overseas product R&D and stimulation of domestic investment in R&D. The author has raised a pertinent question, why don't domestic and foreign pharmaceutical firms have any significant plans to invest in R&D for medicines related to local needs of India? At the same time, the incorporation of the Indian pharmaceutical industry into the emerging international division of labour within the global pharmaceutical industry is already on the horizon and the linked innovation systems are expected to move even further away from the goal of development of medicines needed for the improvements in local health conditions. It is surprising that even developing countries like India remain unable to make use of existing flexibilities in the WTO–IPRs regime for health needs of local people. Therefore, Abrol has suggested that instead of making further wholesale amendments to the IPRs regime, the Indian government should introduce a suitable supporting technological, industrial and health policy package.

This volume consists of research contributions to the core themes of developmental economics that were central to the long-term research interests of Robert Evenson's life. Each contribution adds a new dimension to the existing literature in the area of focus. This book also offers a unique contribution in terms of focusing on the problems from the perspectives of developing countries. Comprehensive in scope and intensive in analysis, this volume provides a fresh perspective on the ongoing debate and will initiate new explorations. As such, we offer this volume as a modest tribute to Robert E. Evenson, his enduring impact through his thoughtful analysis, his care for human welfare and his creative genius.

References

Englander, A.S. and R.E. Evenson. 1993. 'International Growth Linkages Between OECD Countries: An Industry Study', *Memeo*. New Haven, CT: Yale University, Economic Growth Center.

Kuznets, S. 1966. *Modern Economic Growth: Rate, Structure, and Spread*. New Haven: Yale University Press.

Lanjouw, J.O. 1998. 'The Introduction to Pharmaceutical Products Patents in India: Heartless Exploitation of the Poor and Suffering', *NBER Working Paper No. 6366*, Boston: NBER.

Nelson, R.R. 2007. 'The Changing institutional Requirements for Technological and Economic Catch up', *International Journal of technological Learning Innovation and Development*, 1(1): 4–12.

Ruttan, V.W. 2001. *Technology, Growth and Development: An Induced Innovation Perspective*. New York: Oxford University Press.

Singh, L. 2007. 'Innovations, High-Tech Trade and Industrial Development: Theory, Evidence and Policy', in George Mavrotas and Anthony Shorrocks (eds), *Advancing Development: Core Themes in Global Economics*. Houndmills: Palgrave Macmillan, pp. 416–34.

Solow, R.M. 1957. 'Technical Progress and the Aggregate Production Function', *Review of Economics and Statistics*, 39: 312–320.

Szirmai, A. 2009. 'Industrialisation as an Engine of Growth in developing countries 1950–2005', *UNU-MERIT Working Paper No. 2009-010*, Maastricht Economic and Social Research and Training Centre on Innovation and Technology, United Nations University, Maastricht, the Netherlands.

SECTION I

Innovation and Economic Development

1

Education Reforms, Technological Change and Economic Development

Leonardo A. Lanzona, Jr

Introduction and Preview

The rates of enrolment and years of schooling have increased in most countries, mainly a result of successive generations of parents investing in children's education. Nevertheless, while increasing household incomes, shifts in demand for more skilled labour and more classrooms have contributed to some global convergence in education in terms of years of schooling, substantial differences in education exist between rural and urban households and also between males and females, in some countries (Orazem and King, 2008). These discrepancies lead to increasing pressure for improved government policy, and international support, for greater education reforms.

Along with all the increased socio-economic opportunities and welfare that education produces, the established microeconomic link between education policy and growth has led to a focus and an expansion of education programmes in developing countries. Significant resources have been devoted to increase enrolment rates and to reduce dropout rates. Governments have actually devoted widely different percentages of their budgets to education relative to gross national income, at a range of 0.9–11 per cent in 2008 across many countries. These numerous reforms have been attempted to aim at increasing the returns to those investments and at raising quality of schooling (Orazem and King, 2008). These initiatives include the setting of higher eligibility requirements for teachers or increasing the number of textbooks in the hands of students. Remedial programmes also have been attempted to reduce dropout rates. More recently, initiatives have attempted to increase attendance at existing schools through school vouchers or cash transfers conditioned on child enrolment.

Empirical studies suggest that the returns to education (based on Mincer–Becker–Chiswick rate of return) are higher in low-income countries than in high-income countries (Psacharopoulos and Patrinos, 2002). Poor households are able to obtain greater wage returns from additional year of schooling than relatively well-off households. Thus, poverty reduction programmes have recently centred on creating human capital, specifically education, mainly because such programmes are also likely to induce growth or productivity. In many ways, expenditures in schooling have been expected to improve not only efficiency but income equality as well.

Nevertheless, the research studies have shown that the effects of these programmes have not yielded the expected gains in the aggregate (see e.g. Fasih, 2008; Hanuschek, 2001, 2003; Pritchitt, 2001). While decreasing returns to education are not unexpected as the years of schooling increase, the social returns to education have not been substantial in many poor countries, and more recently, have even slackened. Growth figures in several countries have also been noted to be unrelated to continuing investments in education and higher average schooling years. This seems to suggest that while private returns to schooling have remained high,[1] the social benefits, as well as the education externalities intended to spur development, have not been felt (see Krueger and Lindahl, 2001; Psacharopoulos and Patrinos, 2002).

Various accounts have been presented to explain why increased social investments in education have not resulted in substantial economic growth even though the individual returns to education are high. Analyzing data from nearly all developing countries, Pritchett (2001) pointed out three potential and fundamental reasons. The first is the possibility that the demand for educated labour comes, at least in part, from individually profitable yet socially wasteful or unproductive activities. The second reason may be that the expansion of the supply of educated labour when labour demand is stagnant could cause the rate of return to education to fall rapidly. Third, schooling quality may be so degraded that it does not raise cognitive skills or productivity. This could even be consistent with higher private wages if education serves as a signal to employers of some positive characteristics, such as ambition or innate ability.

By analyzing the quality of education reforms in various countries, this chapter will examine why increased social investments in education has not resulted in substantial economic gains at the macroeconomic level while the individual as well as household returns to education are high in most developing countries. Moreover, the chapter presents the argument that this gap is at least partly a result of insufficient aggregate demand for better education: greater expenditures in education—more schools, more teachers and overall increases in the supply of schooling—have not produced the expected productivity gains because of the low demand for higher skills. At the same time, there is a low propensity of people to acquire learning skills in low-income environments where the economic and social conditions depress the formation of skills and depress the returns to education. Under this framework, the worker's choice of occupation and the jobs available are fundamentally determined not only by the community and household characteristics but also by the external (fundamentally structural) conditions such as the education policy, resource constraints, labour market policies, scale of the productive sector and technology.

Low demand for learning can be caused by five major factors. First, the quality of primary education received is too low to justify further investments in higher education. In the discussion of institutional factors, the chapter will highlight the importance of basic education in increasing the productivity and the returns from education. Second, in the face of capital market imperfections, cash constraints prevent the households from using school services. Because of this, economic resources from the government (in the form of subsidies and loan assistance) as well as the private sector (in the form of training) may be needed to increase demand. Third, to the extent that quality of higher education increases the chances of obtaining appropriate employment and increases the pay-off to schooling investments, improving higher education can increase the household's desire to obtain access not only to additional schooling but also to improved quality. In which case, education quality results in higher social returns in the form of better skill composition of workers and a more efficient labour market in matching labour supply with the demand for workers. In other words, the demand for higher education is conditioned by labour market policies that support higher returns and demand for skilled workers. Fourth, education attains its highest rate of return in an economy that is continually adopting newer forms of technology and innovation (Schultz, 1975). If technological improvements complementary to skills are not realized, the demand for higher education will tend to be low. Fifth, educational reforms that do not respond adequately to economic changes are not expected to satisfy the household's demand for learning at school. The skill demand for workers depends on sectoral shifts in the economy, and demand for learning is influenced by the demand for skills. The chapter emphasizes the need for educational reforms to react effectively to the sectoral shifts arising from economic changes. With improved technology, the country's production possibilities expand and the demand for skilled labour is enhanced. Ultimately, the country can move towards high-productivity sectors if educational reforms are successful in forming skills.

This chapter will revolve around these five basic themes. What binds these issues is a conceptual framework that assumes the centrality of household welfare, as opposed to the interests of private firms as the object of development policy. Furthermore, the main focus of this chapter is the main role that technology plays in increasing the demand for education. While the final demand for labour comes from the firms, this demand at a given state of technology is ultimately constrained by the skill levels that households possess. In a technologically challenged country, firms are unlikely to hire skilled labourers. In turn, unless high-quality education and a pool of skilled labourers are available, technological innovation and ultimately education reform are unlikely to be sustainable. The main objective of education reform is then to create conditions for the households to form and increase their demand for skills that will be employed in the labour market.

The major challenge in education reform in many developing countries is how to expand and improve the educational opportunities of their population, especially the less educated and disadvantaged households, in the face of poor economic and social conditions. The widespread fear of losing ground in international competition has generally led to a focus on education as means of improving human capital and enhancing productivity, and creating a resource base necessary for other reforms.

Nevertheless, the attainment of these objectives is dependent on the state's ability to enhance technological progress, which increases the demand for education reforms. In the context of agricultural development, Schultz (1975) argues that households may be 'poor but efficient', and 'efficient but poor'. As noted by Evenson and Pingali (2007), this argument implies that education, particularly agricultural extension, will not induce agricultural development since farmers are already efficient.[2] Farmers are seen to be efficient largely because they have had time to experiment with technological improvements under conditions of slow delivery. However, education can transform agriculture if the technology introduced goes beyond existing local characteristics, hence generating a demand for greater learning. In effect, the success of education reform depends on the country's ability to facilitate the adoption and the access of the technology produced externally.

The chapter aims to determine the factors affecting the success of education reforms. Further, it examines the importance of trade and technological factors in determining this success. It is hypothesized that effective education reforms have to be implemented with technological change, which in turn is enhanced by trade. Moreover, since most of the countries continue to have a substantial agricultural production, much of the analysis has been done in the agricultural sector.

The chapter is organized as follows. The section 'Conceptual Framework of Education Reform and Development' provides a conceptual framework that links skill formation to labour market outcomes and technological change, and includes a brief survey of literature on education and development. The section 'Empirical Model and Findings' offers some empirical evidence of the given framework through a brief examination of the key elements of successful educational and structural reforms as implemented in certain countries. The chapter uses the survey results of World Economic Forum (WEF) in their Global Competitiveness Report. Policy directions are often hard to be concrete because education reforms are country specific. In order to control for country heterogeneity, panel data random effects analysis will be conducted using the data from WEF for two years from 2009 to 2011. The final section presents policy recommendations based on the lessons cited in the previous sections. The study in the end will show how reforms of the education and vocational training systems should be combined with policies related to demand for education, such as investment policies, technology transfer (or R&D) and international trade or, more generally, sectoral policies adjusted to the current global economic environment. Furthermore, the chapter will also discuss how international community may assist in this effort, beyond the usual provision of overseas development assistance.

Conceptual Framework of Education Reform and Development

Demand for quality education and skill formation begins at the household level. Understanding how household decisions are made is then the first in developing the correct education reforms. Drawing from Lanzona (2009), this section uses an economic model of skill formation and summarizes findings of related literature (e.g. Becker and Tomes, 1986; Schultz, 1993; and Heckman, 1999). The determinants of skill formation are viewed as the same factors affecting completed family size and household labour allocation. These operate through a household preference for present and future consumption, children and human capital and through three main constraints. These are:

- the budget constraint that reflects the opportunities and limitations implied by the market prices of goods and services, the wage rate of family members and any non-labour income, and time at the disposal of household members;
- the household technology, which enables it to convert market goods and the time of the family members into basic commodities, including food consumption and human capital; and
- the household's budget that is dependent on the children's future production and income and the income of the resources that were bequeathed by the parents. This last constraint pertains to the intergenerational aspect of household decisions.

Figure 1.1 presents the parent–child dynamics and the general equilibrium model of household decisions. The model assumes that parents make rational choices, which determine the household's welfare and future security and, ultimately, social spillovers that affect the economy's productivity and growth. The possible choice of sending children to school and producing skills is part of these decision-dynamics.[3]

In making decisions, the household is faced with four main elements of the economic environments, namely:

- the invariable genetic structure and background of the family,
- the relatively short-run meso- and macroeconomic environment,

Figure 1.1:
A general model of household decisions

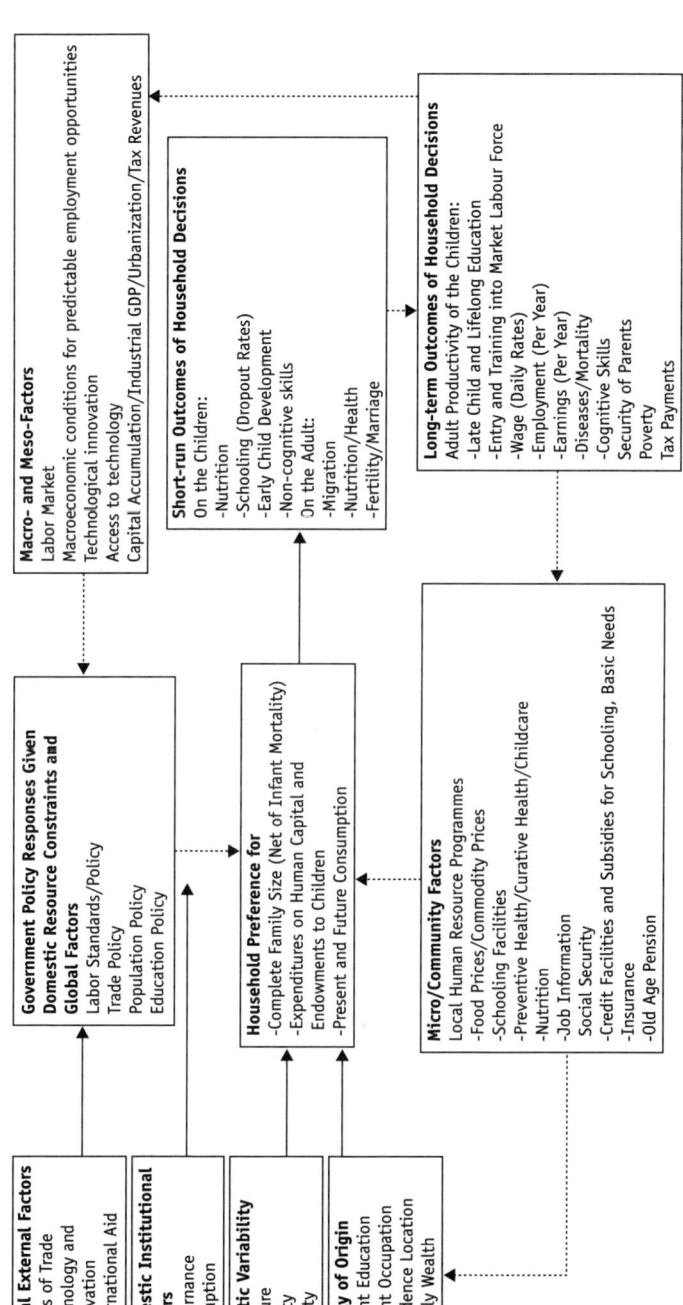

Source: Author.

- the more long-run local (micro) conditions where the family resides and
- the international environment.

Analyses and identification of determinants of skill formation are made difficult, however, because of the cyclical or recursive relationships between outcomes and factors; that is, a choice made today (or in the current period) partly determines the conditions within which the parents have to make decisions tomorrow (or next period). Broken lines show the effects on household choices on the economic and social environment, which can only be partly influenced by the single household. In effect, these broken lines in the diagram (Figure 1.1) show heuristically how final outcomes, such as children's productivity or parents' economic security, reinforce and perpetuate the constraints faced by the household at the local and national levels. The new generation's productivity or parents' security in general will be undermined by the parents' decision to send children to school or to labour market.

Among the outcomes as well as the constraints of various household decisions is education policy. There are number of reasons why the education policy is endogenous to system. While the government influences policy independently, particularly the supply side,[4] its design is somehow mitigated by various (mainly financial) constraints that possibly limit the demand for such policies. A clear case of how education policy may be affected by household decisions can be seen in the way taxes may be collected in the urban and rural areas. Through tax collections, urban areas are more able to finance public education not only because of their higher incomes and the greater concentration of educated individuals. In effect, the returns to education would seem higher in the urban areas since those who remain in the rural areas are often those who have less schooling and have lower productivity while the more educated and motivated individuals reside in the urban areas. Thus, rural households are likely to experience a greater tax burden, relative to their lower incomes, in order to maintain the schools in their communities (Lanzona, 1998).

The supply of education quality can also be affected by the choices made by individual households and the type of supply mechanisms established in the country. For instance, school characteristics, such as infrastructure, basic teaching supplies and number of teachers, depend on the number of resources available to the government and the number of individuals willing to engage in the production of human capital. Other inputs at various levels, such as well-trained teachers and the courses, are also dependent on the income and age composition of the households and society's institutional set-up.

In this framework, household decisions produce two types of outcomes, (early) child development (in the short run) and the lifelong education (in the long run). The former refers to those skills formed at infancy up to the end of primary schooling. This refers mainly to basic education that aims to attain literacy and covers the years when human ability and motivation (non-cognitive skills) are shaped by families and non-institutional environments. The latter denotes long-run skill formation that spans the period from the secondary schooling to post-schooling training, including technical and vocational education. Work experience and skills obtained in the workplace in the form of job search, learning-by-doing and workplace training are also part of the long-term outcomes even if these are not well measured. Post-school learning is an important source of skill formation that accounts for as much as one-third to one-half of all skill formation in a modern economy (Heckman and Lochner, 1999). Because much of this learning takes place in informal settings outside of educational institutions (often on-the-job training) in many developing countries, educational technocrats and politicians who commonly equate skill formation with classroom learning tend to neglect it. Recognizing the importance of informal sources of learning for skill formation can lead to policies that foster skill in a different way.

There are two main features regarding the child development and lifelong education in terms of the formation of skills (see Cunha and Heckman, 2007). First, skills obtained in the early childhood augment the skills learned in the later stages of life. This effect is known as *self-productivity*, and denotes that the skills acquired in one period persist in the future. For instance, the ability to pay attention, a skill learned in the short run, ultimately fosters more vigorous learning of cognitive skills in the long run. Second, the skills acquired at one stage raise the returns of the further skill formation in the subsequent periods. This feature is known as *dynamic complementarity* since the level of skills in a situation of multistage technologies innovation, the levels of skill investment at different periods reinforce each other. This suggests that early investments should be followed by later investments in order to obtain the optimal gains from the early investment.[5] Together, both features denote that skill formation is a lifelong process, and the acquisition of skills is linked further to accessions of other skills.

From a social perspective, self-productivity and dynamic complementarity together suggest that skill formation at later ages is subject to an efficiency–equity trade-off; that is, investments in higher levels of education will produce greater returns if basic education has prepared students for demanding tasks. While disadvantaged and poor individuals may benefit from public expenditures in higher education, the returns from such investments are lower relative to those expenditures spent on high-quality basic education through younger individuals. In deciding whether the state should subsidize later skill formation for poorer households, one has to weigh the gains from this subsidy relative to the gains that can be obtained from an early skill formation subsidy. In other words, disadvantaged individuals who were not able to receive primary schooling are unlikely to acquire productive skills at a later age if they failed to possess basic skills in the early stages of their lives. In effect, because a strong foundation in the early age is necessary for further skill accumulation, human capital investment in the early stages of life is deemed more productive than later age investments. This then calls for greater subsidies in primary education for children.

More importantly, the conceptual framework presented here highlights the crucial role of household decisions in determining the causes and consequences of forming a particular skill. The family can decide to limit skill formation for a number of reasons. For one, the cost of schooling may be too high. While the primary and secondary schooling are supposed to be free in many developing countries, there are other incidental costs such as allowances, transportation and school uniform that may make schooling too expensive. However, a more compelling reason is that returns to education within a particular setting may be too low relative to opportunity costs of skill formation. In other words, future (net) returns of education are viewed to be lower than the wage that is offered in the labour market.

Similar to education, agricultural, labour and trade policies are also determined by conditions affecting the general economy. Policies refer to the government's active response to the main issues and problems that affect the country, thereby possibly enhancing or hindering growth.

However, these are influenced also by prices, incomes, social preferences and other economic conditions, which policymakers use as basis to address issues under various resource constraints. Moreover, the institutions that aim to reduce inefficiencies in the economy or transactions costs shape policies themselves. Policies can point particularly to control the individuals and organizations, which include private orderings, private litigation, regulation and state ownership, and thus can be viewed as points on the institutional possibility frontier, ranked in terms of increasing state powers. In the end, these determine the ability of countries to accumulate, innovate, adopt new technologies and reorganize in the face of technological change (Helpman, 2004).

North (1990) examined the contribution of institutional developments to economic growth throughout history. Using a game analogy to illustrate the difference between institutions and organizations, he defined institutions as the rules of the game, and organizations as the players. The players can consist of groups that are bound by a common interest, such as economic, social, political or educational. The evolving role of these players is influenced by the rules that govern the game. But the players and the conditions they are faced with, in turn, affect the evolution of the rules. Grief (1993) offers a comprehensive view of institutions. He argues for a context-specific analysis of institutions where multiple equilibria are often possible in the formation of institutions. This means that more than one set of institutions can emerge in given circumstances. Examining the context of a given policy helps to clarify why particular institutions and the policies emerged in a particular historical setting and how they could be sustained, because historical path dependence is an important element in the evolution of institutions. In Figure 1.1, it is important to indicate that, as the consequences move from the short run to the long run, the lesser control the households have in the final outcomes. Moreover, as household decisions are aggregated in the long run, the policies and the institutions evolve as a matter of course, and these in turn affect the decisions of the individual households.

While various policies are made by the government, agriculture and trade are crucial to the success of the education reform. Both lead to sector shifts, which require households to form skills in order to develop new forms of production and income generation. In the case of the former, non-farm rural and urban activities become more viable as the surplus generated from agricultural policies allow households to consider other options (see Foster and Rosenzweig, 2008). Technical skills that are more suitable to taking jobs in high-productivity industries may be necessary. In the case of the latter, the consequent changes in the terms of trade cause countries to shift towards sectors where they have either comparative advantage or increasing returns (Helpman and Krugman, 1987). This requires basic skills that allow the workers to be more flexible in the face of changing opportunities.

Another key outcome in this framework is technological innovation. Under autonomy, technological innovation is expected to stagnate as the country's options diminish. In agriculture, for instance, Evenson (2009) points to a technological periodicity without recharge. Faced with a number of technological possibilities, countries under such conditions will undertake a process of selection that is sequence dependent and works generationally with each technological state undergoing an evaluation before the next generational strategy is chosen. As this process develops, given the country's level of skill, certain options will be rejected by households while other potential innovations will prove unpromising. This process follows the theory of induced innovation where the optimal technology is a function of prices, household characteristics and community conditions (see Hayami and Ruttan, 1970). In the case where the R&D process inherent in technical change contain no recharge, technology will grind to a halt before the optimal level of technological determination is reached, thereby limiting the consequences of induced technology.

In contrast, countries that are able to open up their economies to trade are able to experience episodes of technological progress with recharge. Evenson (2009) sees this recharge mechanism in the form of inventions or agricultural commodities that can serve as further parents of recharged invention. In this case, technological determination itself moves, making it parallel to the optimal level of technology. This highlights greatly the value of induced technology.

The framework also considers the important role of global factors, which may consist of terms of trade, as well as technology adoption, development aid. Obviously, such factors should have direct positive effects on national production as these factors not only can change the production possibilities of the country but also can cause certain difficulties if the capacity of the country is unable to take advantage of opportunities that the international market has to offer. For instance, many technologies used by less developed countries (LDCs/the South) are imported from more advanced countries (the North). These technologies are designed to make optimal use of the prevailing factors and conditions in these richer countries (directed technical change). To the extent that these conditions are different in the South, these technologies may not find workers in LDCs suitable, hence causing labour mismatch (Acemoglu and Zilibotti, 2001). It is the function, then, of the government to establish the policies and institutions to mitigate the mismatch and to make sure that technology imported is eventually adapted to local conditions. This will then require new skills and ultimately education content.

Hence, if trade policy then is done properly, this causes a shift in the production possibility frontier, and eventually leads the country to move to a higher level of efficiency. Evenson and Singh (1997) investigated, within endogenous theory framework, the contribution of international spillover R&D capital, in combination with domestic investment and workforce to productivity growth in Asian countries. This suggests that the countries can break away from their low level of efficiency by incorporating greater international technological spillovers into their agricultural and trade policies. Faster growth of international technological spillovers in both agricultural and non-agricultural sectors would be conducive to generate higher-productivity growth in the developing countries. In effect, external factors not only affect the production but policies as well.

In light of this literature review, there are several conditions that need to be established for education and skill training reforms to generate successfully greater and more productive skills—and hence higher social returns—in the long run. First, new skill training enables students to understand and use new technology. If improved technological conditions (especially in agriculture) are not adopted, there are limited returns for higher education. Further as technological improvement eases households' current budget constraints, as the new technology is expected to lead to higher incomes, parents will have more resources at their disposal to invest in their children. On the other hand, a low demand for skilled labour is actually a manifestation of limited technological innovation (see Lanzona, 2009, for the case of the Philippines). The adoption of new technology depends on the presence of R&D, and specialized skills, and unless technology improves, returns to schooling will remain low (Schultz,

1975). The establishment of research and development induces innovation, thereby ensuring the continued demand for quality education.

Second, at lower levels of development, basic education should take precedence over higher levels of skill formation. Educational policy is based on a fundamental belief about the possibility of human development. However, as stated above, the dynamic complementarity of human skill formation implies that compensation for deficient early family environments at a later age is very costly (Cunha and Heckman, 2007).[6] One option available to the disadvantaged adults is technical and vocational education and training (TVET). There are many reasons for the lack of skills, but many European and South East Asian societies have implemented higher-level vocational and technical training, which has been a key element of education policies. This programme led to skilled workers who satisfied the 'manpower demand' of economic growth and modern industrial production and, in the last two decades, helped households to cope with problems of labour market integration and youth unemployment. However, this will require inputs from other schooling levels, including high-quality management schools and substantial participation from the private sector (see Lanzona, 2009, for the case of the Philippines).

Third, improved access to education should be reinforced by other forms of policies that foster skill formation. Search theory points to a number of policies that can lead to unproductive uses of skill that could destroy human capital. In particular, the programmes and policies that favour unskilled workers can cause certain distortions in the labour market and induce skilled workers to search further for better options (Diamond, 1981; Mortensen and Pissiardes, 1994). In effect, such policies make it less profitable to invest in skills (see Shimer, 2007). For European countries, with highly developed welfare state arrangements, high taxation, accompanied by generous social benefits to the unemployed and strong labour market regulations (such as minimum wages for to the semi-skilled and exit restrictions) reduce labour force participation rates, hours worked and employment and thereby lower the utilization rates of human capital (Heckman and Jacobs, 2009). These policies make the earnings of the workers not truly reflective of worker productivity. In addition, there are often weak economic incentives to maintain skills through on-the-job training. Insurance schemes for disability, unemployment and sickness may create important moral hazard problems. The negative effects of such programmes can be offset by policies that encourage greater productivity, such as promotions based on productivity, productivity bonuses and on-the-job training. In the education sector, Hanushek (2003) insisted that disincentives even within the schools themselves could reduce quality of education. In particular, if wages of teachers remain constant, the large variance in teacher quality is an indication that positive incentives have not been forthcoming to good teachers.

Because of emerging skill-based technical changes and heightened globalization of production activities, the prospects of unskilled workers are being threatened, especially in poor economies (see Katz and Autor, 1999). As the relative demand for unskilled workers decreases, increased pressure is placed on the government to protect workers with low skills through labour market regulations or minimum wages. Unfortunately, certain policies that bring about labour market segmentation can increase unemployment or induce skilled workers to enter the informal sector, and creating programmes such unemployment insurance intended to support to unemployed may only reduce total labour demand (see Zenou, 2008, for a developing country setting). Certainly, new forms of protection that are more efficient and cause fewer disincentives will be needed.

Fourth, improved access to schooling should be complemented by greater access to funds and capital, especially at the local and community level. As the productivity of labour increases, the productivity of capital also improves. Without the increase in physical capital stock, labour productivity eventually leads to diminishing returns. Labour needs to be complemented with physical capital so as to increase labour productivity (and wages eventually), and capital stock can only be increased with available funds (at the local or community level) and new technology. Further, once labour productivity increases, further capital increase can be justified.

Nevertheless, it is important to note that capital accumulation should not be used to substitute for long-term increases in labour productivity. Capital should be seen as complementary to labour inputs since skill formation is always grounded on improving the returns to human capital. The problem in this case is not so much the labour markets, but the imperfect credit markets that are unable to evaluate the returns to education, especially for poor households. Because of imperfect credit markets, as often the case in most developing countries, a skills gap, defined as the difference between the skills demanded by firms and the skills supplied by the worker, can ensue. Hence, a redistributive programme is justifiable because it helps to correct these market imperfections (Bardhan and Udry, 1999). Given more access to credit, the households could have properly allocated their resources to provide their children better schooling opportunities and their skills.

Fifth, the conceptual framework suggests that the cross-country relationships between educational policy and the economic activity are affected by substantial heterogeneity in education and labour market institutions. Such heterogeneity in turn can be attributed to the diverse domestic institutional structures. The institutional environment is determined by the legal and administrative framework within which individuals, firms and governments interact to generate income and wealth in the economy. The importance of a solid institutional environment becomes even more important if it addresses effectively the needs of the poor households, especially in periods of financial or food crises. Government attitudes towards markets and freedoms, and the efficiency of its operations, possible sources of frictions that reduce the efficacy of reform: excessive bureaucracy and red tape, overregulation, corruption, dishonesty in dealing with public contracts, lack of transparency and trustworthiness, and the political dependence of the judicial system impose significant economic costs to businesses and hinder the process of economic development.

Another point is that institutions associated with them are path dependent so that what occurred before determines the present and future reforms. Unless a significant break from the past is made, the reforms may have limited effects on skill formation. As already discussed, technological innovation combined with sound policies can be one way of moving away from this low-level equilibrium. Another way of 'breaking away' from weak institutions is to enforce public accountability. This involves the state's ability to reconcile the conflicting demands on education. Public accountability refers to mechanisms through which different groups can voice their concerns and a process by which these concerns are taken into account. According to the World Bank (2008), strengthening the voice of citizens can be achieved in two ways: the first involves improvements in the electoral process itself, which can be complex and time-consuming. The second calls for measures to make public institutions/politicians more accountable to citizens, for example, through carefully designed decentralization, or by making information about resource allocation and education outcomes available to the public. Holding policymakers accountable to citizens has the potential of improving the distribution of education among the population. It also has the potential of bringing about a more rational allocation of resources. Further, it can be an important vehicle for ensuring that education is serving the broader objectives of society.

Given that countries are presently at different stages of development, policies that will work in one country will not necessarily be the same for other countries. In other words, the key factors mentioned earlier will have differential effects on the success of the policies across different countries. For instance, capital build-up may be less productive for a country with poor governance than a country with institutions that are moving beyond factor accumulation and towards technological advancement.

The crucial point is that complementarity in economic programmes, both development and education, is a critical condition for improving social returns of education. Only a coordinated effort that can bring both microeconomic and macroeconomic elements together ensures high returns in education. An interesting case is that of Taiwan's economic development which was spurred by education (Lin, 2003). Since the 1960s, Taiwan has been transformed into one of the few newly industrializing countries in the world. More specifically the contribution of manufacturing to growth was significant as this sector has shown the largest percentage distribution of capital investment among all activities since the 1960s, averaging almost 29 per cent, and 33 per cent in the 1970s. The investment rate was only 14 per cent in 1951 but has risen rapidly since the 1960s, averaging 22 per cent, and 30 per cent in the 1970s. Hence, Taiwan's created growth was sustained in the 1960s, 1970s and 1980s.

What is important here is that during this period, education in Taiwan also expanded correspondingly so that several researches have noted the correlation between the rapid economic growth, globalization and increased education (Lin, 2003; Shen-Keng, 2001). The average number of years of formal education per person of employed people appears to have improved over time, averaging 7.18 years in the 1960s, 7.85 years in the 1970s, 9.28 years in the 1980s and 10.7 years in the 1990s. Thus, the quality of the labour force has also improved. In 1964, the population of workers with a college degree or above was only 3 per cent. The largest population was people who completed a primary school education only (82 per cent). However, the population of workers with a senior high school and college degree has risen steadily over time. By 2000, more than 60 per cent of workers had at least a senior high school degree, and nearly 27 per cent had been awarded a college degree.

It should be pointed out that this was achieved mainly through strong education reforms that affected the households. The Taiwan government first required nine years of mandatory education for boys and girls in 1968 so that the number of graduates from junior high school increased significantly in the 1970s. Spohr (2003) shows that despite the uniformity of intervention, this had affected females more than the males, suggesting shifts in intra-household allocation of resources. The government later introduced a science and technology development programme in 1979 and focused on high-level technologies. Thus, the population of engineering and science majors has grown more rapidly than other majors. Hence, human capital has improved since the 1960s, which in turn sustained the growth of the manufacturing sector.

In direct contrast is the conditional cash transfer programme of various countries that is expected to increase human capital formation in the long run. Programmes like Bolsa Família in Brazil or the Pantawid Pamilyang Pilipino Program in the Philippines have been effective both in increasing school attendance and decreasing dropout rates. However, as in the case of Brazil, the decrease in dropouts has had an unfortunate side effect: it has led to more children falling behind in school (Soares et al., 2007). Such findings confirm that the programme, as a demand-side intervention, is not able, on its own, to have a positive impact on some education outcomes. Namely, it would not necessarily enable disadvantaged children to break the intergenerational transmission of poverty if educational policies did not concomitantly improve the performance of such children while in school. This problem underscores the need to introduce complementary programmes aimed at improving educational quality or providing special attention for underachieving children with poorer parents. For instance, despite the presence of transfers, the role of water security, particularly safe drinking water and sanitation, in bringing about the desired socio-economic changes at the household level, which would increase the opportunity cost of lack of schooling, need to be considered.

In summary, the conceptual framework's main message is the shift towards the household and individual welfare, and away from national economies, as the main objects of development. In the first two decades after World War II, development planners looked to investments in industry and heavy infrastructure as the primary engine of growth, assuming that household and individual welfare would naturally follow. As economic growth proved slow for several countries recently, development planners looked for new ways and structures to define development in terms of the welfare of household and individual members. In this evolving discourse, education has become central in the enhancement of household and individual welfare.

Specifically, the household's decision to send children to school (the formation of skill) is influenced by various factors at the macro- (national), meso- (local and community) and micro- (household) levels. Yet, given the poor technology, inadequate reforms in education and limited resources, the household's decisions themselves can bring about consequences that aggravate this situation. In effect, educational reform and economic development are occurring and reinforcing one another simultaneously, neither one causing the other nor one preceding the other. For the macro-determinants, both are affected by weak economic policies and programmes, resulting in low skills and low employment levels and incomes (and high underemployment), eventually in extreme poverty situations.

The conceptual framework also identified institutional elements that concurrently affect educational reform and economic development. For instance, a weak institutional set-up represented by poor accountability restricts both education and economic performance. With better institutions, social service delivery and safety nets could have cushioned certain sectors of the population from economic risks and vulnerabilities. In addition, the failure to establish strong institutions increases poverty, which in turn raise the probability of lower skills and eventually skill gaps. The inadequacies of the educational facilities and the lack of credit markets also affect the decision to provide children with more skill formation. Most of these macro-factors also operate at the local and community levels and affect the supply of skilled labour.

Furthermore, based on the framework, in order to increase the demand for higher learning and improve the economy's chances for growth, at any given stage of institutional and educational development, three overall factors are necessary. The first are the economic resources, which include not only financial inputs but also capital formation, as well as specialized forms of learning such as TVET. These should ease the constraints brought about not only by lack of income but also by limited preparation needed for the acquisition of further skills. The second are the labour, agricultural and trade policies that should foster and encourage the formation of skills. Unless skilled workers are given adequate returns for their investment, the demand for skills will remain low. Third, innovation is particularly important for economies as skills are needed for extending the frontiers of knowledge through research and development institutes and for integrating and adapting to exogenous technologies, as well as creating new forms of production. As already stated, in environments where innovation is not changing, there is limited demand for skills.

Empirical Model and Findings

In order to substantiate the above framework, it is necessary to empirically define the 'right' educational policy. Given that the content and the provision of education are primarily based on culture, there are different ways of defining the right education policy. Moreover, there are several qualities in the educational system that are deemed to be important, but are not necessarily related to growth. Because of the increasing globalization and competitive pressure, education systems are now required to be more receptive to countries' needs for development. Otherwise, the country will experience limited opportunities for growth and its ability to sustain human capital investments will be severely constrained.

In the absence of microeconomic data, the use of data from the Global Competitiveness Report is proposed to define the school quality for cross-country comparisons. Since 2005, the WEF has surveyed businessmen and economists around the world and derived a global competitiveness index (GCI), a highly comprehensive index for measuring national competitiveness, which captures the microeconomic and macroeconomic foundations of national competitiveness. For this index, competitiveness is defined as the set of institutions, policies and factors that determine the level of productivity of a country. According to the WEF 2009–2010 Report, productivity sets the sustainable level of prosperity that can be earned by an economy. In other words, more competitive economies tend to be able to produce higher levels of income and welfare for their citizens.

The productivity level also determines the rates of return obtained by investments (physical, human and technological) in an economy. Because the rates of return are the fundamental drivers of the growth rates of the economy, a more competitive economy is one that is likely to grow faster in the medium to long run.

This study employs the parts of the WEF Report that pertains to education. Indeed, while there are many reasons why individuals invest in schooling, the main issue here is to determine the society's overall demand for education. To the extent that competitiveness strengthens institutions to sustain growth and creates knowledge spillovers (coming from learning-by-doing) that further increases growth, the information in the Report concerning perceived education quality more than captures private returns but also the success of the reforms undertaken. If the quality of the education is also high, then this implies that the demand for this is also high. Moreover, because these social externalities are largely unobserved, survey data such as the one presented in the WEF Report would seem the next best option.[7]

For the 2009 report, the survey has expanded its scope of completion, achieving this year a record sample of over 12,600 respondents from 133 countries between January and May 2009. For the 2010 report, the survey respondents increased to 13,607 in 139 countries. Among the indicators used in defining this index is the quality of education based on the responses of the businessmen. Higher-quality education and training is considered one of the pillars of productivity, and is seen to be crucial for economies that want to move up the value chain beyond simple production processes and products. The objective of the survey is to determine how economies are able to foster pools of well-educated workers who are able to adapt rapidly to their changing environment. This pillar measures secondary and tertiary enrolment rates as well as the quality of education as assessed by the business community. The extent of staff training is also taken into consideration because of the importance of vocational and continuous on-the-job training—which is neglected in many economies—for ensuring a constant upgrading of workers' skills to the changing needs of the evolving economy.

Businessmen and economists were asked internationally to assess how well does the educational system meets the needs of a competitive economy. The answers range from 1 (which denotes 'not well') to 7 ('very well'). At the same time, a similar question was asked regarding their personal assessment about the quality of the country's primary schools based on how this is able to raise national productivity. Here the answers range from 1 denoting 'poor' to 7, which indicates 'excellence', being among the best in the world.

The education system is made up not only by formal education but also by changes in the content of education and the way this content is presented. The empirical strategy of this chapter is to use the 2009 and 2010 survey results of the WEF to determine correlates for the qualities of education system as well as the determinants of successful reforms, particularly in order to control for the heterogeneity found across several countries. A set of factors that can be used to build up policies for country's education system can be formulated. This set of correlates and possible explanatory variables will be based on the framework discussed in the previous section and the data that is available.

To do this, one can relate these assessments to the per capita gross domestic product (GDPPC) of the countries in order to evaluate the importance of primary education in relation to overall education in affecting productivity. I first cluster the different countries into quartiles, starting with those lower GDP per capita countries. Table 1.1 shows a data summary for these quartiles (of roughly equal sizes). Note that the countries

Table 1.1:
Data summary of international GDP per capita by quartile and by year

Quartile	N	Mean	Standard deviation	Minimum value	Maximum value	Coefficient of variation
A. 2009						
First	33	759.1394	353.074	54.6	1,479.8	0.47
Second	34	3,794.947	1,371.019	1,736.5	6,509	0.36
Third	33	13,256.65	4,894.802	6,579.9	22,595	0.37
Fourth	33	48,409.83	20,143.41	22,997.4	1,13,044	0.42
B. 2010						
First	35	755.9143	327.5664	163	1,560	0.43
Second	35	3,468.257	1,079.357	1,724	5,809	0.31
Third	35	10,821.14	3,902.838	5,824	18,557	0.36
Fourth	34	41,732.06	17,692.19	19,111	1,04,512	0.42

Source: International Monetary Fund, *World Economic Outlook Database*; National sources.

in the upper one-fourth of the quartile have an average GDPPC that is 62 and 55 times higher than the lower one-fourth quartile in 2009 and 2010, respectively. It is also 16 times greater than the second quartile in both years. This suggests extreme income inequality across different clusters and within clusters as seen by coefficient of variation.

Figure 1.2 shows the relationship between the school assessments and the GDPPC quartiles. Four interrelated measures of schooling quality are introduced into the analysis. Apart from the quality of primary schooling and the educational system, the quality of math and science education and the extent of high-quality and specialized training (akin to technical and vocational education and on-the-job training) are also considered. Given the assumption that all countries are engaged in some form of education reform, the higher the assessments for these measures, the more successful these reforms are seen to be. Based on Figures 1.2 and Table 1.2, the technical knowledge and experience provided by math and science education and vocational training is closely associated with the assessments of primary and higher education. This suggests that part of the improvements of the educational system will necessitate the development of a solid math and science curriculum and a highly responsive training programme that can offer specific inputs to the technical demands of the economy.

Figure 1.2:
Average school assessment by level and GDP per capita quartiles

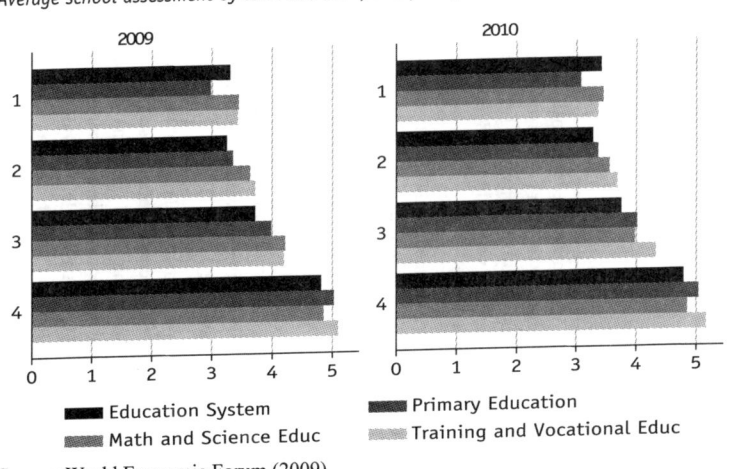

Source: World Economic Forum (2009).

Table 1.2:
Means and standard deviation of correlates of school quality by GDP per capita and year

Variable	Definition	2009				2010			
		First quartile	Second quartile	Third quartile	Fourth quartile	First quartile	Second quartile	Third quartile	Fourth quartile
Higher education Quality	Assessment of educational system in the country in order to meet the needs of a competitive economy? [1 = not well at all; 7 = very well]	2008–09 weighted average	3.31 (0.57)	3.25 (0.75)	3.72 (0.70)	4.81 (0.89)	3.41 (0.62)	3.27 (0.65)	3.75 (0.78)	4.79 (0.84)
Primary education quality	Assessment of the quality of primary schools in your country? [1 = poor; 7 = excellent—among the best in the world]	2008–09 weighted average	2.98 (0.53)	3.34 (0.84)	3.99 (0.91)	5.03 (0.86)	3.07 (0.54)	3.35 (0.86)	4.01 (0.95)	5.04 (0.81)
Math	Assessment on the quality of math and science education in your country's schools? [1 = poor; 7 = excellent—among the best in the world]	2008–09 weighted average	3.44 (0.58)	3.64 (0.92)	4.22 (0.85)	4.85 (0.85)	3.44 (0.57)	3.55 (0.93)	3.98 (1.14)	4.85 (0.83)
q_traing	Assessment of the extent that high-quality, specialized training services (referring to technical and vocational education and continuous on-the-job training) available [1 = not available; 7 = widely available]	2008–09 weighted average	3.43 (0.51)	3.72 (0.61)	4.19 (0.53)	5.10 (0.78)	3.35 (0.57)	3.68 (0.60)	4.32 (0.61)	5.16 (0.78)

	Description								
Primary enrolment rate	Net primary education enrolment rate | 2008. The reported value corresponds to the ratio of children of official school age (as defined by the national education system) who are enrolled in school to the population of the corresponding official school age	79.92 (13.09)	92.42 (4.71)	91.16 (8.76)	96.36 (3.21)	82.62 (12.46)	92.02 (5.44)	93.11 (6.10)	96.60 (2.94)
Secondary enrolment rate	Gross secondary education enrolment rate | 2008. The reported value corresponds to the ratio of total secondary enrolment, regardless of age, to the population of the age group that officially corresponds to the secondary education level	41.37 (21.67)	80.65 (11.41)	91.16 (8.76)	104.56 (12.16)	43.23 (21.12)	77.84 (16.40)	90.23 (8.05)	103.86 (12.30)
Tertiary enrolment rate	Gross tertiary education enrolment rate | 2008. The reported value corresponds to the ratio of total tertiary enrolment, regardless of age, to the population of the age group that officially corresponds to the tertiary education level	7.29 (7.93)	28.69 (13.67)	48.99 (21.10)	56.66 (22.97)	9.69 (12.64)	27.97 (14.61)	50.83 (21.45)	56.16 (22.91)

Source: World Economic Forum (2009).
Note: Figures in parentheses are standard deviation.

Note also that the schooling assessments are significantly correlated with production as measured by the GDPPC. However, it seems that the quality of primary school is more highly associated with increased production. This is particularly in the case of the richer countries where the quality of the primary education is slightly higher than the overall education system.[8] This seems to show that while overall school quality for the poorer countries is related to higher production, its effect may not be as significant as primary education. In contrast, the figure indicates that the quality of primary schools is the factor that seems to be more associated with richer countries and increasing production. This implies that, for the same inputs, investments in basic education have a greater effect on productivity than higher education.

Figure 1.3 shows for the two periods the higher enrolment may ultimately be education quality. To the extent that schooling assessment is positively related to GDPPC, the observed positive relationship between enrolments and GDPPC indicates a similar relationship between enrolment and schooling assessment. This may be verified in Table 1.2 where it can be noted that improvements in schooling assessments are invariably correlated with increases in enrolments at all levels. This implies

Figure 1.3:
Enrollment rates by schooling level and GDP per capita quartiles

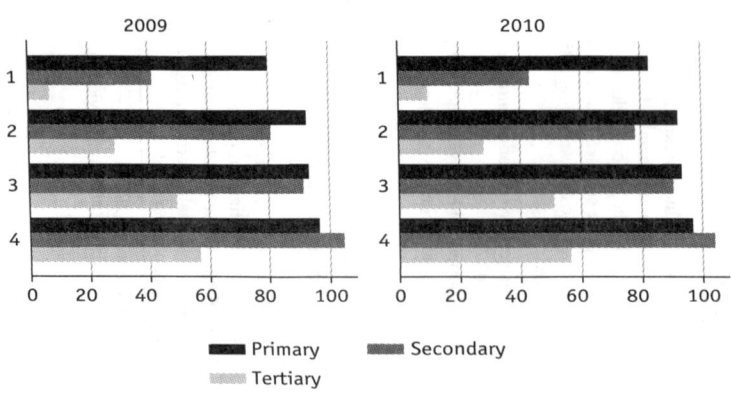

Source: World Economic Forum (2009).
Note: The reported value corresponds to the ratio of total enrolment, regardless of age, to the population of the age group that officially corresponds to the respective education level.

that the overall demand for education (as measured by the enrolment rates) increases with better-quality schooling. Students expect to receive greater incomes if they receive better-quality education, and so the more able students are likely to enrol in schools that provide better education.

Figure 1.4 shows box plots of the results of the two surveys on educational quality. For each group, the box represents the middle half of the assessment (score) distribution lying between the first and third quartiles. The horizontal line near the middle of each box represents the median score of the schools, and the notches represent the range of each score. There is a clear tendency for the scores to rise with GDP per capita, but one can discern other features from the plot. There is a tendency for the dispersion, as measured by the interquartile range of the score, to increase with GDP per capita. This is also accentuated by the upper and lower tails of the score distribution. This implies that the impact of GDPPC on the assessments would likely differ as the former increases as improved economic activity offers more options in the way education may be delivered. In effect, there exist both country-specific and quartile-specific

Figure 1.4:
Assessment of school quality by type and by GDP per capita quartile and by year

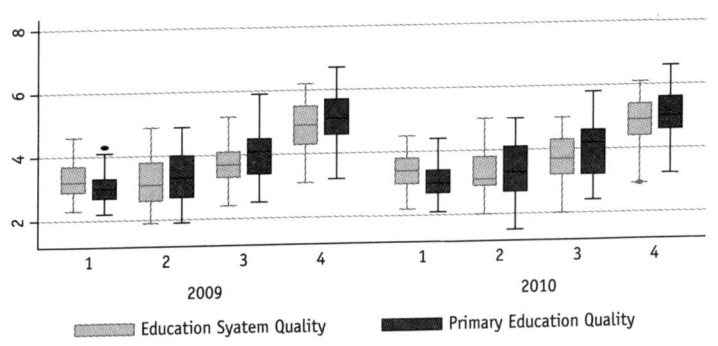

Source: World Economic Forum (2009).
Note: The box plots provide a summary of the distribution of school assessment for the ranked GDP per capita groupings. The upper and lower limits of the boxes represent the first and third quartiles of the assessment. The median for each group is represented by the horizontal line in the middle of each box. The endpoint of each line (whiskers) from the boxes shows either the distance of 1.5 times between the first and third quartiles, or the maximum or minimum values if such values are below this distance. Points either above or below these whiskers are considered outliers.

Figure 1.5:

Assessment of math and science education and training and vocational education by GDP per capita quartile and by year

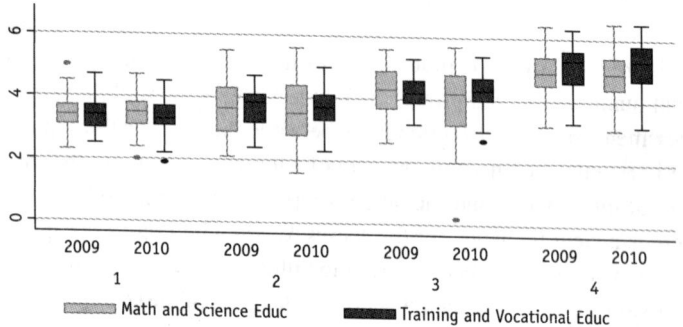

Source: World Economic Forum (2009).
Note: See notes in Figure 1.4.

factors that affect the quality of education. Moreover, given the different observations across time, each quartile-specific factor enters the situation for each period as a single draw. Hence, getting the effect of this variable on the quality assessment will not offer a complete picture than what can be offered by analyzing on the effects using panel data methods. Similar conclusions can be made for math and science education and training and vocational education, as seen in Figure 1.5. One can note also from these two figures that the dispersion across quartiles as well as time is observed for each of these measures of education quality.

The question is what contributes to improve schooling quality?[9] Equivalently, what determines the demand for higher education quality? Based on the data available, I consider core factors discussed in the previous section: the resources available, the labour market, agricultural and trade policies and the state of technology and innovation that foster skills. For resources, apart from a dummy variable to denote the country's associated GDPPC quartile (to capture quartile-specific effects), the percentage of government's budget devoted to agriculture will be used. Increases in these resources necessarily increase the demand for quality education, hence encouraging suppliers to increase its provision. For labour market policies, I use the wage and earnings policy (as this relate to true productivity of the worker) and the rigidity of unemployment as resulting from labour market institutions. If education is expected to increase productivity and improve the chances of employment, the more wages reflect the worker productivity and the less rigid is unemployment, the greater should be the demand for quality education as return for such investments is higher. For agricultural and trade policies, the WEF survey contain some measurement of the effectiveness or the degree of inconvenience of these policies to the population. For innovation and technology, the variables assessing the quality of research institutes, and the extent to which companies invest in R&D, are included in the analysis.

In summary, as indicated in the conceptual framework, these factors are ultimately processed by the households in determining the demand for quality education. Economic resources affect demand in particular because these influence both the financial and institutional constraints that household face in making decisions. Labour market, agricultural and trade policies are also significant since the returns to education are ultimately tied to the wages and the probability of obtaining a job. Finally, technological change is crucial in raising the possibilities of production, especially for under-developed economy. In order to measure the importance of technological innovation, I consider two specifications in the empirical model. The first will be examined without the technological factors, and the second with technological factors. If the effects of agricultural, labour and trade policies are affected with the inclusion of technology, then the value of these policies can be attributed partly to technology.

Table 1.3 shows the list of these variables, its definition, means and standard deviation for each of the GDPPC quartiles and year. It is likely that the relationship of these factors to quality assessments may differ given the significant differences between the means and standard deviations across the four income groups. For countries belonging in the lower GDPPC quartiles, for instance, an increase in the economic resources can possibly not increase quality of schooling if its institutional capacity to offer quality schools is severely impaired. Such institutional capacities are the unmeasured country-specific factors that can affect the success of the reforms.

In order to estimate the success of education reforms, I use a regression analysis, in particular a random effects approach for panel data analysis, in order to control for these unmeasured institutional factors (Greene, 2003). As opposed to the fixed effects model, this presumes that the heterogeneity is also randomly distributed across the cross-sectional

Table 1.3:
Means and standard deviations of explanatory variables for quality assessments by GDP per capita quartiles and year

Variable	Definition	2009				2010			
		First quartile	Second quartile	Third quartile	Fourth quartile	First quartile	Second quartile	Third quartile	Fourth quartile
Exppct	Public expenditures as percentage of GNI | 2006 and 2007. This represents current operating expenditures in education, including wages and salaries and excluding capital investments in buildings and equipment	3.72 (1.94)	3.81 (1.65)	4.66 (1.25)	5.13 (1.40)	3.85 (2.21)	3.93 (1.73)	4.51 (1.11)	4.93 (1.32)
Pay	Extent to which pay in the country is related to productivity [1 = not related to worker productivity; 7 = strongly related to worker productivity] | weighted average	3.65 (0.56)	3.82 (0.57)	4.08 (0.73)	4.36 (0.63)	3.59 (0.61)	3.75 (0.55)	4.01 (0.74)	4.27 (0.61)
Rigid	Rigidity of employment index on a 0–100 (worst) scale | 2009. This index is the average of three sub-indexes: Difficulty of hiring, rigidity of hours and difficulty of firing. The three sub-indexes have several components and all take values between 0 and 100, with higher values indicating more rigid regulation	35.39 (14.17)	35.12 (16.41)	37.23 (16.90)	27.34 (19.74)	30.37 (13.81)	30.17 (17.55)	29.70 (16.51)	22.52 (19.30)

Agri	Assessment of the agricultural policy in the country [1 = excessively burdensome for the economy; 7 = balances the interests of taxpayers, consumers and producers]	weighted average	3.70 (0.55)	3.91 (0.52)	3.82 (0.72)	4.22 (0.65)	3.71 (0.55)	3.88 (0.45)	3.83 (0.66)	4.16 (0.63)
Trade	Extent to which tariff and non-tariff barriers limit the ability of imported goods to compete in the domestic market [1 = strongly limit; 7 = do not limit]	weighted average	4.17 (0.45)	4.36 (0.56)	4.73 (0.73)	5.33 (0.65)	4.20 (0.42)	4.21 (0.54)	4.67 (0.71)	5.35 (0.63)
res_inst	Assessment of the quality of scientific research institutions in your country? [1 = very poor; 7 = the best in their field internationally]	weighted average	3.27 (0.56)	3.35 (0.70)	4.03 (0.55)	4.96 (0.91)	3.10 (0.59)	3.08 (0.69)	3.95 (0.71)	4.93 (0.93)
Invest	Extent to which companies in the country spend on R&D [1 = do not spend on R&D; 7 = spend heavily on R&D]	weighted average	2.77 (0.40)	2.85 (0.50)	3.20 (0.62)	4.31 (0.99)	2.79 (0.40)	2.78 (0.48)	3.21 (0.64)	4.20 (0.99)

Source: World Economic Forum (2009).
Note: Figures in parentheses are standard deviations.

units, not represented by parametric shifts of the regression analysis. For this to be measured however, the approach assumes that these country-specific effects are uncorrelated with the included regression-independent variables. This can be justified if the sampled countries are drawn from the large population. Also, the assumption is feasible since the number of independent variables is limited, and not reflective of the institutions existing in the countries. The drawback of limiting the number of independent results is possibility of inconsistent results, if the assumption of the non-correlation is incorrect.

Table 1.4 shows the results of the random effects estimates. The rows at the bottom of the table show four statistical tests that can support validity of the estimator. The first, overall R^2, indicates how much of the included variables in the model explains for the variation of the education quality measurements. The second, the Wald test, indicates the independent variables are jointly significant in measuring these indicators. The third, the Breusch–Pagan test, indicates the classical regression model (ordinary least squares, OLS) with a single constant term, is inappropriate for these data, thereby rejecting the null hypothesis about the variance of the unmeasured country-specific variables in favour of the random effects model. Finally, the Hausman test shows that except for the estimates on overall education quality, there are indeed individual effects, and suggests that these effects are uncorrelated with the other variables in the model. Overall, given a choice between fixed and random effects models, the latter model is the better choice.

Given these tests, the following points are noteworthy from the estimates on overall education quality. First, proportion of education expenditures to total national income is positively associated with the quality of higher education. The results show that placing more resources in education leads to higher quality as the economy develops and proper institutions are established. Second, from the first regression specification, countries belonging to a higher quartile, denoting greater resources, are seen to have a higher education quality and successful reforms. However, after incorporating the R&D factors for technology, this feature is no longer significant. This suggests that richer countries are able to increase the quality of their education only because there are better research institutes in these places, and the private sector is willing to invest more in R&D. Third, rigidity in employment decreases the score, and better payment

Table 1.4:
Random effects estimates of education quality under two specifications

	(1)				(2)			
	Education system	Primary education	Math and science	Training and vocational	Education system	Primary education	Math and science	Training and vocational
Exppct	0.080*	0.078*	0.037	0.012	0.054*	0.057*	0.013	-0.019
	(2.91)	(2.73)	(1.00)	(0.47)	(2.27)	(2.17)	(0.35)	(1.03)
Quart	0.235*	0.327*	0.241*	0.410*	0.046	0.182*	0.086	0.184*
	(4.85)	(6.50)	(3.63)	(8.82)	(1.00)	(3.59)	(1.21)	(4.92)
Pay	0.326*	0.369*	0.452*	0.173*	0.170*	0.247*	0.331*	-0.021
	(4.85)	(5.42)	(4.63)	(2.73)	(2.79)	(3.76)	(3.38)	(0.40)
Rigid	-0.008*	-0.007*	0.004	-0.004*	-0.008*	-0.007*	0.004	-0.002
	(4.19)	(3.84)	(1.34)	(1.98)	(4.75)	(4.08)	(1.31)	(1.61)
Agri	0.104	0.052	0.323*	0.007	0.074	0.026	0.265*	-0.019
	(1.51)	(0.75)	(3.16)	(0.10)	(1.24)	(0.41)	(2.69)	(0.37)
Trade	0.160*	0.200*	0.112	0.134*	0.081	0.134	0.056	0.040
	(2.43)	(3.00)	(1.15)	(2.09)	(1.40)	(2.16)	(0.59)	(0.79)
res_inst					0.179	0.153	0.294	0.380
					(2.87)	(2.29)	(2.82)	(6.88)
Invest					0.407	0.339	0.117	0.309
					(5.84)	(4.50)	(1.03)	(5.17)

(Table 1.4 Continued)

(Table 1.4 Continued)

	(1)				(2)			
	Education system	Primary education	Math and science	Training and vocational	Education system	Primary education	Math and science	Training and vocational
_cons	0.661	0.299	−0.492	1.848	0.302	−0.075	−0.551	1.345
	(1.72)	(0.75)	(0.91)	(4.97)	(0.91)	(0.21)	(1.06)	(4.96)
R^2 overall	0.4995	0.6082	0.3760	0.5664	0.6589	0.6740	0.4640	0.8329
Wald	143.48	171.14	100.76	151.36	297.20	269.15	134.57	626.54
Breusch–Pagan	109.88	104.65	82.76	101.95	106.80	103.59	77.45	81.62
Hausman	3.91	31.29	37.69	19.33	2.70	65.52	43.54	20.12
N	249	249	249	249	249	249	249	249

Source: Author.
Note: Figures in parenthesis are absolute z-values. * indicates significance at 5 per cent level.

structures that reflect the workers productivity increase it, indicating that poor labour policies that do not foster increased employment and do not provide incentives for productivity after forming skills are bound to decrease school quality. Fourth, more outward-looking trade policies are also expected to increase the demand for quality education. However, like the GDP per capita, this becomes insignificant in case where R&D variables are included. This suggests that trade policies help to encourage more quality education demand and spur success in education reforms if these are used to improve technological innovations. Finally, the absence of superior R&D institutions, and the low private investments in R&D will cause inferior schooling quality. Appropriate conditions are necessary to induce a country to undertake innovation, which increases demand for quality schooling.

The following points are significant in the assessment of primary education. First, the effect of education expenditure as a proportion of the country's total income is significant, suggesting that the effect of increased education budget in countries is crucial. Second, greater access to resources, measured by the GDPPC quartile, translates into higher quality of schooling in all countries. Unlike the case of overall education system, this does not diminish with the inclusion of R&D variables. Third, higher quality of primary schools is also reflected in the labour policy on wages and employment. The more wages reflected the worker's productivity and the less rigid are the employment practices, the higher are the primary schooling assessments. As skills are given a wage premium and as job entries become easier, the more the supply of quality primary education is demanded. Fourth, a less constrained trade policy leads to greater success in the primary school reforms. Like GDPPC, this does not change with the inclusion of the R&D variables. This implies the importance of basic skills to countries that are more inclined to open its markets to the world and are more willing to shift to more viable sectors, thus leading to greater success in the reforms. Finally, the primary schools rate more highly in the presence of high-quality scientific research institutions and greater private investments in these institutes. Better technology emanating from these institutes may have caused increased demand in the country for higher-quality primary education.

In the estimates of math and science education assessments, the following results are significant. First, greater resources as measures by the

GDPPC quartile are associated with higher scores of math and science education. However, when the R&D variables were included, this variable became insignificant. This suggests that resources are only important in improving math and sciences to the extent that these improve the quality of R&D institutes. Second, similar to the assessment of primary schools, the policy of tying wages to worker productivity creates a demand for a better quality of technical education, indicated by math and science education. Third, effective agricultural policies that look to the interests of more sectors in the country are associated with demand for superior math and science education. Technical skills engendered by math and science became more crucial as agricultural development becomes forthcoming, and households shift to rural non-farm activities. Fourth, better R&D seems to have also resulted in improved math and science instruction.

In the case of quality and specialized training for college as well primary school graduates, four points are important. First, better vocational education and training is associated with greater resources. This can suggest that richer countries are able to place a higher emphasis for technical training. Second, labour policies that highlight productivity and less rigidity in employment are also seen to draw better assessments in training. However, these became insignificant with the inclusion of the R&D and technology variables. This suggests a correlation between labour policies and technology, so that the labour policies are effective in affecting training quality only if technology is improved. Third, trade policies lead to a higher demand for better training schools. Finally, the presence of internationally competitive research institutes has also resulted in better-quality training. These results point to the vital role of the R&D and technology in improving vocational education and training and eventually higher education quality.

Summary and Proposed International Responses

In summary, the results seem to indicate the importance of primary education even as the need for higher education becomes evident. As certain sectors become more competitive, and new industries (as opposed to agriculture) are formed with the process of globalization and the subsequent transfer of technology, skill gaps in the manufacturing or service sectors, requiring greater government investments in tertiary and higher education. However, it is necessary that the demand for such programmes on the part of the workers exists, or equivalently the demand for highly skilled workers in the face of technological change should be present. At the same time, labour market policies should offer greater wage premiums to skilled workers in the face of skill-biased technological change. Moreover, expenditures towards improvements in either primary, middle or higher education should be accompanied by proper policies as well. Investments that are not seen as viable and stable programmes will not produce the desired results.

The international community has committed to support these educational reforms. To the extent that these programmes help in achieving Millennium Development Goals (MDGs) and increasing the resources of countries for education quality, the international agencies, particularly the United Nations, are expected to support this overall effort. Efforts to improve higher education should then be aligned with reducing poverty, and not merely to help the countries adjust towards newer technologies.

International development organizations can support or implement several government activities that aim to promote better technology. The crucial role of international agencies is to help introduce new processes and innovations that allow households to have a newer perspective. This is where R&D becomes crucial, a point that Bob Evenson has underlined. While some research programmes are poorly managed and resource constrained, there are various opportunities afforded by better researches and extension programmes. The data used here seem to indicate that policies are more effective if combined with technological innovation. Partial and isolated reforms may lead to some improvements in economic and social performance, but integrated and more comprehensive reforms are likely to be more successful.

Notes

1. Private returns are higher than 'social' returns where the latter is defined on the basis of private and social benefits but relative to total (private plus external) costs. This is because of the public subsidies for education and the fact that typical social rate of return seems to reflect limited social benefits. Part of the issue can be attributed to the definition of social returns to education.

2. This presumes that transaction costs resulting from existing institutions remain constant. Lanzona and Evenson (1997) point out the inefficiencies resulting from the transactions costs, particularly in the labour market.
3. Household preferences and decisions are interrelated and hence represented in the same boxes and in the same way that macro- and meso-factors are also interconnected.
4. In developing countries, education and schools are generally administered by national governments, which have considerable authority over such crucial issues as school facilities, pedagogy, language of instruction, textbooks and other materials and the training of teachers, among many other factors.
5. The term 'complementarity' is used here since both present and future learning are enhanced by past skills. At the same time, the returns from past skill investments are increased with additional investments in further education.
6. It needs to be mentioned that not all labour should be skilled. As the number of skilled workers increases and as the number of unskilled workers declines, the wages of unskilled workers can increase. This may lead to welfare gains for the unskilled workers, as indicated by the experiences of certain countries engaged in globalization (e.g. Lanzona, 2001, in the case of the Philippines).
7. One needs to be conscious however that, as indicated in the framework, the evaluation of the full effects of education still needs to be measured at the household level. Instead of looking at the assessments made by businessmen and economists, the more accurate valuation of education is one based on actual behavioural changes and the distribution of these effects across income classes.
8. Both nonetheless are significant in affecting production. The point here is that production has a greater correlation with primary education than with higher education.
9. The WEF survey on the education system quality pertains not only to management schools but also to high schools, colleges and universities.

References

Acemoglu, D. and F. Zilibotti. 2001. 'Productivity differences', *The Quarterly Journal of Economics*, 116: 563–606.

Bardhan, P. and C. Udry. 1999. *Development Microeconomics*. Oxford: Oxford University Press.

Becker, G. and N. Tomes. 1986. 'Human capital and the rise and fall of families', *Journal of Labor Economics*, 4(3, Part 2): S1–S39.

Cunha, F. and J. Heckman. 2007. 'The technology of skill formation', *American Economic Review*, 97: 31–47.

Diamond, P. 1981. 'Mobility costs, frictional unemployment, and efficiency', *Journal of Political Economy*, 89(4): 798–812.

Evenson, R. 2009. 'Research as Search: Evenson-Kislev and the Green Revolution', in Kimhi, A. and I. Finkelshtain (eds), *The Economics of Natural and Human Resources in Agriculture*. Hauppauge, NY: Nova Science Publishers, Inc.

Evenson, R. and P. Pingali. 2007. 'Overview', in Evenson, R. and P. Pingali (eds), *Handbook of Agricultural Economics, Vol. 3*. Amsterdam: North-Holland, pp. 2253–78.

Evenson, R. and L. Singh. 1997. 'Economic Growth, International Technological Spillovers and Public Policy: Theory and Empirical Evidence from Asia', Yale University Economic Growth Center, Discussion Paper no. 777, New Haven, CT.

Fasih, T. 2008. *Linking Education Policy to Labor Market Outcomes*. Washington, D.C.: World Bank.

Foster, A. and M. Rosenzweig. 2008. 'Rural non-farm employment generation and rural out- migration in South Asia', in T.P. Schultz and J. Strauss (eds), *Handbook of Development Economics, Vol. 4*. Amsterdam: North-Holland.

Greene, W. 2003. *Econometric Analysis,* Fifth Edition. Upper Saddle River, NJ: Prentice-Hall.

Hayami, Y. and V. Ruttan. 1970. 'Factor prices and technical change in agricultural development: The United States and Japan, 1880–1960', *Journal of Political Economy*, 78: 1115–41.

Hanuschek, E. 2001. 'Black-white achievement differences and governmental interventions', *American Economic Review,* 91(2): 24–28.

——— 2003. 'The failure of input-based schooling policies', *The Economic Journal*, 113: F64–F98.

Heckman, J. 1999. 'Policies to Foster Human Capital', NBER Working Paper Series no. 7288, Cambridge, MA, http://www.nber.org/papers/w7288.

Heckman, J. and B. Jacobs. 2009. 'Policies to Destroy and Create Human Capital in Europe', IZA Discussion Paper no. 4680, Bonn, Germany.

Heckman, J. and L. Lochner. 1999. 'Rethinking Education and Training Policy: Understanding the Sources of Skill Formation in a Modern Economy'. Processed paper.

Helpman, E. 2004. *The Mystery of Growth*. Cambridge, MA: The Belknap Press of the University of Harvard Press.

Helpman, E. and P. Krugman. 1987. *Market Structure and Foreign Trade: Increasing Returns, Imperfect Competition, and the International Economy.* Cambridge, MA: MIT Press.

Katz, L. and D. Autor. 1999. 'Changes in the Wage Structure and Wage Inequality', in Ashenfelter, O. and D. Card (eds), *Handbook of Labor Economics, Vol. 3A*. Amsterdam: North-Holland, pp. 1463–1555.

Krueger, A. and M. Lindahl. 2001. 'Education for growth: Why and for whom?', *Journal of Economic Literature*, 39: 1101–1136.

Lanzona, L. 1998. 'Migration, self-selection and earnings in Philippine rural communities', *Journal of Development Economics*, 54: 27–50.

——— 2001. 'An Analysis of Wage Inequality in the Philippines: An Application of the Stolper-Samuelson Theory', in L. Lanzona (ed.), *Labor, HRD and Globalization.* Makati, Philippines: Philippines APEC Study Center Network and Philippines Institute for Development Studies.

——— 2009. *Human Capital and Agricultural Productivity, Mimeo*. Los Banos, Laguna: Southeast Asian Regional Center for Graduate Study and Research in Agriculture (SEARCA).

Lanzona, L. and Evenson, R. 1997. 'The Effects of Transaction Costs on Labor Market Participation and Earnings: Evidence from Rural Philippines Markets', Yale University Economic Growth Center, Discussion Paper no. 790, New Haven, CT.

Lin, T.C. 2003. 'Education, technical progress, and economic growth: The case of Taiwan', *Economics of Education Review,* 22: 213–20.

Mortensen, D. and C. Pissarides. 1994. 'Job creation and job destruction in the theory of unemployment', *Review of Economic Studies*, 61: 397–415.
North, D. 1990. *Institutions, Institutional Change, and Economic Performance*. Cambridge: Cambridge University Press.
Orazem, C. and E. King. 2008. 'Schooling in developing countries: The roles of supply, demand and government policy', in T.P. Schultz and J. Strauss (eds), *Handbook of Development Economics, Vol. 4*. Amsterdam: North-Holland.
Psacharopoulos, F. and H. Patrinos. 2002. 'Returns to Investment in Education—A Further Update', World Bank Policy Research Working Paper no. 2881, World Bank, Washington, D.C.
Pritchett, L. 2001. 'Where has all the education gone?', *The World Bank Economic Review*, 15: 367–91.
Schultz, T.W. 1975. 'The value of the ability to deal with disequilibrium', *Journal of Economic Literature*, 13: 827–46.
Schultz, T.P. 1993. 'Investments in schooling and health of women and men', *Journal of Human Resources*, 28: 694–734.
Shimer, R. 2007. 'Mismatch', *The American Economic Review*, 97(4): 1074–1101.
Shen-Keng, Y. 2001. 'Dilemmas of Education Reform in Taiwan: Internationalization or Localization?' Processed Paper.
Soares, F.V., R. Ribas and R. Osório. 2007. 'Evaluating the Impact of Brazil's BolsaFamília: Cash Transfer Programmes in Comparative Perspective', International Poverty Centre, IPC Evaluation Note no. 1, United Nations Development Programme, Brazil.
Spohr, C. 2003. 'Formal schooling and workforce participation in a rapidly developing economy: Evidence from "compulsory" junior high school in Taiwan', *Journal of Development Economics*, 70: 291–327.
World Bank. 2008. *The Road Not Travelled: Education Reform in the Middle East and North Africa*. Washington, D.C.: World Bank.
World Economic Forum. 2009. *The Global Competitiveness Report 2009–2010*. Geneva, Switzerland: World Economic Forum.
Zenou, Y. 2008. 'Job search and mobility in developing countries: Theory and policy implications', *Journal of Development Economics*, 86: 336–55.

2

Eco-innovation: A Literature Review of the Challenges Facing the Development of Green Technologies

Daniel K.N. Johnson and Kristina M. Lybecker*

Introduction

> Innovation involves attempts to deal with an extended and rapidly advancing scientific frontier, fragmenting markets flung right across the globe, political uncertainties, regulatory instabilities, and a set of competitors who are increasingly coming from unexpected directions. (Tidd, 2006)

Innovation is an inherently challenging activity. It is fraught with uncertainty: uncertainty about the end product of a research process, uncertainty about the reception by the market, uncertainty about the ability to appropriate the returns to research while competitors try to produce similar results and uncertainty about regulatory impacts on the research process and end result. While the remainder of this chapter is devoted to eco-innovation in particular, many of the challenges to innovation more generally (see, for example, Fagerberg et al., 2005; Freeman, 1994; Shavinina, 2006; Stoneman, 1995) are mirrored and exaggerated for the field of eco-innovation.

However, there are also many areas in which the economics of eco-innovation warrants their own attention, separate from the challenges of innovation in general. For example, many of those uncertainties are greater for environmental innovation than in any other field of innovation. After all, the research problems are enormous. Consider the challenges that face(d) the Tesla Motor Company as they developed not just a new

---

* We thank the National Peace Foundation and the United States Chamber of Commerce for their support of this project and David Popp for his generous and insightful comments.

car with new materials and new suppliers but a transcontinental series of fuelling stations to provide quick charging for elective vehicles, a challenge far broader than those faced by their traditional automotive peers like Ford and General Motors (see, for example, Hutter and Starmack, 2013). The valuation by the marketplace is far from certain, as frequently consumers do not have the knowledge or tools to evaluate the environmental impact of an innovation. Witness the surprisingly complex calculations required to evaluate even a rough value of a simpler activity, pre-existing recycling programmes, through a combination of life-cycle impact assessment, life-cycle cost models of the equipment used and contingent valuation of customer appreciation of the environment (Bovea and Vidal, 2004). The problem gets thornier when considering the potential impact of hypothetical future programmes. Even with that knowledge of impact, consumers and producers rarely have the ability to 'value' any given environmental impact, as markets provide little of the information needed to do so. Indeed, most activities related to environmental processes or products encounter externalities, which, by definition, are not incorporated into any market's price without government intervention. A classic example is of course the case of externalities in the production and consumption of petroleum-based products, but after decades of research and debate, there are still arguments about methodology, size and usefulness of all estimates (see Matthews and Lave, 2000, for a good summary). In addition, uncertainty surrounds the pricing of competing as well as complementary goods: solar and wind power innovators are both subjected to exogenous shocks to their planning horizons by policymakers, both domestic and foreign, changing subsidies and incentives (Menz and Vachon, 2006). Further, the appropriability of returns is open to question, as innovators wonder whether their research will be subject to 'public interest' exclusions to patent law, and perhaps subject to compulsory licensing requirements. Examples abound, but consider the compulsory licensing of pharmaceutical products (Love, 2007), in developed and less developed nations alike, such as Merck's Finasteride (used for treatments of prostate cancer and male pattern baldness) or Glaxo's Sumatriptan Succinate (used for treatment of migraines). Finally, the regulatory landscape is variegated and ever changing, such that an innovative environmentally clean process may have enormous value to producers in one location and zero market value in a nearby jurisdiction. Even worse, that situation may easily reverse, or reduce to zero everywhere. One need not look further than renewable energy policy to find examples of temporary policies that reverse themselves or neighbouring jurisdictions with vastly different policies (Menz and Vachon, 2006). Given those uncertainties and the frequent presence of enormous fixed costs in the research and development (R&D) stage, it is a marvel that eco-innovation occurs at all.

On the other hand, it is often uncertainty itself that stimulates innovation. Jaffe et al. (2001) correctly assert that uncertainty about the future rate and direction of technological change is often the largest single source of differences among 'predictions' in global climate change modelling. It may be that fundamental uncertainty which keeps innovators searching for alternative (and frequently better) solutions to environmental challenges. As Tidd (2006) notes, innovation most often takes place in the context of rules, which are clearly understood, but at times the rules are altered to redefine the conditions under which innovation occurs. Sometimes this presents new opportunities, an argument that Porter and van der Linde (1995) champion. Ultimately, uncertainty and changing regulations are factors that both enhance and inhibit eco-innovation, providing policymakers with a critical and challenging role in the process (see Crabb and Johnson, 2010, for a poignant example from automobile emissions regulation).

Policymakers may very well be conflicted about how much structure to provide for innovators, if they truly thrive on some degree of uncertainty. Unfortunately, economics can offer no panacea, no single answer to this question. Aubert (2004) offers a typology of governments at different levels of science and technology (S&T) capacity in Table 2.1, depending upon institutional capabilities, but the appropriate policy response undoubtedly differs by industry, by technological problem and even by time period.

After clarifying our definition of eco-innovation in the section that follows, we summarize the current state of scholarly knowledge regarding five challenges facing eco-innovation: intellectual property rights (IPRs) (e.g. patents), economies of scale, markets and incentives, system complexity and policy choices. We then offer suggestions concerning effective strategies for stimulating eco-innovation, and conclude with policy implications and directions for future research.

Table 2.1:
Innovation systems and policy agendas

Level of institutional and human capital capabilities	Strong institutions (litmus test: business R&D dominate R&D budget) Decision-making horizon: long-term	Limited institutional capabilities (litmus test: large stock of export-driven FDI exists yet national innovation system is virtually irrelevant for business) Decision-making horizon: medium term	Weak or fragile institutions little state activist is possible/ desirable (litmus test: investment climate is poor and volatile) Decision-making horizon: short-term survival
Low ST capabilities Technology adoption		Exports as a springboard agenda: Developing non-traditional exports as entry point for institutional and technology development development Central America (with the exception of Costa Rica) Traditional urban and rural economics in India and China Korea in the 60s Mexico in the 70s Vietnam, Mauritius	Technology basics agenda: Creation of demonstration effect to show that innovation does matter, in particular in health, education, agriculture and crafts. Most of sub-Saharan Africa Most Central Asian states
Medium ST capabilities Technology adoption	*'Turning point' agenda:* a need for transition from global sourcing to proprietary technology		
	Increase in R&D investments Korea, Ireland in the 90s Malaysia India (IT clusters)	Increase in business R&D through recombination of S&T capabilities EU accession countries Chile China, Mexico, Brazil Turkey, South Africa Korea in the 70s and 80s	

(Table 2.1 Continued)

(Table 2.1 Continued)

High ST capa-bilities Technology creation	Innovation leaders agenda: Development of proprietary technology through promotion of innovation clusters Korea, Singapore, Taiwan, Finland, Israel	'Turning point' agenda: Increase in business R&D through recombination of S&T capabilities No country currently fits Russia in the future?	'Embedded autonomy' agenda: Creating a diversity of autonomous business-led innovation organizations (Foundation Chile Agenda) Argentina, Russia, Ukraine, Belarus, Armenia Chile in the 70s

Source: Aubert (2004).

Definition(s) of Environmental Technologies

There are several complementary definitions of environmental technologies, with only slightly different emphases. Shrivastava (1995) offered one of the first clear definitions as:

> production equipment, methods and procedures, product designs, and product delivery mechanisms that conserve energy and natural resources, minimize environmental load of human activities, and protect the natural environment. They include both hardware, such as pollution control equipment, ecological measurement instrumentation, and cleaner production technologies. They also include operating methods, such as waste management practices (materials recycling, waste exchange), and conservation-oriented work arrangements (car pooling, flextime), used to conserve and enhance nature.

Shrivastava considers five thematic approaches for eco-innovation: design for disassembly (e.g. production with an eye towards waste reduction via simpler reusing and recycling), manufacturing for the environment (e.g. innovative cleaner technology using fewer inputs or reducing emissions), total quality environmental management (i.e. adopting a total systems approach to design and manufacturing), industrial ecosystems (i.e. creating inter-organizational linkages like waste exchanges or symbiotic firms) and technology assessment (including technology transfer to areas of greatest marginal impact).

Alternatively, Rennings (1998) cites a German government definition for environmental innovation (short: eco-innovation) as 'all measures of relevant actors ... which a) develop new ideas, behavior, products and processes, apply or introduce them and b) which contribute to a reduction of environmental burdens or to ecologically specified sustainability targets'. Rennings suggests that social and institutional innovations are particularly important in this arena of innovation, but are neglected by neoclassical economics. As such, he calls for a combination of thought with evolutionary economics to develop new theory for eco-innovation, required due to a combination of factors: their double externality (costs and benefits for those who are not decision-makers in either the production of environmentally related goods or the creation of the knowledge needed to change that production in content or process), the importance of the regulatory framework in this sector and the importance of social/institutional changes as part of the innovative activity. On the other hand, Murphy and Gouldson (2000) argue that organizational innovations usually do no more than facilitate the implementation of process and product environmental innovations. Accordingly, they do not merit anything more than equal consideration alongside more traditional forms of technological change.

Finally, Bernauer et al. (2006) follow Organisation for Economic Co-operation and Development (OECD) (1997) in defining environmental innovations as 'all innovations that have a beneficial effect on the environment regardless of whether this effect was the main objective of the innovation, including process, product, and organizational innovations'. They focus primarily on explanations of product and process innovations, defining process innovations as improvements in the production process resulting in reduced environmental impacts. Bernauer et al. argue that the primary environmental impact of many products stems from their use (e.g. emissions by vehicles) and disposal (e.g. heavy metals in batteries) rather than their production process, so product innovations aim at reducing environmental impacts during a product's entire life cycle. Regardless of which of these definitions one chooses, the following sections are equally useful in the identification and analysis of challenges, all currently faced by environmental innovators.

The Challenges of Eco-innovation

This section focuses on five themes that affect the development of eco-innovations: IPRs (e.g. patents), economies of scale, markets and incentives, system complexity and policy choices. For a more general approach, Kemp (1997) is a good, if dated, resource dedicated to explaining the literature's models of environmental innovation and diffusion, including empirically based case studies.

Role of Patents

One of the first themes that arises in most discussions surrounding innovation is also one of the most provocative and emotionally charged topics, that of intellectual property (IP), specifically patents. As an innovator's legal right to exclude others from an activity, they present a double-edged sword: without some guarantee of repayment for the risk and financial sacrifice of the research process, little innovation will occur, but too great an exclusion right may hamper follow-on innovation or may extract inappropriately large monopoly rents from the consumer. This situation is exacerbated when the innovation is suitable for or desperately needed by developing nations and their impoverished citizens. Naturally, the decision about what repayment for risk and investment is 'appropriate' usually depends upon whether one takes the perspective of the producer or the consumer.

Leading up to the 2009 Copenhagen Summit on Climate Change, the Copenhagen Economics study (Copenhagen Economics A/S and The IPR Company ApS, 2009) examined the claim frequently made by developing nations that strong IPRs on carbon abatement technologies hinder developing countries' greenhouse gas abatement efforts. The study finds that IPRs do not constitute as significant a barrier as claimed since a variety of technologies exist for reducing emissions. Based on the cost per unit of carbon emission reduction, IPR-protected technologies are not necessarily more costly than unprotected alternatives. The study notes that the expense of some innovative carbon abatement technologies stems from the immaturity of the technology rather than patent protection. Moreover, the study finds that while there is a small number of emerging market economies that account for the majority of patents protected in the sample (99.4 per cent), there is a much larger number of low-income nations that protect very few patents (0.6 per cent of the total sample). Given that patents are virtually non-existent for these technologies in most developing countries, it is difficult to argue that IPR protection is a significant barrier to technology transfer.

Further, Copenhagen Economics A/S and The IPR Company ApS (2009) presents some intriguing information about the IPR ownership of environmental technologies. Between 1998 and 2008, a sizeable share of the IPRs for eco-innovations in emerging economies were owned by firms within those economies themselves, rather than by firms in industrialized nations. In their words,

> The patent count on the relevant technologies covered by this study has indeed increased rapidly. Globally, some 215,000 patent applications were filed worldwide over this period 1998–2008, including some 22,000 in developing countries—out of which about 7,400 were actually owned by developing country residents. When the last four years of the period are compared to the first four years, the global patent count increased by 120%, but by nearly 550% in developing countries. Solar energy and fuel cell patents account for 80% of the count and for most of the growth as well, followed by wind energy as a distant third. (Copenhagen, 2009)

In biomass, 17.60 per cent of patents were owned by firms in emerging nations; in fuel cells, 21.80 per cent; in geothermal, 12.40 per cent; in ocean-energy technologies, 39.40 per cent; in solar, 39.80 per cent; in waste management, 48.40 per cent; and in wind energy, 58.60 per cent. No single country has market dominance in any of the technologies studied, where the largest market shares are held by China (38 per cent of solar energy patents) and Japan (28 per cent of fuel-cell patents). The authors conclude that the price of carbon abatement technology is not driven up by monopoly power since significant competition exists within and between eco-technology markets.

There are additional clearly documented reasons to encourage strong patent law in particular. Levin et al. (1987) provide the landmark survey of US firms on the importance of IPRs, a study that launched dozens in its wake. They found that the value of patents and other forms of protection varies across industries and across innovations and though patents were important, secrecy, lead times and learning curve advantages were all considered more effective. The study also confirmed substantial inter-industry variation in the evaluation of different appropriability mechanisms, suggesting that the impact of policy changes should be assessed at the industry level. For example, a longer patent life would have little impact on innovation in the airline industry, but the effect would be a significant force for additional innovation in the pharmaceutical industry. Beyond industry differences, the value of patents differs across product and process innovations. Notably, for new processes, patents were generally rated the least effective of the appropriability mechanisms. This distinction is important when one considers that pollution can be reduced through either end-of-pipe treatments (e.g. pollution control products) or changes in production processes. Assuming no third-party involvement in the modifications, process changes are less likely to be patented. This point is especially salient when considering whether changes in IP protection would matter for environmental technologies.

Most firms indicate that IPRs are essential to the profitability of commercial research, so in its absence they simply will not commit R&D funding to the market in question. That leaves policymakers with a choice of whether to perform the research themselves as a public policy initiative (an option explored below), or to develop patent law that will carefully tread the line or sufficient returns to risky research while protecting consumers and encouraging subsequent research.

Gallini and Scotchmer (2001) describe this challenge succinctly in the language of economists. Naturally, for economists, efficiency is achieved when the deadweight loss of taxation is minimized, meaning that the unintended distortions to behaviour are at their lowest:

> With intellectual property, projects are funded out of monopoly profits. Monopoly pricing is equivalent to taxing a single market, which is generally thought to impose greater deadweight loss than the broad-based taxation that generates general revenue. Thus, to justify intellectual property, there must be some type of asymmetric information about the costs and benefits of research programs. (Gallini and Scotchmer, 2001)

Indeed, there usually is asymmetric information about research programmes, which is the entire reason to bestow IPRs on firms in exchange for public revelation of their research insights. It is this trade, of profits in return for information, which constitutes the heart of any IPR system. As such, policymakers carefully tailor the many dimensions of patents, and scholars analyze and critique the efficiency and equity of existing policy as well as proposed changes.

The potential elements of an effective patent system have been explored in a large and rich literature. Hopenhayn and Mitchell (2001) correctly state that a patent is defined by its length, its breadth and the fees associated with its origination and renewal. Each of these elements has been theoretically modelled, analyzed and evaluated in numerous economic studies including Gallini (1992), Gallini and Scotchmer (2001), Green and Scotchmer (1995), Scotchmer (1996, 1999), Yiannaka and Fulton (2001) and Yiannaka and Fulton (2003). Given that patent design is essential to manipulating the incentives that drive innovation, this work is relevant to the development and dissemination of environmental innovation, but does not specifically address the unique elements of eco-innovation that distinguish it from other innovation.

In designing an optimal patent system to enforce, policymakers must pay particular attention to technologies that build, one generation to the next generation. Some fields of innovation such as biotechnology, information sciences and technology and environmental technology face the challenge of sequential, or cumulative, innovation. The challenge of sequential innovation occurs when innovation happens in separable stages (Green and Scotchmer, 1995), with the original innovation verging on pure science and having little commercial value but relying on the follow-on stages for the social values based on practical applications. Scotchmer (1991) explores this concern, noting the difficulty in rewarding early innovators for the technological foundations they develop, while also allowing for the reward of subsequent innovators who improve and extend the original technology to new applications. In Scotchmer (1996), she argues that patents are not necessary for the development of second-generation products and that original innovators would collect a larger share of profit if second-generation products are not patentable. She notes that this problem is 'particularly acute when the entire commercial value is contained in the applications facilitated by the basic research, and when the basic innovation has no commercial value on its own', such as may easily be true for environmental innovations. The first innovator may have insufficient incentive to invest, leading to a potentially large role for public research in basic science.

Some of these same concerns surrounding cumulative innovation are empirically examined in Cohen (2005). Focusing on the pro-patent movement since 1980 in the United States, Cohen explores the claim that the growth of patenting in upstream innovations may constrain critical follow-on research. While the empirical basis for these concerns is limited, he does find evidence that patenting stimulates innovation. Similarly, Cahoy and Glenna (2009) collect evidence on patenting around biofuels, to evaluate the concern that IPRs are stifling sequential innovation. They evaluate the evidence that thickets, or the tragedy of the anti-commons, might occur, finding 231 biofuel patents spread among 72 owners. Ownership is therefore markedly less concentrated than genetically modified (GM) patents as a whole (biofuels show 33 per cent by the top three owners, as opposed to 85 per cent by top three owners for GM corn, 70 per cent for non-corn GM). Private ordering, or the collaboration between firms to reach goals of market value, is predicted to occur as it has done in GM more generally, with private contracts against the background of public enforcement as a key example. Vertical consolidation, joint ventures, cross-licensing, patent pooling and standard setting are all examples of the private ordering that could solve the potential anti-commons problem. Cahoy and Glenna believe that a private solution will be more likely in case with a limited number of patents, significant R&D barriers to entry, complementary infrastructure and technology and long-term market potential. In short, they conclude that the likelihood of a private solution for biofuels is extremely high.

Along this line of thought, it is important to draw a distinction between the roles that patents play in the pharmaceutical sector as compared to the eco-innovation sector. While the underlying principle is the same (to accord a limited degree of market power, limited by both time and the entrance of competitors, in return for the research and creative process as well as the public sharing of information sufficient to replicate the innovation), Barton (2007) summarizes a key difference particularly well. In the case of eco-innovation, most fundamental technologies have long ago been absorbed into popular technical knowledge, having been off-patent in many cases for decades. Thus, current patents primarily protect amendments or improvements to those fundamentals, encouraging competition among alternative models that serve the same general purpose. Pharmaceutical compounds, on the other hand, are usually the result of completely new biochemical research, facing much higher research costs to recover and fewer short-run competitors in the IPR-protected marketplace. In summary, while patents serve to encourage innovators in both sectors, it would be extremely simplistic (and probably dangerously incorrect) to make an argument about eco-innovation on the basis of conclusions made for pharmaceuticals.

Mandel (2005) presents a very readable review of the role of patent law in encouraging eco-innovations in particular, discussing the possibility of alterations to patent law for environmental technologies in particular. He considers and dismisses the merits of two possible changes: extended patent terms on eco-innovations (which should alter incentives only marginally) and accelerating patent prosecution (which already exists on special terms in the United States and is rarely used). Instead, he suggests a patent reward system in parallel with the current legal framework, giving extra value to eco-innovations out of the public purse, to align social value with private market values. Naturally, this poses the administrative challenge of valuing innovations without market signals, but it poses little risk. The United States already has a patent reward system in place for atomic energy innovations (42 U.S.C. § 2181. 81, 42 U.S.C. § 2187). Mandel considers and responds to four major criticisms of patent rewards:

1. Rewards fail to incentivize commercialization (but that could be a separate issue, under government authority in this case).
2. Rewards based on marginal costs do not compensate for fixed costs (but in this case, we could calculate social value rather than marginal private value).
3. Rewards do not screen out invalid patents (but the evaluative board could summarily reject awards).
4. Rewards are costly to administer (which may be true, but we have a template in place. Costs are in payment and administration, and should pale next to costs of under-provision of eco-innovations, not to mention that they can be offset against other costs like environmental protection and mitigation costs).

In short, the current legal system is entirely compatible with supplementary efforts to encourage specific types of innovation. There need not be any special consideration of eco-innovation within IP law, but encouragement could take the form of public funding to augment the incentives provided by market forces, thus minimizing the opportunity for distortion of private economic activity.

In the specific case of developing nations, Park and Lippoldt (2008) empirically analyze the impact of strengthened IPRs in the developing world on local innovation and technology transfer, and discover a positive relationship in both cases. Strengthened IPRs are significantly and positively associated with: developing country patent applications and expenditure on R&D as a share of GDP, inward foreign direct investment (FDI), merchandise imports, service imports and the inflow of high-tech products. Case studies of Brazil, Russia, India and China reinforce the finding that technology transfer and FDI are among the most important factors contributing to the development of indigenous technological capacity, a result confirmed by Kanwar (2007) in subsequent work. Utilizing country-level data for 44 developing and developed countries, between 1981 and 2000, Kanwar analyzes the relationship between R&D investment (a proxy for innovation) and an index of patent protection (a proxy for IP protection), concluding that the strength of IPRs exerts a strong, positive impact on innovation.

Given the importance of IPRs to innovation, it is essential to learn more about the differences that exist in IP regimes between developing and industrialized nations. Kanwar and Evenson (2001, 2009) examine the claim that the technology-haves (developed nations) provide relatively stronger IP protection, while the technology-have-nots (developing nations) opt for weaker protection. Utilizing cross-national data for 1981–2000, the studies find only weak support for this claim, noting that weak IP protection is more likely due to the lack of financial resources and human capital, and their inward-looking trade orientation.

In conclusion, there is virtually unanimous consent among economists that strong IPRs are an essential prerequisite to the development of environmental technologies. The dissenting voices (e.g. Hutchinson, 2006) make the valid claim that patent law increases the cost of technology acquisition by consumers or intermediary producers, but do not explain how technology arrives more cheaply by another means. Given that innovation is costly and risky, there is quite simply no alternative to IPRs proposed in the literature that will adequately encourage eco-innovation.

Given that IPRs are necessary, there are potential alterations that we should consider to make IPRs work more effectively for eco-innovation in particular. Both financial awards and the clearer distinction between primary research and cumulative/application research could be avenues for policy consideration.

Role of Economies of Scale

For most innovative products and processes, one of the challenges is how to scale up production in order to lower costs. In other words, most innovations are subject to economies of scale (or increasing returns to scale), in which higher levels of output are associated with lower per-unit costs. Eco-innovations are no exception, and evidence of this output–cost relationship is documented universally (see, for example, Cowan, 1990; Cowan and Gunby, 1996; Cowan and Hulten, 1996; Cowan and Kline, 1996; Islas, 1997; Kemp, 1997). Kemp (1997) reviews the case of chlorofluorocarbons (CFCs) and the shift by DuPont to less destructive hydrofluorocarbons (HCFCs). DuPont estimated the fixed cost of retooling production at roughly $1.25 billion, in order to produce an HCFC product that in the end still has a downward sloping average cost curve (decreasing per unit costs as quantity produced increased), but is at every level more expensive per unit than the CFC-11 that it replaced. While full-cost curve information is held privately by DuPont, they did reveal some startling cost differences. Production of 5 kilotons would cost $5 per unit for CFC-11 but $25 per unit for the more environmentally responsible HCFC. If customers balked at those prices, DuPont would presumably sell less product, which made the cost differences worse. In fact, a reasonably market-competitive price of $10 per ton was reached for CFC-11 at production levels of only 2.5 kilotons, while that cost level was not reached for HCFC until production of 25 kilotons.

The same is true for learning curves, which is another way of relating scale of production to costs, over time as opposed to simultaneously. Evidence on the ability to lower costs for eco-innovation as more units are produced can be found in Joskow and Rozanski (1979), Zimmerman (1982), Sharp and Price (1990), Lester and McCabe (1993), Neij (1997), Grübler and Messner (1999) and Grübler et al. (1999).

Friedman (2006) quotes General Electric's chief executive officer (CEO) Jeffrey Immelt as noting 'the big energy players are not going to make a multibillion-dollar, forty-year bet on a fifteen-minute market signal'. In short, they need a promise of a long-term market for whatever they develop. To illustrate, he cites the case study of First Solar, an Ohio company that produces exclusively in Germany because they guaranteed the buy-back price of solar energy from consumers for 20 years after installation.

The prevalence of economies of scale and significant learning curves may also be linked to another characteristic of successful eco-innovation firms, specifically firm size and resources. Baylis et al. (1998) argue that environmental activities go along with a higher amount of financial and human resources, which is why larger firms have better opportunities and abilities to reduce environmental impacts. Several empirical studies show that, in general, firm size has a positive influence on environmental innovation (e.g. Arimura et al., 2007; Cleff and Rennings, 1999; Rehfeld et al., 2006).

In a similar vein, Berrone et al. (2007) propose that firms with more available resources, or organizational slack, have a better ability to evaluate outside influences and to adapt to internal pressures. They use an unbalanced panel of 340 firms drawn from the 20 most polluting industries, firms listed in Compustat 1997–2001 and firms listing more than 40 patents in the United States. Utilizing regression analysis, they find that larger, more R&D-intensive, capital-intensive and Environmental Protection Agency (EPA)-litigated firms have more eco-innovations. Interestingly, Cainelli et al. (2007) use an Italian census of 773 firms 1993–95 to test the opposite causality, namely, the impact of eco-innovation strategies on employment, turnover and labour productivity. They find a negative relationship on employment and turnover, with no significant effect on productivity.

Friedman (2006) argues that the US role in eco-innovation development is to provide the upfront investment, just as it did for personal computers (PCs), digital videodiscs (DVDs) and iPods. Then the global community can draw on India's low service costs and China's low manufacturing costs to produce at a scale and price that will make it accessible to all. Without massive investment in the development phase, we simply chip away at a large problem with a small tool.

Clearly there are several effects at play, effects that change over time. Teece (1998) points out that 'with increasing returns, that which is ahead tends to stay ahead ... mechanisms of positive feedback reinforce the winners and challenge the losers'. On the other hand, global markets have become increasingly liberalized, so restrictions on knowledge transfers have evaporated. Given this, firms are no longer able to earn extra-normal returns by capitalizing on trade restrictions. While lower transportation costs have facilitated large-scale production, competition has increased and information about market opportunities diffuses virtually instantaneously. Paradoxically, this might mean that competition keeps individual innovators from profiting much from their work, because competitive forces slow any one firm from reaching sufficient scale to achieve minimum economic size.

The challenge of achieving this efficient scale and reducing per-unit production costs is critical to the success of most innovative products and processes. Economic evidence indicates that successful firms are those that more quickly achieve economies of scale and quickly move along technological learning curves. Though there is no guarantee, larger firms with more resources seem better poised to exploit the output–cost relationship.

Role of Markets and Incentives

Innovation is the response of market-based firms to profit potential and other market-based incentives. The evidence is overwhelming on this point (see Mansfield, 1977; Mowery and Rosenberg, 1979 for early evidence but the literature sprawls outward from there). Within the immense body of work on incentives and innovation, there is a branch of literature dedicated to the empirical testing of 'induced innovation', or the suggestion that higher prices lead consumers to search for alternatives, at least partly in the form of new products and processes.

Kemp and Soete (1990) point out that there are several factors auguring against environmental innovation. On the supply side, technological opportunity and appropriability affect this field of innovation in a fashion similar to other fields of innovation. As the development process proceeds, the amount of profits innovators reap from innovations will lead to environmental innovations. On the demand side, innovation faces much higher hurdles here. First, there are problems related to knowledge and information, including who is responsible for costs and how to price damage. Second, there is uncertainty about actual costs, consumer values and policy platforms now and in the future. Third, many eco-innovations are processes in nature, but aim to market to the end consumer without necessarily lowering costs, making them a strange commodity. In addition, the market is complicated by competing technologies (e.g. fossil fuels) subject to negative externalities in which the user does not bear the full cost of the good. Further, the public goods nature of environmental technology prevents the user (and the innovator) from fully capturing the benefits of the innovation. Daniels and Johnson (2014) seems to confirm this point, showing minimal induced innovation in the oil and gas sector in terms of exploration technologies in response to anticipated price changes.

Popp (2006a) examines government subsidies for innovation in the context of addressing the two market failures that characterize green technology. The first is the public goods nature of knowledge, which leads to knowledge spillovers. This is where policies such as IPRs protection and R&D subsidies play an important role. Popp finds that while R&D subsidies do lead to increases in climate-friendly R&D, they address only the public good problem. Notably, this market failure characterizes all forms of innovation. The second market failure, environmental externalities, is unique to environmental innovation. The market does not reward, or allow for complete appropriation of the benefits of, innovations that increase costs but reduce pollution. Since the environmental externality problem is not addressed, there are no additional incentives to adopt the new technologies. In this case, environmental regulation provides the incentives for innovation. As such, policies that directly impact the environmental externality result in greater gains in terms of both atmospheric temperature and economic welfare. This illustrates the importance of putting environmental policy in place as a first step, and it also demonstrates that expectations about future policy are a key component of the uncertainty surrounding eco-innovation.

In a similar vein, Arrow et al. (2004) refer to three reasons that natural resources may be underpriced: unclear property rights, externalities and government subsidies. They refer to the 1992 World Development Report by the World Bank that showed that in 29 of 32 less developed countries (LDCs) surveyed, subsidies had caused the price of electricity, water and fossil fuels to fall below cost (not even including externality costs). They similarly report that the International Energy Agency (1999) 'has estimated that in India, China and the Russian Federation, full-cost pricing would reduce energy consumption by 7, 9 and 16 percent, respectively ... where most of the departure from social cost pricing is attributed to energy subsidies'.

Nevertheless, there is strong evidence of a statistical linkage between energy prices and the development of environmental technology. Newell et al. (1999) tested the effect of energy prices on innovation in home appliances, while controlling for regulatory effects. Popp (2002) is a seminal piece linking environmentally related innovations in industrial energy-using equipment to energy prices, controlling for the supply of available knowledge. Perhaps the most striking of Popp's findings is the speed at which innovation responds to incentives. Popp (2005) enumerates the lessons as follows: innovation responds quickly to incentives; innovation in a given field experiences diminishing returns over time; the social returns to environmental research are high; and the type of policy used affects the nature of new innovations.

As evidence, consider the evidence in the automotive industry. Both prices and regulation affect fuel efficiency (Atkinson and Halvorsen, 1984; Berry et al., 1996; Crabb and Johnson, 2010; Goldberg, 1998; Goodman, 1983; Greene, 1990; Ohta and Griliches, 1986; Pakes, 1993; Wilcox, 1984). The results are case sensitive, and method dependent, but all show separate effects of both factors. Most find that much of the improvement over time was autonomous or exogenous, with very strong effects of both price and policy. For example, Crabb and Johnson (2010) describe policy as a ratchet, to keep the impacts of price on innovation from backsliding during periods of lower energy prices. They calculate that a $5-per-barrel increase in the price of crude oil (roughly 12 cents a gallon for gas at the pump) translates into a 4 per cent increase in granted patents dealing with energy efficiency in automobiles (36 per year in the United States).

On the other hand, Jaffe et al. (2002) point out that it is more difficult to test induced innovation in eco-innovation because the prices are frequently not explicit, but rather shadow prices felt differentially by each industry. They briefly review the large literature on the impact of environmental regulation on productivity, a result that is clearly case specific. Nordhaus and Boyer (2000) calibrate Nordhaus' previous Dynamic Integrated model of Climate and the Economy (DICE) model of the economics of climate change, and finds that induced innovation has very little effect on emissions. This is largely due to the nature of his model that features a fixed factor production function, though partially due to the fact that new innovation completely crowds out innovation in other sectors. Nordhaus' paper is the most extreme negative result in the induced innovation literature. Bernauer et al. (2006) showcase nine studies on markets for green products as examples.

There is a strong theme in the literature encouraging policymakers to help to create markets for eco-innovation, although there are some minor disagreements on how that should be done. Cahoy and Glenna (2009) encourage the use of the Coase theorem, to enable private solutions to pursue eco-innovation. They suggest encouraging information dissemination as early and fully as possible, to avoid duplicative research and to maximize collaborative potential. Regulation probably will not accomplish this, so they encourage federal incentives (e.g. in alternative energy via agriculture) to be tied to information disclosure. The suggestion parallels the disclosure in the pharmaceutical industry via the Orange Book, tied to incentives under the Hatch–Waxman Act.

Beyond federal incentives, there is a large literature on the role of federal regulations. Porter and van der Linde (1995) recommend strong regulation, which will itself create new markets for environmental technology. Their article provoked a deluge of commentary and exploration into the conditions under which increased regulation could or could not stimulate greater profits (e.g. Bonato and Schmutzler, 2000; Jaffe et al., 1995; Mohr, 2002; Roediger-Schluga, 2004; Schmutzler, 2001; Sinclair-Desgagne and Gabel, 1997). While possible, the mathematical conditions are unlikely, but could indeed easily divert research activity to a more desirable end goal in eco-innovation compared to their current goals elsewhere in industry. For example, Bonato and Schumutzler (2000) build a theoretical model to test the Porter hypothesis and find that it holds only under fairly strict mathematical (although theoretically possible) conditions. Environmental regulation might lead to cost-saving innovation if (a) the fixed costs of innovation are lower than compliance plus production, or (b) spillover effects make innovation strategically a bad idea for the firm but a good idea for society, or (c) regulation helps to fix incentive problems between managers and owners or (d) regulation helps to clear information flow. However, the compliance costs must be low and there must be initial underinvestment for the arguments to hold in a mathematical model.

Desrochers (2008) places the Porter hypothesis within the larger framework of a literature on the incentive to create by-products for profit out of industrial waste. He argues that the profit motive generates the activity, not regulation. Regulation might simply help to set the property rights in place for a new market to develop more easily. The three variants of the Porter hypothesis presented in Jaffe and Palmer (1997) are empirically tested in Lanoie et al. (2007):

- Weak version: environmental innovations will be stimulated by environmental regulation.
- Narrow version: flexible environmental policy regimes give firms greater incentive to innovate than prescriptive regulations.
- Strong version: properly designed regulation may induce cost-saving innovation that more than compensates for the cost of compliance.

The authors find qualified to strong support for each with data from 4,200 facilities across seven OECD nations. The greatest support emerges for the weak variant. Most significantly, environmental policy induces innovation (as proxied by R&D expenditures).

Additional evidence comes from Costantini and Crespi (2007) who test the hypothesis that stronger environmental regulation creates a comparative advantage in those nations in the production of eco-innovation. They posit that the 'pollution haven' hypothesis is the opposite, where low-regulation areas become low-cost producers of all goods (but what about comparative advantage?). Using a 1996–2004 sample of 20 OECD exporting nations and their trade flows with 148 importing nations, they use a gravity model augmented with environmental policy variables to test the impact of regulation on trade flows in goods related to energy and energy savings alone. Environmental policy is proxied by carbon dioxide (CO_2) emissions, current environmental protection expenditures of both the public and the private sectors, the percentage of revenues from environmental taxes on total revenues and public investments on environmental protection. Innovation is alternatively measured as the number of patents in the energy sector, the number of total patents from residents or the percentage of R&D expenditures. The model general results show the expected signs and significance.

Cahoy and Glenna (2009) consider the impact of the imperfectly competitive agribusiness industry and the consolidation of market power there in a few firms. The danger is that the horizontal integration is 'necessary' for economies of scale, but reduces competition and limits the gains flowing back to the small-scale producer. The same may occur in eco-innovation, particularly since there are efficiency advantages to local production (rather than shipping biomass to a central facility for energy conversion, thereby losing the energy content advantage over traditional fuels). In short, the oligopolistic nature of the distribution or even production system must be taken into account when forming policy.

There is also a theme in the literature questioning whether the returns to R&D are sufficient to encourage eco-innovation. Since knowledge has positive spillovers (benefits to those who bore none of the cost of acquisition), economists usually conclude that the amount of R&D provided by private markets will be lower than the socially optimal level. That is, if we all paid what research is truly worth to us, more would be provided by the firms involved, evidenced by Hall (1996) and Jones and Williams (1998). However, Goolsbee (1998) provides a convincing counterargument, namely, that although the social return to R&D is higher than the private return, thus warranting public investment, the supply of researchers is inelastic so an increase in public funding often serves as a return to human capital investment rather than as a spur to innovation. In fact, public funding may crowd out private investment.

Friedman (2006) suggests that there are two kinds of innovation, namely, the big laboratory moments and the smaller adaptation moments. The United States focuses on the first, neglecting the crucial role of the second. The second is enhanced by quicker and more widespread diffusion, along with regulatory incentives to adopt at large scale. He argues that one role of the government should be to fund basic research, since from 100 lines of inquiry only one might merit commercialization. That one might be commercialized by venture capital, but initial funding has to be done using basic science without a profit motive.

Unfortunately, it is unclear that markets for eco-innovations will develop on their own. Roberts (1996) shows that demographics explain less ecologically conscious consumer behaviour now than it did in the past. Instead, it is an attitudinal emphasis that matters, with the belief in environmental impact of individual behaviour explaining most effectively whether consumers buy eco-innovative products or not. Straughan and Roberts (1999) confirm that result with college students. This opacity makes it extremely difficult for potential innovators to gauge the size, depth or even location of their potential market before they engage in costly product research.

Finally, it is unclear that eco-innovation is beneficial to a firm. As noted earlier, Cainelli et al. (2007) use an Italian census of 773 firms 1993–95 to carefully estimate the impact of eco-innovation strategies on employment, turnover and labour productivity. They find a negative relationship on employment and turnover, with no significant effect on productivity. Mazzanti et al. (2008) use the same firms to confirm a negative link between environmental motivations and growth in employment. Mazzanti et al. (2009) follow up using a larger sample of 61,219 Italian manufacturing firms 2000–04, with results showing a trade-off of lower environmentally efficiency in the recent past which allows slightly faster growth in the short and medium term, although there are some possibly complicated nonlinearities in that relationship involving policy types.

A great deal of clarifying empirical work could still be done here. There is a small literature that attempts to link patent grants with stock market valuation (see Pakes, 1985, for an early example or Johnson and Scowcroft, 2013 for a recent example and a relevant literature review), but we are unaware of any work specific to eco-innovation. Similarly, there is work to evaluate the characteristics of patents that contribute to their value at auction (e.g. Sneed and Johnson, 2009), but nothing specific to environmental technologies.

In sum, the greatest potential for propelling innovation is frequently found in market forces and incentives. Uncertainty, externalities and subsidies to competing goods undoubtedly hinder the process, but the motivation provided by potential profit is undeniable. Economic studies show that the innovative process may be enhanced (or inhibited) by appropriate government incentives or regulations. Given the spillovers associated with eco-innovation and the public goods nature of these technologies, there is a role for government intervention in order to spur an increase in environmental innovation.

Role of System Complexity

System complexity is a serious problem for any agency to consider in environmental policymaking. Not only is the modelling of the economic–environmental system complex, but each policy decision has both direct and indirect impacts on multiple sectors of the economy. As a simple example, Goolsbee (1998) and Jaffe et al. (2001) raise the question of the elasticity of supply of R&D inputs, so that if innovation is pursued actively in environmental areas, there may be a deterioration of innovation in other fields.

Models of the economic–environmental system abound, and none are simple. The DICE (Nordhaus, 1994), arguably the first and simplest of the models, includes 74 variables, most of which vary over time and with the state of the model's development, and 32 equations, many of which dynamically fix the relationships between constraints and objectives. Extensions of the DICE model (e.g. in Nordhaus and Boyer, 2000) run to 36 pages of appendices outlining the programming of the constraints and functions. Further extensions or alternatives, each excelling in one particular nuance of theory or another, can be found in a spectrum of sources (Bosetti et al., 2006a, 2006b; Buonanno et al., 2003; Gerlagh and van der Zwaan, 2003; Nordhaus and Boyer, 2000). As a result, it would be futile here to try to summarize the (frequently fragile) results.

Karl et al. (2005) examined 13 case studies of eco-innovation in Italy, and they found that the challenges vary considerably by sector. In all cases, they point to the underlying challenge of coordination/cooperation between firms, as innovation invariably means spillovers to other agents. They find very positive effects attributable to intermediary organizations, to the degree of trust between agents and to specific policy initiatives that facilitated cooperation and information exchange.

Uncertainty, as outlined in the introduction, clearly contributes to the complexity of the situation. Tidd (2006) points out that technological discontinuity leads to extremely challenging problems that emerge in a complex system. Innovative and production players, as well as regulators and government agencies, may be faced with a new environment or uncertainty stemming from the extent of system complexity. Chandrashekar and Basvarajappa (2001) propose that creating a network of working relationships, consisting of industry, academia and government entities, across industries with key technological inputs is one of the greatest challenges to technology policy. However, they note that creating networks in different key industries with technology inputs may accelerate the process of change in these industries through organizational, institutional and personal relationships that ameliorate the risk and disruption associated with technological change.

Interestingly, less developed nations may have an advantage in eco-innovation where their systems are less developed. Larson (2006) analyzes fuel-cell technology, suggesting that it may be adopted in less developed nations before developed nations, since the need is greater in the absence of a well-developed power distribution grid. In the absence of an existing reliable power infrastructure, fuel cells could follow the path of cell phones, leapfrogging the challenges of fixed distribution lines and moving ahead to a decentralized model. As the least capital-intensive system (compared to stringing transmission lines from a central generation plant to remote areas), it may very well win based purely on lower upfront costs.

Along similar lines, less developed nations may also have the advantage of less techno-economic certainty, resulting in less resistance to new complex eco-innovative systems, since they have fewer effective institutionalized systems currently in place for energy provision, manufacturing, waste disposal and other environmentally-sensitive sectors. Craig and Moores (2006) look at survey responses from 278 Australian family-owned businesses and find that higher techno-economic uncertainty and better information flows both work in favour of innovation.

Role of Policy

The literature is unanimous in asserting that policy clearly has a tremendous role to play in the support (or stifling) of eco-innovation. Press (2007) provides a good review of the literature on the impact of regulation on environmental protection, but also on competitiveness and innovation and capital movements. There are several key themes upon which virtually all economists agree.

First, there is a clear portfolio of policy alternatives to stimulate innovation or energy-related investment including taxes, subsidies, permits and standards/regulations. Unfortunately, Requate (1998) shows that comparing taxes and permits depends critically on the parameters, so the social preference on policy should be situation specific. Montero (2000) finds that standards and taxes yield higher incentives for R&D in a Cournot environment (where firms compete based on quantity), but yield the worst results in a Bertrand market (where firms compete based on price). Parry (1998) presents model simulations showing that the welfare gain of policies encouraging innovation may be limited, depending on the type of spillover externalities. In short, economists agree that the details of the policy matter more than the overall degree or intent of the policy (Bernauer et al., 2006; Cleff and Rennings, 1999; Frondel et al., 2004; Jaffe et al., 2004; Johnstone et al., 2008; Kemp, 1997; Rehfeld et al., 2006).

Second, there is strong evidence that regulatory policies can be very effective. For example, Popp (2001, 2003) empirically shows that regulations requiring plants to install sulphur dioxide (SO_2) scrubbers created an incentive for eco-innovation, distinctly different from the pre-regulation period. While patenting rates fell, the nature of innovation was shifted to pollution-limiting goals from other private cost-reducing goals. Rehfeld et al. (2006) analyze firm-level data in the EU to show that the certification of environmental management systems has a significantly positive effect on environmental product innovations. Jaffe and Palmer (1997) find a less inspirational result, namely, that lagged environmental compliance expenditures have a significant positive effect on R&D expenditures, but they do not find that successful patent applications are related to compliance costs. Berrone et al. (2007) confirm unsurprisingly that in the United States, firms with a history of more litigation by the EPA have more eco-innovations, holding other factors equal. Work by a host of others (e.g. Bonifant et al., 1995; Brunnermeier and Cohen, 2003; Johnstone et al., 2008; Shrivastava, 1995) all document the empirical effects of environmental regulation. However, the costs are substantial: Brunnermeier and Cohen (2003) conclude that firms spend $170–185 billion per year complying with environmental regulations, up 50 per cent from 1990.

Arimura et al. (2007) use 4,200 firm-level observations across the OECD to study the propensity for firms to do environmentally related R&D. In a simple Tobit estimation, they find that subjective perception of the stringency of environmental regulation is a strong predictor of environmentally related R&D. Firms with an environmental accounting system, or access to technical assistance programmes, are likewise more likely to do more R&D. There is also a strong nation specific, and perhaps cultural, effect. In a similar study drawing on the same data, Lanoie et al. (2007) find that environmental policy induces innovation.

A worthwhile and readable account of this sensitivity to location and culture is presented by Calef and Goble (2005) as they compare the policy encouragement during the 1990s of electric vehicle diffusion by California and France. They argue that California's stringent regulation spurred the development of innovative hybrid and fuel-cell vehicles more effectively than the French approach, calling it 'technology-forcing'. On the other hand, there is mixed evidence about the impact of regulatory policy on automobile fuel efficiency standards (see Crabb and Johnson, 2010, for a recent review and evidence).

In a larger empirical study on the impact of regulation, Cleff and Rennings (1999) used the Mannheim Innovation Panel survey of 2,264 companies with follow-up for 929 eco-innovative firms in the survey. The data are only summarized, but generally show that a combination of regulation and public image drove the majority of innovations, rather than stated green goals. These data would be very interesting to analyze again with a more rigorous model and statistical toolkit.

Popp (2010) correctly concludes that increases in the stringency of a regulatory standard has two competing effects, 'a direct effect of increasing costs, which increases the incentives to invest in R&D in order to develop cost-saving pollution abatement methods; and an indirect effect of reducing product output, which reduces the incentive to engage in R&D'. Others, like Carraro and Siniscalco (1994), show that innovation subsidies could be used instead of pollution abatement taxes, achieving the same goal without the output drag of a tax or regulation.

However, the evidence on the effectiveness of financial policy measures (taxes and subsidies) is ambiguous at best, and negative in many cases. Kemp (1997) argues that they have been largely ineffective, a result largely upheld by Hall and van Reenen (2000) through their extensive review of the literature. Carraro and Soubeyran (1996) found an R&D subsidy preferable to an emissions tax only if the output contractions induced by the tax are small or if the government finds output contractions desirable for other reasons. Katsoulacos and Xepapadeas (1995) confirmed that a simultaneous tax on pollution emissions and subsidy to environmental

R&D may work best of all in overcoming the joint market failure (negative externality from pollution and positive externality or spillover effects of R&D). Goulder and Mathai (2000) and Jaffe et al. (2001) note that in many cases, in the presence of eco-innovation subsidies, the optimal rate of pollution taxes will be lower, even though the desired level of abatement is higher, resulting in a worse government budget situation.

Johnstone et al. (2008) studied the impact of environmental policies on renewable energy technology using panel data from 26 countries from 1978 to 2003, finding that public policy has a significant impact on the development of new technologies. Utilizing a composite policy variable (including production tax credits, mandatory production quotas, differentiated tariff systems and tradable certificates), the authors find that the policy instruments are statistically significant for all renewable energy sources. Notably, the study found that only tax incentives have wide influence on renewable energy innovation.

Tradable permits for pollution have almost universal support from economists, as they encourage the efficient distribution of costs from a regulatory measure. By requiring and enforcing a maximum aggregate amount of pollutant, such markets ensure that standards are upheld. By encouraging firms to trade permits, markets distribute benefits to firms capable of low-cost reductions in emissions, thus encouraging the aggregate reduction of pollution in a least-cost manner. Laffont and Tirole (1996) utilize a theoretical model to reorganize tradable permits markets with an attached futures market to encourage eco-innovation for the next generation of standards. In a related proposal, Driesen (2003) suggests an 'Environmental Competition Statute' to authorize those who achieve low emissions to collect the cost of achieving low emissions plus a premium from competitors with higher emissions.

Third, the policy may serve to create a market for previously uncertain or ill-defined environmental commodities. Porter and van der Linde (1995) presented this forcefully as the assertion that environmental regulation can be good for industry by pointing out, or even forcing, unseen opportunities, and Desrochers (2008) restates it as part of a larger literature on the incentive to create by-products out of industrial waste. Ricci (2004) builds a model in which environmental regulation can improve productivity and economic growth, that is, supporting the Porter hypothesis via an increased productivity of inputs, better education, economies of scale in abatement, expectations of a better environment encouraging greater household savings and therefore cheaper investment, and stimulated overall R&D because it is a clean activity.

Bonato and Schmutzler (2000) build a theoretical model and find that the Porter hypothesis holds only under fairly strict mathematical (although theoretically possible) conditions. Environmental regulation might lead to cost-saving innovation if (a) the fixed costs of innovation are lower than compliance plus production, or (b) spillover effects make innovation strategically a bad idea for the firm but a good idea for society, or (c) regulation helps to fix incentive problems between managers and owners or (d) regulation helps to clear information flow. However, the compliance costs must be low and there must be initial underinvestment for the arguments to hold in a mathematical model.

Jaffe et al. (2001) summarize five ways in which regulation may focus the ability of markets to create profit from environmental concerns: they focus attention on the issue, they create information useful in developing solutions, they reduce uncertainty, they create first-mover advantages if other regions follow in regulation and they create pressure to overcome inertia by polluters. The empirical evidence is thin on each side of this hypothesis.

Fourth, current policymakers are frequently unable to muster the political will to enact legislation that is pro-environmental innovation. Arrow et al. (2004) point to government subsidies as one of the three most important threats to eco-innovation. In a similar vein, Arrow et al. (2004) refer to three things where natural resources may be underpriced: unclear property rights, externalities and government subsidies. The World Bank Report (1992), on the basis of 29 LDCs surveyed, concluded that subsidies extended by the LDCs had caused the price of electricity, water and fossil fuels to fall below the cost (not even including externality costs). A similar report was also given by the International Energy Agency (1999), which "has estimated that in India, China and the Russian Federation, full-cost pricing would reduce energy consumption by 7, 9 and 16 percent, respectively ... where most of the departure from social cost pricing is attributed to energy subsidies". Friedman (2006) argues that China has one political advantage over the United States, that it can make decisions against special interests and all bureaucratic obstacles or worries about voter backlash and simply order a change. If the United States could do that for one day, to institute responsible regulations, standards, education, infrastructure and prices, he argues, then our system would make sure that they are enforced via legal action if necessary.

Fifth, heterogeneity may be a desirable attribute in policy. Adler (2002) advocates 'competitive federalism as a promising alternative to rigid, inefficient national regulation and regimentation'. He asserts that many environmental issues are local or regional in nature, so require local knowledge and solutions. He advocates a national policy of 'ecological forbearance', where states would petition the EPA for waivers of particular requirements in order to pursue state-level innovation and experimentation. Rather than rely on a patchwork system of prescriptive policies which may slow innovation and impose non-trivial costs, a new system might encourage (or at least permit) states to deviate from the national norm in pursuit of better solutions. Kemp (1997) even goes so far as to argue that uncertainty about regulation may spur innovation, and in fact simultaneous innovation and regulation discussions may offer the best path. While both are fairly dramatic positions for mainstream economists to espouse, most would indeed agree that a variety of approaches is wise, in order to encourage the widest possible base of knowledge to tackle problems common to us all. The comprehensive survey of alternative energy technologies by Hoffert et al. (2002) concludes that a portfolio approach for simultaneous development is the only realistic strategy for success.

Strategies for Effective Innovation

There are several key lessons then for the effective policymaker. First, economics has effective tools, many of them highly mathematical models that are difficult to summarize in prose, which permit policymakers to simulate policy choices and their effects on the environment and related innovation. None are perfect, nor do they pretend to be. However, they all work from a common set of assumptions that empirical studies have proven to be reasonable. Naturally, the future is unpredictable and innovation is doubly so by its creative nature. A direct comparison of their predictions would be a valuable contribution to the literature, but include at least the following: Kaufmann (1997), Messner (1997), Goulder and

Schneider (1999), Goulder and Mathai (2000), Van der Zwaan et al. (2002), Buonnano et al. (2003), Bosetti et al. (2006a, 2006b), Popp (2004, 2006b). Keeping in mind that results are often sensitive not only to the underlying parameters but also to specifications of the functions in the model (see Soderholm and Sundqvist, 2007, on the fragility of modelling results based on alternative specifications of a learning function), parallel analyses are warranted.

Second, we have pointed to five themes that resonate with all economists as challenges to eco-innovation: IPRs, economies of scale, markets and incentives, complex systems and policy. Economists are virtually unanimous in pointing to IPRs as an essential precondition, not a constraint to innovation, but there are some suggestions in the literature about ways in which the existing system could be improved for the stimulation of eco-innovation specifically. Economies of scale are undeniably important, raising the question of how to present a sufficiently sized market to drive costs per unit down to levels permitting widespread access. The market has a clear role to play, but when the systemic effects are complex and uncertainty is high, the role of policy becomes disproportionate.

Aubert (2004) identifies a number of features that characterize successful innovation in less developed nations in particular: motivated individuals or groups, assistance from foreign partners in financing or market networks, support from local politicians to overcome administrative barriers and concentration in a well-defined locality or industry. Market forces will shape some of these factors, but there is a clear role for effective policymaking in order to foster these conditions and facilitate environmental innovation in the nations that most require, and perhaps can least afford, these technologies.

Conclusions and Policy Implications

The challenge to policymakers is one of balance: encouraging competition while guaranteeing a large market for minimum economic scale, reducing uncertainty about future resource prices while keeping alternatives open, offering rights of exclusion to IP holders while not curtailing the ability of sequential innovators to build upon past successes, promoting social goals while respecting market pressures. This is no doubt complicated by the policy distortions and market failures that characterize the markets for competing and complementary goods. The exercise is one of structuring the future but permitting innovators to creatively fill in the frame and to build out in unpredictable directions. The unenviable challenge requires flexibility and vigilance by policymakers, but the challenge is only commensurate with the stakes.

References

Adler, J.H. January 2002. *Let Fifty Flowers Bloom: Transforming the States into Laboratories of Environmental Policy*. The Federalism Project/American Enterprise Institute.
Arimura, T.H., A. Hibiki and N. Johnstone. 2007. 'An empirical study of environmental R&D: What encourages facilities to be environmentally innovative?' in Johnstone, N. (ed.), *Environmental Policy and Corporate Behaviour*. Cheltenham, UK: Edward Elgar.
Arrow, K., P. Dasgupta, L. Goulder, G. Daily, P. Ehrlich, G. Heal, S. Levin, K.-G. Maler, S. Schneider, D. Starrett and B. Walker. 2004. 'Are we consuming too much?', *Journal of Economic Perspectives*, 18(3): 147–72.
Atkinson, S.E. and R. Halvorsen. 1984. 'A new hedonic technique for estimating attribute demand: An application to the demand for automobile fuel efficiency', *Review of Economics and Statistics*, 66: 417–26.
Aubert, J.-E. 2004. *Promoting Innovation in Developing Countries: A Conceptual Framework*. World Bank Institute.
Barton, J.H. December 2007. 'Intellectual Property and Access to Clean Energy Technologies in Developing Countries', ICTSD Programme on IPRs and Sustainable Development, Selected Issue Briefs, no. 2.
Baylis, R., L. Connell and A. Flynn. 1998. 'Company size, environmental regulation and ecological modernization—Further analysis at the level of the firm', *Business Strategy and the Environment*, 7(5): 285–96.
Bernauer, T., S. Engels, D. Kammerer and J. Sejas Nogareda. 2006. 'Explaining Green Innovation: Ten Years after Porter's Win-Win Proposition: How to Study the Effects of Regulation on Corporate Environmental Innovation?' Working Paper No. 17, ETH Zurich and University of Zurich.
Berrone, P., L. Gelabert, A. Fosfuri and L.R. Gomez-Mejia. 2007. 'Can Institutional Forces Create Competitive Advantage? An Empirical Examination of Environmental Innovation', IESE Business School Working Paper no. 723.
Berry, S., S. Kortum and A. Pakes. 1996. 'Environmental change and hedonic cost functions for automobiles', *Proceedings of the National Academy of Sciences*, 93: 12731–12738.
Bonato, D. and A. Schmutzler. 2000. 'When do firms benefit from environmental regulations? A simple microeconomic approach to the porter controversy', *Schweizerische Zeitschrift fur Volkswirtschaft und Statistik*, 136(4): 513–30.
Bonifant, B.C., M.B. Arnold and F.J. Long. 1995. 'Gaining competitive advantage through environmental investments', *Business Horizons*, 38(4): 37–47.

Bosetti, V., C. Carraro, M. Galeotti, E. Massetti and M. Tavoni. 2006a. 'WITCH: A world induced technical change hybrid model', *Energy Journal*, 27: 13–37.
Bosetti, V., C. Carraro and M. Galeotti. 2006b. 'The dynamics of carbon and energy intensity in a model of endogenous technical change', *Energy Journal*, 27(1): 191–205.
Bovea, M.D. and R. Vidal. 2004. 'Increasing product value by integrating environmental impact, costs and customer valuation', *Resources, Conversation and Recycling*, 41(2): 133–46.
Brunnermeier, S.B. and M.A. Cohen. 2003. 'Determinants of environmental innovation in US manufacturing industries', *Journal of Environmental Economics and Management*, 45(2): 278–93.
Buonanno, P., C. Carraro and M. Galeotti. 2003. 'Endogenous induced technical change and the costs of Kyoto', *Resource and Energy Economics*, 25(1): 11–34.
Cahoy, D.R. and L. Glenna. 2009. 'Private ordering and public energy innovation policy', *Florida State University Law Review*, 36(3): 415–58.
Cainelli, G., M. Mazzanti and R. Zoboli. 2007. 'Environmentally-oriented innovative strategies and firm performances in services: Micro-evidence from Italy', *Nota di Lavoro*, 104.
Calef, D. and R. Goble. 2005. 'The Allure of Technology: How France and California Promoted Electric Vehicles to Reduce Urban Air Pollution', FEEM Working Paper no. 07.05.
Carraro, C. and A. Soubeyran. 1996. 'Environmental Taxation and Employment in a Multi-Sector General Equilibrium Model', in Carraro, C. and D. Siniscalco (eds), *Environmental Fiscal Reform and Unemployment*. London: Kluwer Academic Publishers.
Carraro, C. and D. Siniscalco. 1994. 'Environmental policy reconsidered: The role of technology innovation', *European Economic Review*, 38: 545–55.
Chandrashekar, S. and K.P. Basvarajappa. 2001. 'Technological innovation and economic development: Choices and challenges for India', *Economic & Political Weekly*, 36(34): 3238–45.
Cleff, T. and K. Rennings. 1999. 'Determinants of environmental product and process innovation', *European Environment*, 9(5): 191–201.
Cohen, W.M. 2005. 'Patents and appropriation: Concerns and evidence', *Journal of Technology Transfer*, 30(1/2): 57–71.
Copenhagen Economics A/S and The IPR Company ApS. 2009. 'Are IPRs a Barrier to the Transfer of Climate Change Technology?', Commissioned by the European Commission (DG Trade), 19 January.
Costantini, V. and F. Crespi. 2007. 'Environmental Regulation and the Export Dynamics of Energy Technologies', FEEM Working Paper 53.2007.
Cowan, R. 1990. 'Nuclear power reactors: A study in technological lock-in', *Journal of Economic History*, 50: 541–67.
Cowan, R. and D. Kline. 1996. *The Implications of Potential 'Lock-In' in Markets for Renewable Energy*. Golden, CO: National Renewable Energy Laboratory.
Cowan, R., and P. Gunby. 1996. 'Sprayed to death: Path dependence, lock-in and pest control strategies', *Economic Journal*, 106: 521–42.
Cowan, R. and S. Hulten. 1996. 'Escaping lock-in: The case of the electric vehicle', *Technological Forecasting and Social Change*, 53: 61–79.
Crabb, J. and D.K.N. Johnson. 2010. 'Fueling innovation: The impact of oil prices and CAFÉ standards on energy-efficient automotive technology', *Energy Journal*, 31(1): 199–216.

Craig, J.B.L. and K.A. Moores. 2006. '10-Year longitudinal investigation of strategy, systems, and environment on innovation in family firms', *Family Business Review* 19(1): 1–10.
Daniels, B. and D.K.N. Johnson. 2014. 'More Where That Came From: Induced Innovation in the American Oil and Gas Sectors', Colorado College Department of Economics and Business Working Paper 2014–03, July 2014.
Desrochers, P. 2008. 'Did the invisible hand need a regulatory glove to develop a green thumb? Some historical perspective on market incentives, win-win innovations and the Porter hypothesis', *Environmental and Resource Economics*, 41(4): 519–39.
Driesen, D.M. 2003. *Economic Dynamics of Environmental Law*. Cambridge, MA: MIT Press.
Fagerberg, J., D.C. Mowery and R.R. Nelson. 2005. *The Oxford Handbook of Innovation*. Oxford: Oxford University Press.
Freeman, C. 1994. 'The economics of technical change', *Cambridge Journal of Economics*, 18: 463–514.
Friedman, T.L. 2006. *Hot, Flat and Crowded*. New York: Farrar, Straus and Giroux.
Frondel, M., J. Horbach and K. Rennings. 2004. 'End-of-Pipe or Cleaner Production? An Empirical Comparison of Environmental Innovation Decisions Across OECD Countries', ZEW Discussion Paper no. 04–82, Mannheim: Center for European Economic Research.
Gallini, N.T. 1992. 'Patent policy and costly imitation', *Rand Journal of Economics*, 23(1): 52–63.
Gallini, N.T. and S. Scotchmer. 2001. 'Intellectual Property: When is it the Best Incentive System?' Institute of Business and Economic Research, Department of Economics, University of California, Berkeley, Paper E01'303.
Goldberg, P.K. 1998. 'The effects of the corporate average fuel efficiency standards in the U.S.', *Journal of Industrial Economics*, 46: 1–3.
Goodman, A.C. 1983. 'Willingness to pay for car efficiency: A hedonic price approach', *Journal of Transport Economics*, 17: 247–66.
Goolsbee, A. 1998. 'Does government R&D policy mainly benefit scientists and engineers?', *American Economic Review*, 88(2): 299–302.
Goulder, L.H. and K. Mathai. 2000. 'Optimal CO_2 abatement in the presence of induced technological change', *Journal of Environmental Economics and Management*, 39: 1–38.
Goulder, L.H. and S.H. Schneider. 1999. 'Induced technological change and the attractiveness of CO_2 emissions abatement', *Resource and Energy Economics*, 21(3–4): 211–53.
Green, J.R. and S. Scotchmer. 1995. 'On the division of profit in sequential innovation', *Rand Journal of Economics*, 26(1): 20–33.
Greene, D.L. 1990. 'CAFE or price? An analysis of the effects of federal fuel economy regulations and gasoline price on new car MPG, 1978–89', *Energy Journal*, 11(3): 37–57.
Grübler, A. and S. Messner. 1999. 'Technological change and the timing of mitigation measures', *Energy Economics*, 20: 495–512.
Grübler, A., N. Nakicenovic and D.G. Victor. 1999. 'Dynamics of energy technologies and global change', *Energy Policy*, 27: 247–80.
Hall, B. 1996. 'The private and social returns to research and development', in Smith, B. and C. Barfield (eds), *Technology, R&D, and the Economy, Brookings*. Washington, D.C.: Brookings Institution and the American Enterprise Institute, pp. 140–83.
Hall, B. and J. van Reenen. 2000. 'How effective are fiscal incentives for R&D? A review of the evidence', *Research Policy*, 29: 449–69.

Hoffert, M.I., K. Caldeira, G. Benford, D.R. Criswell, C. Green, H. Herzog, A.K. Jain, H.S. Kheshgi, K.S. Lackner, J.S. Lewis, H.D. Lightfoot, W. Manheimer, J.C. Mankins, M.E. Mauel, L.J. Perkins, M.E. Schlesinger, T. Volk and T.M.L. Wigley. 2002. 'Advanced technology paths to global climate stability: Energy for a Greenhouse planet', *Science*, 298(1): 981–86.

Hopenhayn, H.A. and M.F. Mitchell. 2001. 'Innovation variety and patent breadth', *Rand Journal of Economics*, 32(1): 152–66.

Hutchinson, C. 2006. 'Does TRIPS facilitate or impede climate change technology transfer into developing countries?', *University of Ottawa Law and Technology Journal*, 3(2): 517–37.

Hutter, C. and T. Starmack. 2013. 'Tesla Roadster: The New Standard of Electric Automobiles', Unpublished working paper at University of Pittsburgh, Accessed from: http://136.142.82.187/eng12/history/spring2013/pdf/3096.pdf on 11 August 2014.

International Energy Agency. 1999. *World Energy Outlook—1999 Insights. Looking at Energy Subsidies: Getting the Price Right*. London: IEA Publications.

Islas, J. 1997. 'Getting round the lock-in in electricity generating systems: The example of the gas turbine', *Research Policy*, 26: 49–66.

Jaffe, A.B. and K.L. Palmer. 1997. 'Environmental regulation and innovation—A panel data study', *The Review of Economics and Statistics*, 79(4): 610–19.

Jaffe, A.B., S. Peterson, P. Portney and R. Stavins. 1995. 'Environmental regulation and the competitiveness of US manufacturing: What does the evidence tell us?', *Journal of Economic Literature*, 33: 132–63.

Jaffe, A.B., R.G. Newell and R.N. Stavins, 2001, 'Technological Change and the Environment', KSG Working Paper 00–002.

——— 2002. 'Technological change and the environment', in Maler, K.G. and J. Vincent (eds), *Handbook of Environmental Economics*. Amsterdam: North-Holland/Elsevier Science.

——— 2004. 'Technology policy for energy and the environment', *Innovation Policy and the Economy*, 4: 35–68.

Johnson, D.K.N. and S. Scowcroft. 2013. 'The Importance of Being Steve: An Econometric Analysis of the Contribution of Steve Jobs's Patents to Apple's Stock Market Valuation', Colorado College Department of Economics and Business Working Paper 2013–01.

Johnstone, N., I. Hascic, L. Clavel and F. Marical. undated. 'Renewable Energy Policies and Technological Innovation: Empirical Evidence Based on Patent Counts', Working Paper.

Johnstone, N., I. Hascic and D. Popp. 2008. 'Renewable Energy Policies and Technological Innovation: Evidence Based on Patent Counts', NBER Working Paper no. 13760.

Jones, C.I. and J.C. Williams. 1998. 'Measuring the social return to R&D', *Quarterly Journal of Economics*, 113(4): 1119–35.

Joskow, P.L. and G.A. Rozanski. 1979. 'The effects of learning by doing on nuclear plant operating reliability', *Review of Economics and Statistics*, 61: 161–68.

Kanwar, S. 2007. 'Business enterprise R&D, technological change, and intellectual property protection', *Economic Letters*, 96: 120–26.

Kanwar, S. and R.E. Evenson. 2001. 'Does Intellectual Property Protection Spur Technological Change?' Yale University Economic Growth Center Discussion Paper no. 831.

Kanwar, S. and R. Evenson. 2009. 'On the strength of intellectual property protection that nations provide', *Journal of Development Economics*, 90(1): 50–56.

Kaufmann, R.K. 1997. 'Assessing the DICE model: Uncertainty associated with the emission and retention of Greenhouse gases', *Climatic Change*, 35(4): 435–48.

Karl, H., A. Möller, X. Matus, E. Grande and R. Kaiser. 2005. 'Environmental Innovations: Institutional Impacts on Co-operations for Sustainable Development', FEEM Working Paper no. 58.05.

Katsoulacos, Y. and A. Xepapadeas. 1995. 'Environmental policy under oligopoly with endogenous market structure', *Scandinavian Journal of Economics*, 97(3): 411–20.

Kemp, R. 1997. *Environmental Policy and Technical Change—A Comparison of the Technological Impact of Policy Instruments*. Cheltenham, UK; Brookfield, US: Edward Elgar.

Kemp, R. and L. Soete. 1990. 'Inside the "Green Box": On the economics of technological change and the environment', in Freeman, C. and L. Soete (eds), *New Explorations in the Economics of Technological Change*. London: Pinter.

Laffont, J. and J. Tirole. 1996. 'Pollution permits and compliance strategies', *Journal of Public Economics*, 62: 85–125.

Lanoie, P., J. Laurent-Luccetti, N. Johnstone and S. Ambec. September 2007. 'Environmental policy, innovation and performance: New insights on the Porter hypothesis', *Cirano*, Scientific Series, 2007s-19.

Larson, A. 2006. 'Fuel Cell Technology and Market Opportunities', Darden Case no. UVA-ENT-0016. Available at SSRN.

Lester, R.K. and M.J. McCabe. 1993. 'The effect of industrial structure on learning by doing in nuclear power plant operation', *Rand Journal of Economics*, 24: 418–38.

Levin, R.C., A.K. Klevorick, R.R. Nelson and S.G. Winter. 1987. 'Appropriating the returns from industrial research and development', *Brookings Papers on Economic Activity*, 3: 783–820.

Love, J.P. 2007. 'Recent examples of the use of compulsory licenses on patents', *Knowledge Ecology International Research Note*, 2(1).

Mandel, G. 2005. 'Promoting Environmental Innovation with Intellectual Property Innovation: A New Basis for Patent Rewards', *Journal of Science, Technology, and Environmental Law's Symposium Paper*, 24(1): 51–69.

Mansfield, E. 1977. 'Social and private rates of return from industrial innovations', *Quarterly Journal of Economics*, 91: 221–40.

Matthews, H.S. and L.B. Lave. 2000. 'Applications of environmental valuation for determining externality costs', *Environmental Science and Technology*, 34(8): 1390–95.

Mazzanti, M., G. Cainelli and R. Zoboli. 2008. 'Environmentally-Oriented Innovative Strategies and Firm Performances in Services—Micro-Evidence from Italy', FEEM Working Paper no. 104.2007.

——— 2009. 'The Relationship between Environmental Efficiency and Manufacturing Firm's Growth', FEEM Working Paper 258.

Menz, F.C. and S. Vachon. 2006. 'The effectiveness of different policy regimes for promoting wind power: Evidence from the states', *Energy Policy*, 34(14): 1786–96.

Messner, S. 1997. 'Endogenized technological learning in an energy systems model', *Journal of Evolutionary Economics*, 7(3): 291–313.

Mohr, R.D. 2002. 'Technical change, external economies, and the Porter hypothesis', *Journal of Environmental Economics and Management*, 43: 158–68.

Montero, J.-P. 2000. 'Optimal design of a phase-in emissions trading program', *Journal of Public Economics*, 75(2): 273–91.

Mowery, D.C. and N. Rosenberg. 1979. 'The influence of market demand upon innovation—A critical review of some empirical studies', *Research Policy*, 8: 102–53.

Murphy, J. and A. Gouldson. 2000. 'Environmental policy and industrial innovation—Integrating environment and economy through ecological modernisation', *Geoforum*, 31(1): 33–44.

Neij, L. 1997. 'Use of experience curves to analyze the prospects for diffusion and adoption of renewable energy technology', *Energy Policy*, 23: 1099–1107.

Newell, R.G., A.B. Jaffe and R.N. Stavins. 1999. 'The induced innovation hypothesis and energy-saving technological change', *The Quarterly Journal of Economics*, 114(3): 941–75.

Nordhaus, W.D. 1994. *Managing the Global Commons: The Economics of the Greenhouse Effect*. Cambridge, MA: MIT Press.

Nordhaus, W.D. and J. Boyer. 2000. *Warming the World: Economic Models of Global Warming*. Cambridge, MA: MIT Press.

OECD. 1997. *The Oslo Manual—Proposed Guidelines for Collecting and Interpreting Technological Innovation Data*. Paris.

Ohta, M. and Z. Griliches. 1986. 'Automobile prices and quality: Did the gasoline price increases change consumer tastes in the U.S.?', *Journal of Business & Economic Statistics*, 4: 187–98.

Pakes, A. 1985. "On patents, R&D, and the stock market rate of return', *Journal of Political Economy*, 93: 390–409.

Pakes, A., S. Berry and J.A. Levinsohn. 1993. 'Applications and limitations of some recent advances in empirical industrial organization: Prices indexes and the analysis of environmental change', *American Economic Review*, 83: 240–46.

Park, W.G. and D.C. Lippoldt. 2008. 'Technology Transfer and the Economic Implications of the Strengthening of Intellectual Property Rights in Developing Countries', OECD Trade Policy Working Papers 62.

Parry, I.W.H. 1998. 'Pollution regulation and the efficiency gains from technological innovation', *Journal of Regulatory Economics*, 14: 229–54.

Popp, D.C. 2001. 'The effect of new technology on energy consumption', *Resource and Energy Economics*, 23(3): 215–39.

——— 2002. 'Induced innovation and energy prices', *American Economic Review*, 92: 160–80.

——— 2003. 'Pollution control innovations and the Clean Air Act of 1990', *Journal of Policy Analysis and Management*, 22(4): 641–60.

——— 2004. 'ENTICE: Endogenous technological change in the DICE model of global warming', *Journal of Environmental Economics and Management*, 48(1): 742–68.

——— 2005. 'Lessons from patents: Using patents to measure technological change in environmental models', *Ecological Economics*, 54: 209–26.

——— 2006a. 'R&D subsidies and climate policy: Is there a "free lunch"?', *Climate Change*, 77: 311–41.

——— 2006b. 'ENTICE-BR: The effects of backstop technology R&D on climate policy models', *Energy Economics*, 28: 188–222.

——— 2010. 'Innovation and climate policy', *Annual Review of Resource Economics*, 2: 275–98.

Porter, M.E. and C. van der Linde. 1995. 'Toward a new conception of the environment–competitiveness relationship', *Journal of Economic Perspectives*, 9(4): 97–118.

Press, D. 2007. 'Industry, environmental policy, and environmental outcomes', *Annual Review of Environment and Resources*, 32: 317–44.
Rehfeld, K.M., K. Rennings and A. Ziegler. 2006. 'Integrated product policy and environmental product innovations—An empirical analysis', *Ecological Economics*, 61(1): 91–100.
Rennings, K. 1998. 'Towards a Theory and Policy of Eco-Innovation—Neoclassical and (Co-)Evolutionary Perspectives', ZEW Discussion Paper 98–24.
Requate, T. 1998. 'Incentives to innovate under emission taxes and tradeable permits', *European Journal of Political Economy*, 14: 139–65.
Ricci, F. 2004. 'Channels of Transmission of Environmental Policy to Economic Growth: A Survey of the Theory', FEEM Working Paper 52.04.
Roberts, J.A. 1996. 'Green consumers in the 1990s: Profile and implications for advertising', *Journal of Business Research*, 36: 217–31.
Roediger-Schluga, T. 2004. *The Porter Hypothesis and the Economic Consequences of Environmental Regulation*. Northampton, MA: Edward Elgar Publishing.
Schmutzler, A. 2001. 'Environmental regulations and managerial myopia', *Environmental and Resource Economics*, 18: 87–100.
Scotchmer, S. 1991. 'Standing on the shoulders of giants: Cumulative research and the patent law', *Journal of Economic Perspectives*, 5(1): 29–41.
——— 1996. 'Protecting early innovators: Should second-generation products be patentable?', *Rand Journal of Economics*, 27(2): 322–31.
——— 1999. 'Cumulative innovation in theory and practice', Goldman School of Public Policy Working Paper 240, University of California, Berkeley.
Sharp, J.A. and D.H.R. Price. 1990. 'Experience curve models in the electricity supply industry', *International Journal of Forecasting*, 6: 531–40.
Shavinina, L. 2006. *International Handbook on Innovation*. Oxford: Elsevier.
Shrivastava, P. 1995. 'Environmental technologies and competitive advantage', *Strategic Management Journal*, 16: 183–200.
Sinclair-Desgagne, B. and H.L. Gabel. 1997. 'Environmental auditing in management systems and public policy', *Journal of Environmental Economics and Management*, 33(3): 331–46.
Sneed, K.A. and D.K.N. Johnson. 2009. 'Selling ideas: The determinants of patent value in an auction environment', *R&D Management*, 39(1): 87–94.
Soderholm, P. and T. Sundqvist. 2007. 'Empirical challenges in the use of learning curves for assessing the economic prospects of renewable energy technologies', *Renewable Energy: An International Journal*, 32(15): 2559–78.
Stoneman, P. 1995. *Handbook of the Economics of Innovation and Technological Change*. Oxford: Wiley-Blackwell.
Straughan, R.D. and J.A. Roberts. 1999. 'Environmental segmentation alternatives: A look at green consumer behavior in the new millennium', *Journal of Consumer Marketing*, 16(6): 558–75.
Teece, D.J. 1998. 'Capturing value from knowledge assets: The new economy, markets for know-how, and intangible assets', *California Management Review*, 40(3): 55–79.
Tidd, J. 2006. 'Innovation Models', Imperial College London, Tanaka Business School, Discussion Paper 1.
Van der Zwaan, B.C.C., R. Gerlagh, G. Klaassen and L. Schrattenholzer. 2002. 'Endogenous technological change in climate change modeling', *Energy Economics*, 24(1): 1–19.

Wilcox, J. 1984. 'Automobile fuel efficiency: Measurement and explanation', *Economic Inquiry*, 22: 375–85.

Yiannaka, A. and M. Fulton. 2001. 'Strategic Patent Breadth for Drastic Product Innovations', Working Paper, American Agricultural Economics Association Annual Meeting, Chicago.

—— 2003. 'Strategic Patent Breadth and Entry Deterrence with Drastic Product Innovations', Working Paper, First Biennial Conference of the Food System Research Group, Madison, WI.

Zimmerman, M.B. 1982. 'Learning effects and the commercialization of new energy technologies: The case of nuclear power', *Bell Journal of Economics*, 13: 297–310.

3

Social Inclusion and Institutional Innovations: Working towards a Policy-theoretical Framework

*M.A. Oommen**

Social inclusion has occupied an important place in contemporary development discourse (see Rodgers et al., 1995; Kabeer, 2006 among others). It is projected as the major development goal by the planners and policymakers of India. The three-volume 12th Five-Year Plan (2012–17) draft with the subtitle 'Faster More Inclusive and Sustainable Growth' approved by the National Development Council on 27 December 2012 has added a new chapter captioned *Social Inclusion*. This is expected to be different from 'a purely corporate-led growth strategy' (Planning Commission, 2012:5)]. Even so, the term is neither defined nor strategized. The underlying assumption of this chapter is that the existing institutional architecture to be truly inclusive needs reform and that in any redesigning exercise decentralized democratic institutions have a place of significance. Ever since the United Nations/Centre for Development Studies' (CDS) (1975) study identified the relatively high social achievements of the regional economy of Kerala, with low per capita income, Kerala has attracted considerable attention from development thinkers the world over. In the development literature, Kerala's development experience came to be widely characterized as a 'Model' (more as an *expost facto* generalization of a historically evolved experience than as an exemplar). A high human development without a sharp rural–urban divide or obvious gender disparity, an extremely respectable food and nutritional security, a wide social security system, a very high female–male ratio, fertility decline way below replacement rate achieved as a matter of

---

* This chapter is written in fond memories of Robert Evenson with whom I worked in the Yale University as a Senior Fulbright Scholar during the academic year 1974–75. I am thankful to D. Shyjan who read the first draft of this chapter and offered comments.

choice, collapse of an oppressive caste structure (with not only untouchability but also with unapproachability and unseeability as its dominant character) and imparting self-respect to the 'lower' caste/strata people, abolition of feudal relations of production in the agrarian sector, a multi-religious, multicultural population living in symbiotic relationships and so on, indeed, are great social achievements in human history (see Robin Jeffrey, 1992, 2001; Dreze and Sen, 1995 and Oommen, 1999; among others). The 'big bang' People's Plan Campaign (PPC) launched in August 1996 backed up by the decision to devolve 35–40 per cent of the state plan outlay was in the best traditions of Kerala's public action (see Isaac and Franke, 2000). Besides the capability demonstrated in mobilizing people and resources for augmenting productive base and the great potential for improving the quality of public services, the most important contribution of the people's plan has been the introduction of a new methodology for decentralized planning which has widened the avenues of people's participation. Indeed, participating in social choice is a valued capability and freedom (see Sen, 1999). The purpose of this chapter is to explore the possibilities of building the contours of a policy-theoretical frame for a meaningful social inclusion based on the institutional innovations in decentralized governance initiated by the Kerala state since the mid-1990s.

Social Inclusion, Inclusive Growth and Inclusive Development: Some Conceptual Issues

At the outset, it is important to clarify the concepts of social inclusion, inclusive growth and inclusive development because of the confusing explanations centring around the terms. No theoretical postulate or policy option can be built on ambiguous concepts. We prefer the term social inclusion rather than inclusive growth which is used profusely in the 11th and 12th Plan documents. In 'inclusive growth', the major emphasis is on expanding the size of national income and is projected as a normative framework for development, whereas a more fundamental and rational approach is to focus on the lives that people live and the freedom they enjoy. Economic growth is only a means to improve the quality of human well-being. It is not an end in itself. Looking back, the concept of *Hind Swaraj* of Mahatma Gandhi could be envisioned as a goal towards inclusive human development. But this is too idealistic with almost impossible demands on transforming human nature and values. The 'reverse engineering' involved is indeed too formidable.

For the purpose of this chapter, we prefer to use the concept of social inclusion rather than inclusive growth and inclusive development where the context warrants the use. Social inclusion implies that no one should be left out or kept out including the social misfits. We may elaborate the concept more specifically. First, unless you have the exchange entitlements to participate in the market (e.g. income, employment, assets, social security, credit and so on), you will be kept out. Charity can help, but that is antithetical to all genuine ideas of inclusion. Social arrangements for the public provision of education, health care, food entitlements, social net provision and so on assume significance here. They are the barometers of the progress and sincerity of a society. Second, social inclusion is not a binary model in which some are 'in' and others are out. Inclusion has no meaning unless it is done to enhance human dignity and freedom. In other words, inclusion should be on equitable terms. The classic example of inequitable social inclusion is the caste system of India. Institutional architecture and social arrangements for provision of basic needs and all arrangements for building capabilities have to be suitably reformed and modified to promote social inclusion. It is assumed that Panchayati Raj Institutions (PRIs) following the 73rd Constitutional Amendment have a special role in promoting social inclusion as part of an integrated design. Third, growing economic inequalities are an expression of social exclusion. In a situation where growth keeps increasing and poverty is reduced, but at the same time faces widening inequalities, there can be no equitable social inclusion (Sen, 2000). The opportunities offered by the market in the social production of goods and services should not be appropriated by a small minority. The institutional arrangements must provide equitable social and economic opportunities. Fourth, inclusive development should be participatory. This means increased efforts are needed to include the disadvantaged people in the process of decision-making (e.g. decentralized planning in Kerala) as well as in the process of social production. Fifth, relevant policy formulations related to social inclusion will have to recognize the diversity of exclusions which result in capability deprivations that range from enforced exclusion like slavery and attached labour to credit market exclusion, gender-related exclusions, lack of educational facilities, health care and the like (see Sen, 2000).

The Government of India (GoI) has never precisely defined or clearly and comprehensively strategized its policy regarding the provision of entitlements to include the excluded. Sen's capability and freedom perspective has a great potential to be used as a normative framework to evaluate development practices. This is in sharp contrast to the unidimensional measure of growth of income or consumption or industrialization and the like even when they are declared to be pro-poor.[1] Without expanding the capabilities and freedom of individuals, there can be no social mobility and social inclusion. Bhagwati and Panagariya (2012) take a somewhat different view and argue that what is important for India is to promote economic growth through more and better economic reforms. They do not address the question of social inclusion as their argument is that more and rapid growth can take care of that. Sen in his various writings argues that people differ in their ability to convert goods into valuable achievements due to personal and locational factors, besides the social arrangements in the provision of education, health care, social security, democratic rights, participatory institutions, ruling central values such as the treatment of women and so on. Democratic decentralized governance if well designed and strategized can help in the process of social inclusion considerably.

A fundamental flaw in the approach of the GoI about inclusive growth has been its treatment of inclusiveness and growth as disparate entities. Presenting growth and inclusiveness as co-equal development objectives, M.S. Ahluwalia (2011) observes:

> The Eleventh Plan aimed at delivering faster and more inclusive growth and it is logical to assess performance against this dual objectives [p. 88]

and goes on to say that,

> it is difficult to assess inclusiveness than performance on growth. [p. 89]

Treating inclusiveness and growth as separate entities and again maintaining that as 'logical' policy choice rules out structural transformations, institutional reforms, market failures, redistribution strategies, envisioning sustained capability enhancing policies and so on, besides upholding the trickle-down approach. By increasing the number of monitorable variables of inclusiveness to 27 (health, education, sanitation, etc.) as is done by the

Planning Commission, you cannot define social inclusion meaningfully. Increasing the number of target variables to assess departmental progress cannot be inclusive growth or development by any reckoning.

Not only that but Ahluwalia (2011) also speaks of the 'systemic reforms' since 1991 based on a wider play of market forces, gradual liberalization of the financial sector and the opening up of the economy to world trade and capital flows (Ahluwalia, 2011). But the moot question is whether 'the cumulative effect of the systemic reforms' has promoted social inclusion, in the sense in which we have postulated it. Presumably, the more important question is: Why have the planners and policymakers of India failed to think of corresponding systemic reforms for a growth process that includes those excluded or likely to be marginalized on the wayside of growth?

Logically and factually, rising inequalities are an inherent by-product of the growth process in any market-mediated economy particularly with deep disparities in initial endowments.[2] As Ifzal Ali (2007) argues, inequality has to be strongly contained through relevant policies to promote an inclusive democracy. The GoI does not have a deliberate inequality reduction strategy.[3] Besides the lack of positive redistributive strategies, the poor progress and appalling inequities in the access to health services and education are now very well documented (Baru Rama et al., 2010; William Joe et al., 2008; Sonalde Desai and Amaresh Dubey, 2011, among others). That large inequities in health and access to health services continue to persist and have even widened across states, between rural and urban areas and within communities in the India of faster growth, is well documented by five scholars (most of them health and community medicine experts) (Baru Rama et al., 2010). This is a powerful evidence to show that the goal of inclusive growth, leave alone social inclusion, remains a far cry.

The approach of the Asian Development Bank to inclusive growth seems to be far more meaningful than that of the GoI as is clear from the following definition:

> Growth is inclusive when it allows all members of society to participate in, and contribute to the growth process on an equal footing regardless of their individual circumstances. (Ali and Zhuang, 2007:10)

This definition is positive and implicitly envisages a widespread employment pattern that grows faster than the rate of growth. However, a growth process that works towards expanding equitable employment opportunities must spell out the technological paradigm underlying the growth process, define the policy choices relating to the composition of commodity flow, the structure and pattern of initial endowments, the social arrangements for enhancing the capabilities of individuals and the freedom needed to achieve the capabilities and so on. Apart from the protective support given to the traditional industries such as khadi, handicrafts, coir, cashew and the like, India is for a capital-intensive technological paradigm. The latest numbers regarding employment from the 66th-round National Sample Survey (NSS) data (2009–10) reveal a 'jobless growth' story as regards the rate of growth in employment which shows a dramatic deceleration in the rate of employment generation to 0.8 per cent per annum during the quinquennium ending 2009–10 as against 2.7 per cent during the previous quinquennium (see Chandrasekhar and Jayati Ghosh, 2011).

In sum, the concepts inclusive growth and inclusive development that frequently appear in GoI's policy documents and often used interchangeably need precise definition and clear policy strategization. It was because of this ambiguity that we have tried to broadly define social inclusion which envisages an equitable and inclusive development.

Innovative Institutions as Change Agents

In this section, we examine the potential and possibilities of institutional innovations. Long back John Rawls pointed out:

> Justice is the first virtue of social institutions, as truth is of systems of thought. A theory however elegant and economical must be rejected or revised if it is untrue; likewise laws and institutions no matter how efficient and well-arranged must be reformed or abolished if they are unjust. (Rawls, 1971:3).

Many of India's political institutions were borrowed or adopted from the West and surely need reform to suit her growing needs and challenges. For example, as Amartya Sen (1981) has demonstrated, it was not the lack of food availability or production, but institutional failure that caused the great famines of Bengal, Ethiopia, Sahel and so on in recent times (Sen, 1981). Relevant institutional innovation is continuously needed in strategizing development based on meaningful social inclusion. The 73rd/74th Constitutional Amendments envisaged a welcome decentralization reform in Indian federal polity.

For the purpose of this chapter, innovation refers to the use of new ideas or new ways of doing things. Quite often, innovation is associated with high-tech and is considered to be motivated by profit. For such innovators, consumers and markets are the target field. The institutional innovations that we postulate, however, are social innovations motivated by the goal of social well-being. Innovations need not be about brand new ideas. It is as much about developing and adapting or refashioning existing ideas. But the application and adaption should yield enduring social value. The nature and quality of institutional architecture is crucial to any transformative project.

In socially oriented institutional innovations (e.g. the participatory mechanisms developed as part of the People's Plan and PRIs in Kerala), how challenges are framed is important in determining the kind of innovation one achieves. When you formulate problems and challenges keeping the needs of the micro-unit, individual, family, community and the like, the chances of making radical institutional innovations increase. Here, people are not treated as consumers or customers, but as agents and participants. People are also not conceived as the recipients or beneficiaries of the so-called development programmes but as agents enacting change as we have already noted. Capacity and motivation to act depend on relationships and mutual trust and support or what may be called social capital a la Robert Putnam.[4] Considerable personal relations, community networks and cooperation are fostered in a genuine democratic society. Individuals achieve additional functioning through a direct connection with another person and in recent literature are referred to as 'external capabilities'.[5] The range of external capabilities can be significantly widened by information and communications technology (ICT). Capabilities of individuals and groups are enhanced because of networks. Communication facilitated by ICT (e.g. mobile phones) is an important instrument in building capabilities and expanding freedoms. Getting connected is a way of reducing the isolation of the poor people and fostering inclusiveness.

In what follows, we try to explain the role and potential of the local governments (LGs) in India with particular reference to the innovative practices Kerala has initiated since mid-1996.

PRIs' Claims for Social Inclusion

The institutional reforms following the 73rd/74th Constitutional Amendments which provide the legal framework for decentralization in Indian federal polity bear tremendous potential for strategizing social inclusion. The 11th Plan noted that it is 'absolutely critical for the inclusiveness of our growth process' that the 3.2 million elected representatives in the PRIs are 'fully involved in planning, implementing and supervising the delivery of the essential public services' (Planning Commission, 2007:23). While involving people's representatives is important, empowering the PRIs to function as integral agencies of social inclusion however requires an altogether different approach and strategy.

The two Amendments seek to usher in 'Institutions of self-government' at the local level endowed with the tasks of 'planning for economic development and social justice' (Articles 243 G and 243 W). This mandate to deliver social justice is evidently a call to include the traditionally excluded, minorities as well as women and all those marginalized in the scheme of planning for economic and human development. The institution of *gram sabha*, the assembly of voters mandated to deliberate on development priorities, budgets, audit reports, etc., 50 per cent reservation for women and representation for the historically marginalized communities on population basis, the establishment of the state election commission to ensure fair elections every 5 years, the creation of the State Finance Commission (SFC) to rationalize state- and sub-state-level vertical and horizontal fiscal imbalances and so on are important steps towards enhancing the quality of participatory governance and inclusive development. A report of the World Bank that critically studied the decentralization process in India notes that PRIs as political institutions can play a critical role in strengthening voice and representation, access and inclusion of the marginalized, and that they can be effective only if they are able to broaden the democratic base by tackling political exclusion and strengthening the 'synapse' between communities and governments for better local governance (World Bank, 2006:78).

The institution of *gram sabha*[6] which is direct democracy par excellence has few parallels in the world. Equally important are the 50 per cent reservation for women (although currently only a few states follow this, it is going to be mandatory following a Constitutional Amendment that is underway), representations to the traditionally marginalized classes and the various other reforms which were virtually efforts to create an inclusive, responsive and responsible democracy at the local level in India. These are somewhat different from the theory and practice of local democracy in the West. In the fiscal federalism literature produced and popularized mostly by the Western scholars[7], the role of LGs (municipalities alone matter to them) is confined to providing certain basic services to the residents (customers of these services) in a local jurisdiction at the cheapest possible price (taxes, user charges, etc.) (see Oommen, 2005). This is far removed from the conceptualization underlying Part IX and Part IX A of the Indian Constitution. Poor performance of decentralization in some states does not deny the intrinsic value and instrumental role of decentralization reforms in enhancing human well-being and democratic practice.

In the contemporary world, the ruling political paradigm of state governance is representative democracy characterized by balloting and majority rule. Strictly speaking, it has no compulsion to empower citizens, generate high or direct participation in government affairs or ensure social justice. The economic philosophy underlying representative democracy could be characterized as market-mediated economic growth (Dunning, 1993) which is the enemy of inclusive development as it believes in the natural trickling down of the benefits of growth. However, meaningful democracy should contribute to real social choice (production process and distribution should also reflect this) of the people, by the people, for the people as we have been told for years now.

The Local Government Reform Initiatives of Kerala

While the 73rd/74th Constitutional Amendments provide for certain basic institutional reforms, the details of making them operational were left to the individual states. Kerala is one state that radically reformed the LGs. In this section, we outline the salient features of these reforms that were made to widen the avenues of citizens' participation in formulating and implementing the LG plan especially at the panchayat levels.

Unlike other states, Kerala launched its decentralized planning and governance initiative in a campaign mode called 'People's Plan Campaign' (PPC) widely characterized as a 'big bang' approach. The campaign mode was launched because it helped to facilitate people's involvement in the new process of local planning. The first major decision of the Left Democratic Front (LDF) Government that came to power in July 1996 was to transfer 35–40 per cent of the state plan outlay to the LGs. This discretionary budgetary fund made available to the LGs was a quantum jump compared to the past. Instead of the traditional wisdom to lay down the procedures and rules first and then create the structures and instruments, the PPC adopted a learning-by-doing approach. Although conceptually a bottom-up planning exercise, a lot of trial and error and guidelines from above were but inevitable and the State Planning Board (SPB) virtually acted as the nodal agency. However, it is important to note that to begin with, the setting out of state-wide theatrical processions on the theme of power to the people called *Janadhikara Kalajathas* prepared the ground and created the ambience and awareness to launch and keep alive the campaign mode.

After the momentous decision to devolve a sizeable proportion of the state plan outlay to the LGs, almost immediately a Committee on Decentralisation of Powers (1996–99) (popularly called Sen Committee after the name of its first chairperson Subrato Sen) worked out the scheme of devolution of functions, funds and functionaries and the enabling legal framework for democratic decentralization. Although subject to guidelines and accountability norms, there was considerable budgetary autonomy to these governments whose functional domains (broken up into activities and sub-activities) also have been mapped out in due course. More than a decade has passed since then and the institutionalization process is still underway.

The first step in the methodology of planning launched was the special *gram sabha* meet to identify the needs and priorities of the community. Each electoral ward presided over by the ward member (elected representative) constitutes a *gram sabha* in Kerala. There are as many *gram sabha*s as there are wards which number 16,680 today. Besides that there are 5,002 urban wards, making a total of 21,682 wards all of which constitute the citizen assemblies. In fact, an important contribution of PPC was the convening of the special *gram sabha* for ventilating the felt needs of local citizens and to bring *gram sabha* into the centre stage of decentralized planning. Under the PPC regime (circa 1996–2000), considerable preparation by way of elaborate publicity and pamphlet distribution was done to enlist more people's participation and to make it more purposive. The special *gram sabha* is generally divided into subject-wise committees to make the deliberative process more participatory. In the initial years during the 9th Plan, this was preceded by 'transect walks',[8] resource mapping and similar such exercises to make the *gram sabha* meetings fruitful and to prevent such meetings from becoming a cataloguing of wish lists. Trained local facilitators helped to catalyze the process during the early periods. Direct beneficiaries of programme if any were selected by subsequent *gram sabha*s on the basis of well-defined criteria.

Once the 'felt needs' were identified and evaluated, the second stage was the preparation of a *Development Report* for each panchayat based on a scientific assessment of the human and material resources of the area, an exercise which was buttressed through several participatory mechanisms. This *Development Report* prepared for all village panchayats and municipalities under the campaign dispensation was a basic document for local planning during the 9th Five-Year Plan. Unfortunately, so far it is not yet updated or revised. The Development Reports were discussed in what has come to be called Development Seminars which organize sectoral task forces (subsequently rechristened working groups) whose responsibility is to prepare projects. In the words of Isaac and Franke, one important purpose was 'to organise development seminars in every gram panchayat and municipality to discuss the (development) reports so that the ward-wise development dialogue initiated at the grama sabhas could be scaled up to a higher level of the village' (Isaac and Franke, 2000:105). That under the PPC regime nearly 5 lakh people consisting of *gram sabha* members, political party leaders, civil servants, local experts and so on participated in the Development Seminars (see SPB, 2000) underscores the potential for deliberative problem solving and participatory planning in a decentralized democratic set-up. Needless to say, this does not mean that all the Development Seminars were of uniform success.

The third is the projectization stage which requires technical skill of a high order. This is the task of the working groups which were designed to be an ideal mix of stakeholders like peoples' representatives, bureaucrats and experts. Surely public participation does not automatically ensure the technical skill required. It is broadly understood that the projects should have both backward and forward linkages (Isaac and Franke, 2000:128). During the PPC regime, a 300-page *Hand book on project preparation*[9] was made available to all concerned so as to not let the project preparation become a 'blind man's bluff'. More importantly, special attention was given to the preparation of special plans for women (called Women Component Plan), for schedule castes (SCs) and scheduled tribes (STs)[10] and that too by a working group presided over by an SC/ST member. Once a shelf of projects[11] was prepared, the actual choice was constrained by the availability of resources.

The first three stages have meaning only when the proposals get concretized into a local plan. The fourth stage is related to the determination of the actual size of the annual plan. The state plan share is given as a grant-in-aid to LGs to meet the bottom-up plan they prepare. The plan grants (subsequently called development funds) as designed during the PPC and continued later have three components: special component plan (SCP) or funds earmarked for the development of SCs, tribal sub-plan (TSP) or funds for the development of STs and the general sector funds. The horizontal distribution of the plan grant (development fund) till 2010–11 was on the basis of a composite formula based on population, area and backwardness with undue weight (ranging from 65 to 75 per cent) on population. The fourth State Finance Commission (SFC-IV) changed this and recommended a formula which gives weight to population, area, tax effort and poverty measured in terms of five deprivation criteria viz. families living in huts, families without latrine, families whose access to drinking water is more than 300 m away from residence, families without electricity and landless households (for more on this, see section 'PRIs' Claims for Social Inclusion'). This came into effect from 1 April 2011 and has imparted more equity into the scheme of devolution. Despite the best efforts even under the PPC, the plan preparation at the level of village panchayats and municipalities moved haltingly. While the responses of some of the elected representatives, chairpersons and mayors were positive, those of several others were lukewarm although everybody had to fall in line. Actually, the transparent and participatory methodology was alien to the prevailing pattern of decision-making and definitely antithetical to the tradition of the contractor, politician, engineer, bureaucrat nexus that shaped the 'development culture' in the past. It was in this circumstance, several directives had to be given from above.

The task of integrating or rather meaningfully coordinating the village panchayat plan with that of the block and district panchayats, virtually a new dimension as far as Kerala is concerned, constitutes the fifth stage.

A lot of duplication was inevitable especially because each tier had the freedom to plan for its functional domain. The multitude of centrally sponsored schemes (CSSs) implemented through the LGs in the past also had to be made a viable component of the three-tier system. For managing the system in such a circumstance, Kerala abolished the District Rural Development Agency (DRDA) created by the central government to implement the CSS and merged it with the district panchayats. The CSSs were not only out of sync with the approach and strategies of the PPC but also seldom met the specific needs of a state that has made significant progress in education, health care, road connectivity and so on. To make the CSSs a viable and meaningful part of the state's decentralized planning, the state government's matching contribution towards CSSs was routed through the block panchayat as against the prevailing DRDA in order to give them autonomy to use the state share as per the norms of PPC. Moreover, plan funds could be used to supplement CSSs where necessary. With the traditional block development officer acting as the block panchayat secretary, the operational significance of many of the schemes was lost. The block plans, expected to avoid all duplications at the village level and dovetailing projects of inter-village panchayat relevance, had to be discussed at Development Seminars before they could be approved.

It is the constitutionally mandated duty of district panchayats to have a consolidated plan for the district as a whole. Despite the fact that urban areas do not come under the functional domain of district panchayats, as per the Constitutional mandate district planning will have to take into account the plans of municipalities and corporations. The overall planning as well as the consolidation as mandated by Article 243ZD of the Constitution is the responsibility of the constitutionally created institution of the District Planning Committee (DPC). The administrative sanction has to be given by the DPC. The technical support to the district plan is made by the District Level Expert Committee (DLEC) and scrutiny of village panchayat schemes by the Block Level Expert Committee (BLEC).

An important problem the PPC had to encounter in its quest for a methodology of local-level planning which became most apparent almost at the last phase was the need to have a body that would evaluate the technical and financial feasibility of projects. The creation of the voluntary technical corps (VTC), later on called the technical advisory group (TAG) was in response to this need. Retired engineers, professionals, unemployed technical experts and so on were encouraged to register themselves as volunteers to appraise the projects of LGs and offer suggestions for improvement. The TAG in course of time became an appraisal team of officials and non-officials at the block and district levels. During the 12th Plan, the role of TAG was deliberately deemphasized. During this time, over 35,000 persons, mostly retired persons, registered as VTC members which is a major indicator of popular participation.

The success of participatory planning can be seen by its capacity to reduce (rooting out completely would be impossible) corruption and progressively clean the system at the local level. What has the decentralization methodology done about corruption? The Public Works Department (PWD) has been notorious for corruption. The growing construction activities on roads, buildings, irrigation and so on even at the local level—village, block and district—have been huge as more funds have devolved. This has further accelerated resulting in an unholy nexus between contractors, civil servants, technical personnel and politicians leading to huge leakage of funds which increases the cost of public services. In the words of Isaac and Franke:

> Contractors made big profits, which were shared with politicians who connived to grant the work, engineers who gave technical sanction and monitored the work, as well as clerical staff who approved payment of bills. To contain corruption, innumerable checks and balances and adhoc rules had been introduced over time. But these myriad rules and regulations trapped honest persons in a web of delays and approvals that discouraged them from participating. (Isaac and Franke, 2000:237)

The Beneficiary Committee (BC) system was PPC's answer to tide over this situation and introduce greater transparency and accountability in public works. If the beneficiaries come forward (the assumption was that they had the time, energy and expertise) to manage the projects that benefit them and the technical and administrative system lend them a supporting hand, ideally there is no better alternative. Although over 30,000 BCs were formed during the initial phase, most of them succumbed to the continuing pressures and fights from the vested interests including engineers, officials and politicians. However, that nearly 25 per cent cases successfully completed projects demonstrates the tremendous potential of BCs to function as a viable participatory mechanism (Pillai, 2000). In course of time, the BC system fell by the wayside and the contractor system has staged a comeback.

Another agency widely known as Kudumbashree (KDS) formally inaugurated in May 1998 and launched in April 1999 (fully operationalized by 2003), functions as a viable participatory structure in the decentralized governance of the state. It is 'a women-oriented anti-poverty programme' cast in a mission mode under the leadership and patronage of panchayats and municipalities. As the mission statement puts it:

> To eradicate absolute poverty in ten years through concerted community action under the leadership of Local Self Governments, by facilitating organisation of the poor combining self-help with demand led convergence of available services and resources to tackle the multiple dimensions and manifestations of poverty, holistically.

KDS through its microfinance self-help group structure and through its many-faceted activities taken up through LGs over these years seeks to empower poor women who were identified using a nine-point criterion. The empowering process and activities of KDS are worked through a three-tiered community-based organization (CBO) referred to as community development society (CDS). The multi-pronged approach focuses on human resource development, community health, child welfare, basic human needs, rehabilitation of the destitute (The Ashraya Project of village panchayats), lease-land farming, housing and promotion of micro enterprises, besides its microfinance activities. The CDS which was functioning long before it was changed into KDS, continues to function as part of the participatory planning process. It is a fact that the KDS which has over these years functioned as a sub-system of the municipalities and panchayats not only integrated the various anti-poverty programmes but also functioned as their delivery system and participatory mechanism. The strength of the CDS system is largely because it is functionally linked to the LGs (see Oommen, 2007 for more details).

Towards More Institutional Effectiveness: Some Suggestions

The institutional reforms briefly outlined above and the actual practice have a wide gap. In this section, we present some suggestions or correctives in the light of what happened in the state.[12] Most importantly, the three Fs—functions (broken-up into activities and sub-activities based on the principle of subsidiarity avoiding overlapping and concurrency), funds (which should match the functions transferred) and functionaries (adequate for the tasks assigned)—should be properly aligned for any local governance apparatus that is responsive and accountable. This strategy implies considerable expenditure discretion, a large proportion of untied transfers along with own source revenue (the transfer system should incentivize revenue efforts) and the functionaries deployed should be answerable to the LGs. In many states, this is not happening. Even in key sectors, such as primary education, primary health, minor irrigation, crop production, etc., which are the exclusive functional domain of PRIs, the departments dominate. In Kerala also, the system needs considerable improvement. The accounting and accountability arrangements in particular need drastic reform. Budget formulation and financial reporting in vogue despite ongoing reforms continue to be unreliable and irregular and need to be radically improved to facilitate accountability, monitoring and evaluation. Also, the pattern of spending should be evenly spread throughout the year. The prevailing pattern of spending in Kerala shows a heavy bunching of development expenditure during January–March.[13] Incidentally, this is indicative of the extant corruption which has to be reduced if not rooted out.

Again, the *gram sabha* should be made the kingpin of inclusive governance in the villages of India. Kerala's decentralized performance was at its best when citizen participation in *gram sabha*s and Development Seminars was high. However, as the Government of Kerala (GoK) (2009) points out, when *gram sabha*

> becomes a routine affair and a meeting place of beneficiaries or benefit seekers or when attendance gets fudged an important democratic tool stands discredited and the bureaucracy comes back with a vengeance. The gram sabha/ward sabha meetings which could have been developed into a forum for consensus building or throwing up constructive ideas for development purposes lose these significance. (GoK, 2009:45)

The colossal failure of the BC system and the continuation of the endemic leakage of public resources through artificial escalation of project estimates, total disregard to quality specification in the use of materials and so on are documented in the report. In the words of the committee, 'Apparently nobody, seems to be worried about the continuation of the contract system and the archaic public works manual which have facilitated and legitimized corruption in the state' (GoK, 2009:46).

Certainly, participatory methodology needs rethinking and refashioning. Making participatory institutions always answerable to bureaucracy (obeying the rules of game and rule of law notwithstanding) is permitting them to recapture power. Even so, a mix-up of participation with administration has to be avoided. While participation of people has no direct role such as where a doctor, engineer, planner and other actors are directly involved, it is certainly for the people to design the destiny and direction of their future transformation.

This again drives home the need for strong systemic change and deep political commitment without which any worthwhile participating and inclusive governance can be made.

Sustainability of PRIs depends a great deal, surely not in romanticizing their achievements but in fighting those vested interests that work against them, rallying forces that make democratic practice a way of life and above all in consistently building the capabilities of the citizens.

Inclusion of the Excluded: Towards Relevant Postulates

In sum, it is clear that the philosophy and institutional structures underlying the 73rd/74th Constitutional Amendments bear tremendous potential to promote inclusive development. The approach of GoI, however, as explained in section 'Social Inclusion, Inclusive Growth and Inclusive Development: Some Conceptual Issues' and section 'PRIs' Claims for Social Inclusion', envisages inclusiveness and growth/development as two disparate boxes and does not work towards social inclusion. Evidently, this approach needs correction and the PRIs offer great possibilities. That policy choices for the inclusion of the excluded are very great in a decentralized regime is well exemplified in the design of the transfer of development funds from the state to the sub-state level in Kerala followed by the SFC-IV of the state.[14] An improved version of the SFC-IV recommendations is discussed below (see also GoK, 2011; Chapter VIII).

A comprehensive below poverty line (BPL) survey based on a census of households in all village panchayats, municipalities and corporations of

Kerala undertaken by the Commissionerate of Rural Development, GoK, in 2009 using school teachers as investigators (to ensure quality of data), which SFC-IV processed yielded a wealth of information (generating relevant data is essential for inclusive policy initiatives). From the BPL survey data, SFC-IV constructed a deprivation index[15] and made use of it to identify the vulnerable communities viz. SC, ST, fisherfolk (admittedly a vulnerable group like SCs or STs) and other categories in every gram panchayat (GP). Besides the deprivation index, a ratio index was also constructed to rank every village panchayat as regards the deprivation of these social groups compared to the total situation in the state as a whole. That this could be done for every LG holds out the possibilities for mapping out a strategy for the inclusion of the excluded.[16]

The demarcating deprivation line was drawn based on the data of the BPL survey, which used two schedules, form A and form B. Form A was used for every household. It contained questions related to 11 variables.[17] The social group classification of the household was also generated through this form. Form B was canvassed in a house only if all the 11 characteristics of form A were reported nil. This schedule contained information on the housing conditions, health status, land holdings, availability of other household basic amenities and so on. The main economic activity (occupation) of the families is also generated from this form. From this data, four sets of information are obtained for SC, ST, fisherfolk and general population. The details of SC, ST and general population are obtained from the social groups in form A.[18]

From the data, it is possible to generate eight to ten variables that can indicate human deprivations. Out of this, SFC-IV used five equally important variables for constructing the deprivation index. They are given below:

1. Housing status: Percentage of families living in huts and dilapidated houses.
2. Sanitation: Percentage of families reported as without latrine, kayal latrine, latrine without tank, pit latrine and common latrine.
3. Drinking water: Percentage of families whose access to drinking water source is more than 300 m away from dwelling place and relying on pond, river and private distributors for drinking water.
4. Electricity: Percentage of families without electricity connection for household purpose.
5. Landholding: Percentage of landless families.

In order to compare different social groups under consideration, and also to identify the vulnerable GPs, a composite index is constructed comprising all these five variables[19] and used them for the actual allocation of development funds and also for a definite resource allocation transfer arrangement to help the excluded viz. SC, ST, fisherfolk and the economically backward sections in the most vulnerable GPs identified on the ratio index basis.

The formula for calculating the index of deprivation used was similar to that of human poverty index employed by the United Nations Development Programme (UNDP) and is given below:

$$ID = [\tfrac{1}{5}(d_1^\alpha + d_2^\alpha + d_3^\alpha + d_4^\alpha + d_5^\alpha)]^{1/\alpha}$$

where ID is index of deprivation; d_1, d_2, d_3, d_4 and d_5 are the percentage of families who lack the five basic amenities in a given GP; α is a weight parameter. If $\alpha = 1$, the ID is the average of its indicators. As the α increases, greater weight is given to the indicators in which there is the most deprivation. Similar to the human poverty index used by UNDP and deprivation index used by Kerala Human Development Report (HDR) (2005), a value of $\alpha = 3$ for computing the ID was used. Using this index, first all the GPs were ranked so as to identify the vulnerable panchayats and then the relative position of SC, ST and fisherfolk with that of the general population was found out. Apart from the power weight of 3 for α, additional weights can be assigned to the variables. This enables one to give weights depending on the priority of the variables (on this, see Oommen and Shyjan, 2014). These illustrative cases actually employed by SFC-IV of Kerala are mentioned to show the potential of the tool in mapping out the excluded in the state (by each LG) which is a necessary condition for any strategy for social inclusion.

Conclusion

Innovative institutional architecture is a sine qua non for any sustainable transformative practice that works towards social inclusion. The 73rd/74th Constitutional Amendments have provided the necessary conditions for helping inclusive decentralized development process by creating a set of institutions such as the *gram sabha*, DPC, SFC, state election commission,

50 per cent reservations for women, representation on population basis for the traditionally marginalized and so on in the LGs which are mandated to deliver 'economic development and social justice' and usher in 'institutions of self-government' at the local level. A lot of sufficient conditions also need to be fulfilled to put these into an integral part of the everyday practice of local democracy. The decentralized planning exercise of Kerala although launched on a campaign mode has been stabilized through a multistage process of participatory structures that have demonstrated the possibilities of inclusive development. True, there is a yawning gap between theory and practice and it has to be rectified. We have also shown that through a census of households in the village panchayats, municipalities and corporations, it is possible to identify all the vulnerable households through universally valid deprivation criteria. That you can physically map out the socially and economically excluded is no pipe dream. Through continuing adaptations and innovative actions, it is possible to make local democracy a live instrument for local development. Probably there is no other institution that can be fashioned to increase the access and inclusion of the marginalized and strengthen the 'synapse' between communities and governments for better local governance.

Notes

1. Terms like pro-poor growth have very little in common with the inclusive development that strategizes the inclusion of the excluded through relevant policy choices.
2. For an elaborate empirical evidence covering a period of nearly 300 years of economic growth, one may read Piketty (2014). It is also useful to read Stiglitz (2012) and Wilkinson and Pickett (2010) in this context.
3. The following observation of Ahluwalia (2011), a leading policy strategist of India, is indicative of the indifferent approach to inequality.

    It is sometimes argued that inequality should not matter as long as the poor are getting better off and it is probably true that a rapid rate of improvement in incomes for the poor may make them willing to accept some increase in inequality. However, large increases in inequality accompanying only modest improvements in the levels of living of the poor are unlikely to be acceptable. (p. 89).

    Are the poor to be treated as a distant somebody or cartoon characters to be commented upon by the elites and scholars?

4. See, for example, D. Robert Putnam (1994, 2001)
5. For further information on this, see http://www.ophi.org.uk/pubs/OPHI.WP8.pdf.
6. It is interesting to recall that the United Progressive Alliance (UPA) government in their National Common Minimum Programme declared (May 2004) that 'the Gram Sabha is empowered to emerge as the foundation of Panchayati Raj'. This is a major declaration of development goal.
7. For review of literature on this, see Oommen (2005).
8. Resource appraisal team consisting of experts in agriculture, soil and so on along with local people, local teachers, etc., walk through the length and breadth of the village to gather basic data. This practice was done only during the initial process of the PPC, although the special *gram sabha* has continued since then in the preparation of annual plans.
9. It is important to note that on almost every major aspect of decentralization, there was a handbook to help the planning process.
10. SCs refer to the so-called low castes popularly called *dalits* and STs refer to *adivasis* or tribal communities, both listed in separate schedules of the Indian Constitution.
11. Preparing a fewer number of sound projects could be a better alternative to multiplying projects that are an ensemble of good, bad and indifferent types.
12. For an elaborate critique of decentralized participatory governance and development in Kerala, see *Report of the Committee for Evaluation of Decentralised Planning and Development,* 2009 (GoK, 2009).
13. It goes as high as 83.6 per cent for municipalities, 66 per cent for GPs, 59.9 per cent for block panchayats and 68.4 per cent for district panchayats. (This is based on a survey conducted by the SFC-IV and covers the period 2004–09 and made available to the author.)
14. The author of this chapter was the chairman of SFC-IV and wrote the chapter entitled, Devolution and the Inclusion of the Excluded (Chapter VIII).
15. For the devolution of plan grants to panchayats and municipalities also, a formula with 30 per cent weightage to devolution index was followed (along with 50 per cent for population, 10 per cent for area and 10 per cent for tax effort).
16. This exercise is attempted in another paper by the author (see Oommen and Shyjan, 2014).
17. Families with employees in government sector (Class I to Class IV), employees in private/quasi-government/aided sector, employees in co-operative sector, pensioners in government/service, pensioners in quasi-government/aided institutions, pensioners in co-operative institutions, regular salaried in public/private institutions, house structure higher than 1,000 sq ft, four-wheeler motor vehicle for private purpose, non-resident Indians (NRIs) with land possession higher than 1 acre (except STs).
18. It may be noted that since the social groupings given in 'form A' are as SC, ST, minorities and others, it is difficult to get fisherfolk as a separate category from form A. Therefore, the fisherfolk group can be arrived at only by using form B, where the occupational details are furnished. The group fisherfolk includes only those families whose main source of income is fishing either marine or inland. Hence, the 'general' group includes fisherfolk and minorities also.
19. The extra variables left out are households with school dropouts, households with members above 65 years of age, households with disabled members, unmarried women with children, etc.

References

Ahluwalia, M.S. 2011. 'Prospects and policy challenges in the Twelfth Plan', *Economic and Political Weekly*, 46(21): May 21–May 27.
Ali, I. 2007. 'Inequality and the imperative for inclusive growth in Asia', ERD Working Paper, Manila: Asian Development Bank.
Ali, I. and Z. Juzhong. 2007. 'Inclusive growth toward a prosperous Asia: Policy implications', ERD Working Paper Series No.97, Asian Development Bank. (Downloaded from the website: http://www.adb.org/Documents/ERD/Working_Papers/WP097.pdf)
Baru, R., A. Acharya, S. Acharya, A.K. Shivakumar and K. Nagraj. September 2010. 'Inequities in access to health services in India: Caste, class and region', *Economic and Political Weekly*, XLV(38).
Bhagwati, J. and A. Panagariya. 2012. *India's Tryst with Destiny Debunking Myths that Undermine Progress and Addressing New Challenges*. India: HarperCollins Publishers.
Chandrasekhar, C.P. and J. Ghosh. 2011. 'Latest employment trends from the NSSO', *The Hindu Business Line*.
Desai, S. and A. Dubey. 2011. 'Caste in 21st century India: Competing narratives', *Economic and Political Weekly*, March 21.
Dreze, J. and A. Sen (eds). 1995. *India, Economic Development and Social Opportunity*. New Delhi: Oxford University Press.
Dunning. J. 1993. *Democracy: The Unfinished Journey, 508 BC to AD 1993*. New York: Oxford University Press.
Government of Kerala (GoK). 2005. *Human Development Report*. Thiruvananthapuram: State Planning Board.
——— 2009. *Report of the Committee for Evaluation of Decentralised Planning and Development*. Thiruvananthapuram: State Planning Board.
——— 2011. *Report of the Fourth State Finance Commission, Kerala, Part I*. Thiruvananthapuram: State Planning Board.
Isaac Thomas, T.M. and R. Franke. 2000. *Local Democracy and Development: Peoples Campaign for Decentralised Planning in Kerala*. New Delhi: LeftWord.
Jeffrey, R. 1992. *Politics, Women and Well-being How Kerala Became: 'A Model'*. New Delhi: Oxford University Press.
Joe, W., U.S. Mishra and S. Navaneetham. 2008. 'Health inequality in India: Evidence From NFHS-3', *Economic and Political Weekly*, August 2.
Kabeer, N. 2006. 'Poverty, social exclusion and the MDGs: The challenge of "durable inequalities" in the Asian context', *IDS Bulletin*, 37(3): 64–78.
Oommen, M.A. (ed.) 1999. *Kerala's Development Experience Vol. I & Vol. II*. New Delhi: Concept.
——— 2005. 'Rural fiscal decentralisation in India: A brief review of literature', in L.C. Jain (ed.), *Decentralisation and Local Governance*. New Delhi: Orient Longman.
——— 2007. *Report of Kudumbashree of Kerala: An Appraisal*. Thiruvananthapuram: State Planning Board.
Oommen, M.A. and Shyjan. 2014. *Local Governments and the Inclusion of the Excluded: Towards a Strategic Methodology with Empirical Illustration*. Working Paper No. 458, Centre for Development Studies, Thiruvananthapuram.
Pillai, A.R.V. 2000. 'Beneficiary Committees: An Experiment under People's Planning', Paper presented at the International Seminar on Democratic Decentralisation, Thiruvananthapuram, 23–27 May.

Piketty, T. 2014. *Capital in the Twenty-First Century*, translated by Arthur Goldhammer. Cambridge, MA: Harvard University Press.

Planning Commission. 2007. *Eleventh Five Year Plan*. New Delhi: Government of India.

——— 2012. *Twelfth Five Year Plan*. New Delhi: Government of India.

Putnam, R.D. 1994. *Making Democracy Work: Civic Traditions in Modern Italy*. Princeton: Princeton University Press.

——— 2001. *Bowling Alone: The Collapse and Revival of American Community*. New York: Simon & Schuster.

Rawls, J. 1971. *A Theory of Justice*. Oxford: Clarendon Press.

Rodgers, G., R.D. Lee and W.B. Arthur (ed.). 1995. *Economics of Changing Age Distributions in Developed Countries (International Studies in Demography)*. Oxford: Clarendon Press.

Sen, A. 1973, 1998. *On Economic Inequality*. Oxford: Oxford India Paperbacks, Oxford University Press.

——— 1981, 1999. *Poverty and Famines*. Oxford: Oxford University Press.

——— 1999. *Development as Freedom*. New Delhi: Oxford University Press.

——— 2000. 'Social Exclusion: Concept, Application, and Scrutiny', ADB Social Development Paper No. 1, Asian Development Bank, Manila, Philippines.

State Planning Board. 2000. *Economic Review*, Thiruvananthapuram: Government of Kerala.

Stiglitz, E.J. 2012. *The Price of Inequality*. London: Allen Lane.

UN/CDS. 1975. *Poverty, Unemployment and Development Policy: A Case Study of Selected Issues with Reference to Kerala*. Trivandrum: Centre for Development Studies.

Wilkinson, R. and K. Pickett. 2010. *The Spirit Level—Why Equality Is Better for Everyone*. London: Penguin Books.

World Bank. 2006. 'India: Inclusive Growth and Service Delivery: Building on India's Success: Development Policy Review', Report No. 34580-IN, Washington: The World Bank.

SECTION II

Technological Progress and Agricultural Development

4

Measuring Public Agricultural Research Capital and Its Impact on State Agricultural Productivity in the United States

Wallace E. Huffman

Introduction

A half-century ago, Schultz (1953, pp. 109–11) argued that modern science is supported mainly for the fruits that it bears, measured in terms of new techniques. Furthermore, he proposed that pure science and its contribution to society are closely interrelated, and advances in science and technology require investments of real resources. These resources consist largely of scientists' effort, complementary inputs of assistants, laboratories and equipment such as computers and the use of the existing stocks of knowledge. Schultz also argued that new techniques are a type of input that entrepreneurs would pay to obtain because they increase expected productivity or output per unit of input of an enterprise.

Although Schultz saw that basic and applied research in the sciences contribute to advances in agricultural technologies, organized research is not undertaken by farm firms but primarily by public agencies—the State Agricultural Experiment Stations (SAESs) and the Agricultural Research Service (ARS) of the US Department of Agriculture (USDA) and large private sector firms. The reason that farm firms do not undertake organized research is the large fixed costs, long gestation periods and very specialized talent needed to successfully undertake research, relative to farm sales, and the public goods nature of the discoveries, that is, benefits tend to be non-rival for many discoveries (Cornes and Sandler, 1996; Khanna et al., 1994). In contrast, in private industry most of the research is undertaken in large firms or corporations, and this research focuses on innovations in products and processes that are protected by patents, copyrights or trade secrets, thereby having the potential to enhance innovators' future profits but not easily spilling over to other firms.

Schultz also argued that the competitive structure of agriculture is conducive to the introduction and adoption of new technology. Competition among farmers spurs them on to be more aggressive than they otherwise would be. However, most new agricultural technologies are geo-climate sensitive—responding to climate, soils and topography—as well as responding to the local ecosystems (Alston et al., 2010; FAO, 1978–81; Huffman and Evenson, 2006, p. 271; USDA, 1957, 2006). New technologies reduce the expected cost of production to farmers in a particular geo-climatic region where they are best suited. Some of these farmers will be early adopters, as with hybrid corn (Griliches, 1960), and others will adopt later. When this technology is successful, it gives early adopters a competitive advantage and higher profits than would otherwise occur. As the successes of new technologies are observed in an area, more farmers will try the technology and many will eventually adopt it (Wozniak, 1984). Diffusion of a new technology takes place when a large share of the farmers in a geo-climatic region adopts the technology.

Griliches is well known for his pioneering research in the field of productivity and economic analysis in which he employed econometric techniques to link agricultural productivity or output to past investments in agricultural research and extension in regional or national data. At the heart of his research was the idea that technical change was a major source of productivity growth, and that technical change was the result of a productive economic activity—discovery and innovation, which are funded by the public and private sectors (Griliches, 1998, p. 1). Hence, knowledge and knowledge generation are the primary sources of productivity growth in the long run. However, research capital is a form of intangible capital, creating major challenges in how to measure it well (Griliches, 1998), and measuring it is considerably more challenging than measuring physical capital, which has its own set of tough issues (Ball et al., 2002; Jorgenson et al., 2005).

The objective of this chapter is to describe a methodology for measuring public agricultural research investments and capital at the state level and reporting new estimates of the impact of public agricultural research on state agricultural productivity. The chapter uses the USDA's official state agricultural productivity statistics for the 48 contiguous states over

1970–2004 (USDA, 2009). The research revises and updates measures of public agricultural research capital over 1998–2004 and presents new estimates of the impact of public agricultural research capital on state agricultural productivity relative to the Huffman and Evenson (2006a) paper. The emphasis in this chapter is on estimating the impact of within-state research capital and spill-in research capital on state agricultural productivity in a simple model without drawing on the issues of alternative agricultural research funding mechanisms. The chapter unfolds in the following five sections.

The Model of Productivity

A model of state agricultural productivity incorporates the following considerations. Assume that agriculture of a given state can be adequately summarized by an aggregate production function

$$Y_t = F(X_t, K_t, \mu_t) \qquad (1)$$

where Y_t is an index of farm outputs in time period t for farms in a given state; $F(\ )$ is the algebraic form for the production function; X_t is an index of quality-adjusted conventional inputs of farm capital services, labour, chemicals, energy, other intermediate materials and land in t; K_t is the stock of knowledge in t for a given state and μ_t represents all other factors affecting the conversion of conventional inputs into agricultural output in period t for a given state. Under special conditions, total factor productivity (TFP) relationship is written as

$$\ln(\text{TFP}_t) = \ln(Y/X)_t = G[W(\mathbf{B})R_t, t, \nu_t], \qquad (2)$$

where $G(\ )$ represents the productivity function. Public agricultural research capital in t is defined as a vector of current and lagged values of real agricultural research investments

$$W(\mathbf{B})R_t = w_0 R_t + w_1 R_{t-1} + w_2 R_{t-2} + w_3 R_{t-3} + w_4 R_{t-4}$$
$$+ w_5 R_{t-5} + \ldots + w_m R_{t-m} \qquad (3)$$

Moreover, in creating the research capital, the weights on current and past real investments sum to one (i.e. $\Sigma w_\ell = 1$). Other variables in Equation (2) are a time trend t and v, which represent other factors that affect agricultural productivity, including weather shocks.

One plausible relationship between knowledge capital and research investments is

$$K_t = [W(\mathbf{B})R_t]^\eta \exp(\alpha + ct + v_t) \tag{4}$$

Substituting Equation (4) into Equation (2), we obtain an econometric model of agricultural productivity:

$$\ln(\text{TFP})_t = \alpha + \eta \ln[W(\mathbf{B})R_t] + ct + v_t \tag{5}$$

An estimate of η represents the impact elasticity of public agricultural research capital on agricultural productivity. Shocks (v_t) are expected to spill over from one period to the next, which causes first-order autocorrelation or $v_t = \rho v_{t-1} + \varepsilon_t$. For the stochastic process to be stationary, ε_t must be distributed with zero mean, constant variance and $|\rho| < 1$. By incorporating trend into the econometric model of productivity and using an estimator that takes account of first-order autocorrelation, the model is likely to be covariance stationary (Enders, 2010; Greene, 2003).

Measuring Public Agricultural Research Capital

Over the past half-century, advances in constructing the measures of agricultural research capital have been slow. Griliches (1964) and Evenson (1967, 1980) were pioneers in this effort. In addition, Huffman and Evenson (1993) and Alston et al. (2010) built upon this base.

Early Measures and Their Performance

Griliches (1964) undertook the first econometric analysis of the impact of public agricultural research and extension on agricultural output at the state level.[1] His R&D variable was a bundle of public agricultural research and extension capital and created from the undeflated sum of total public expenditures on research and extension within a state during the previous year and five years earlier. His R&D variable had some shortcomings. It was too broad in the sense that it used all expenditures on public agricultural research and extension as the appropriate measure of investment in technology and information, irrespective of whether they could plausibly impact agricultural output or productivity. The measure was too narrow in the sense that it ignored the very significant intramural agricultural research activity of the USDA in the states (Huffman and Evenson, 2006a). In addition, his timing weights were crude, and no allowance was made for spillin/spillover effects of research discoveries in one state on agricultural productivity in other states.

Given these limitations, his estimate of the regression coefficient for the natural log of public agricultural research and extension capital per farm was 0.059. The estimate implied a marginal product of 13 dollars of output per year for an additional dollar of public agricultural research and extension expenditure per year and a 65 per cent social rate of return.

Evenson (1967) was the first to experiment with timing weights in constructing a measure of public agricultural research capital. He adopted Schultz's (1953) hypothesis that research discoveries of the USDA and SAESs cannot be simply expressed in terms of a small number of important 'breakthroughs', but rather as a continuous stream of advances that impact agricultural productivity at the state and national level. In addition, Evenson (1967) argued that public agricultural research provides important knowledge needed to enhance the quality of inputs—fertilizers, pesticides, feed, seed, breeds of livestock, etc.[2]

In addition, Evenson suggested that a lag exists between the investment in the effort of scientists and additions to knowledge. Second, a lag exists from the discovery of new knowledge to the development and marketing of new technologies by farmers. Third, it takes time for farmers to adopt new and profitable technologies. Fourth, most knowledge eventually depreciates or becomes obsolete. He summarized all of these effects in a knowledge or technology production function: $R_t = W(L)Z_t + C(L)\mu_t$, where $W(L)$ is a lag operator providing timing weights for current and past values of real research investments Z_t and $C(L)\mu_t$ is a distributed lag of error terms (μ_t). Then, the stock of existing knowledge (or technology) is defined as

$$K_t^* = R_t + (1-\delta) K_{t-1}^* \qquad (6)$$

where δ is the depreciation rate on the stock of knowledge (or technologies), $0 \leq \delta \leq 1$. For example, a discovery might enable a new commercial pesticide, but pests evolve over time to their new environment and develop resistance to the pesticide, which is a type of depreciation of past discoveries. In addition, depreciation of knowledge can occur when existing knowledge (technology) is replaced by new knowledge (better technologies), sometimes called 'creative destruction'. The net result of Evenson efforts was a knowledge production function of the form

$$K_t^* = F(L)W(L)Z_t + F(L)C(L)\mu_t.$$

The second term reflects the impacts of random shocks to the stock of knowledge (μ_t).

Evenson converted USDA and SAES research expenditures into real terms by using a price index of salaries of university professors. He then followed a suggestion by Griliches to impose a symmetric inverted V-lag structure on past real public agricultural research investments in order to derive a measure of public agricultural research capital. Using national aggregate data, he experimented with the lag length between investments in public agricultural research and changes in agricultural productivity.[3] In his preferred regressions, he obtained a mean lag of 6–7.5 years. He also tested the hypothesis that research discoveries of the SAESs and USDA impacted agricultural productivity similarly. His thinking was that USDA research during the study period might be more intensely engaged in basic and less intensely engaged in applied research than the research of the SAESs. If this were true, the mean lag length would be longer for the USDA's research. However, he could not reject the null hypothesis.

Evenson's (1967) methodology for constructing a measure of public agricultural research capital was more sophisticated than that of Griliches (1964), but his decision to use national aggregate data, rather than state Census data, in his agricultural productivity research greatly limited the potential for new discoveries, and his empirical results were weak. Thus, the 1960s' research on agricultural productivity by Griliches and Evenson was only partially successful.

Evenson (1980), which is not widely available, was the first study to investigate the impact of public agricultural research on agricultural productivity using annual state aggregate data. Evenson used unpublished

SAES records as the source of annual investments in public agricultural research.[4] From these data, he was able to distinguish 24 research commodity categories—22 were denoted 'applied' research categories and two were denoted 'basic' research categories. Six categories related to livestock research are: five 'applied' categories—beef, dairy, hogs, poultry and sheep—and one 'basic' livestock research category for livestock research that was not directly linked to any of the five specific types of livestock. Similarly, he distinguished 17 categories for 'applied' crop research (on barley, corn and sorghum, cotton, flax, forestry and forest products, fruits, hay, oats, peanuts, potatoes, rice, soybeans, sugar beets, sugar cane, tobacco, vegetables and wheat), and one 'basic' crop research category for crops research that was not directly linked to any single one of the 17 detailed categories.

Although the state and national governments, which are geopolitical units, collect taxes and allocate the tax revenues to investments in intramural research of the USDA and SAESs and other things, public agricultural research undertaken in any state seems unlikely to impact all farmers in that state uniformly, except possibly for tiny states such as Rhode Island. The main reason is that the geo-climate conditions are not homogeneous across most states. Recognizing this fact, Evenson prorated the applied research expenditures—investments of a state across sub-regions of the state based upon farmers' commodity revenue shares for these sub-regions as determined in county data from the *Census of Agriculture*. He then used the number of applied agricultural research commodities receiving positive research funding to convert applied research investments into a per commodity basis.

To operationalize his geo-climate idea and to be able to construct spatial weights, Evenson (1980) chose the geo-climatic region (and sub-region) structure laid out in the 1957 *Agricultural Handbook on Soils*, pp. 451–627.[5] In this handbook, Barnes (pp. 452–455) argues that US soils and climates follow definite regional patterns. Differences across regions are due to latitude, elevation and worldwide movement of air masses, and major differences in soils across regions result from the climate under which the soils developed, the parent materials from which the soils developed and the slope and drainage potential. Across regions of the United States, major differences exist in climates and soils, and they greatly affect the opportunities and constrain the choices of local farmers,

Figure 4.1:
US agricultural geo-climatic regions and sub-regions

Source: Adapted from US Department of Agriculture (1957) and Evenson (1996).

especially for crop production. For example, consider the Midlands Feed Region, Region 6 (see Figure 4.1). This region combines productive soils, gentle relief and moderate rainfall over a wide area. Summers are warm, but winters are cold. In this region, a very large share of rural land is in farmland, producing feed grains, oilseeds and hay and beef, swine and dairy. Although Alston (2002) and Alston et al. (2010) suggest that these geo-climatic regions are composed of state/geopolitical units that lie in close proximity one to another, we see that Region 6 does not follow strict state boundaries. Moreover, it stretches from western Nebraska to northeastern Ohio, which is roughly 800 miles, and from southern Wisconsin to northern Missouri. This is not a small region.

In addition, consider Region 13, the grazing-irrigated region. This region and all of the other geo-climate regions (Figure 4.1) do not follow strict state boundaries. Region 13 covers a very large geographic area that stretches from Idaho and western Montana in the northern United States to New Mexico and western Texas in southwestern United States, which is a little over 1,000 miles, and from central Colorado to eastern California. Yet, Alston (2002) and Alston et al. (2010) argue that California, Florida,

Georgia and New York produce fruit, and hence, are similar. However, the climate, parent material of the soils, topography and drainage are quite different, and this translates into different growing and harvesting seasons and types of endemic pests.[6] Furthermore, these states do not compete much in the seasonal market for fresh fruit, because they either supply different types of fruit or supply during a different season of the year.

To represent timing and spatial weights, Evenson (1980) chose the following structure:

$$R(a, b, c)_t = AR(a, b, c)_t + \alpha_0 SA(a, b, c)_t \text{ or } AR(a, b, c)_t + \beta_0 SRA(a, b, c)_t \tag{7}$$

where $AR(a, b, c)_t$ is the within-state applied public agricultural research stock, $SA(a, b, c)_t$ is the stock of applied public agricultural research spilling in from similar sub-regions to a given state and $SRA(a, b, c)_t$ is the stock of applied agricultural research spilling in from the same geo-climatic region (which includes the sub-regions). For timing weights, Evenson (1980) limited himself to the generalized trapezoidal shape (see Figure 4.2 for one version). In Equation (7), a is the number of years over which there are rising weights, b denotes the years over which the timing weights are at a peak and constant and c denotes the length of the following period over which the timing weights are declining to zero. For example, $R(7, 8, 15)_t$ denotes a research stock variable created using trapezoidal timing weights that start at zero in period t, then increase linearly for the

Figure 4.2:
Public agricultural research timing weights: Trapezoidal pattern

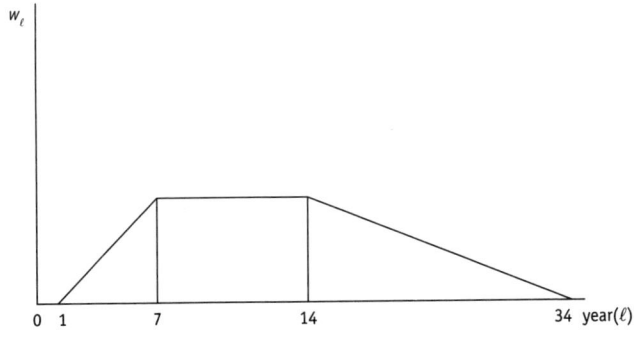

Source: Author.

next 6 years, are constant at maximal contribution for the following 8 years and then decline linearly over the following 15 years. Hence, there is an early gestation period where weights are zero or increasing and a 15-year period and at the end where depreciation is occurring. At year 29, the net contribution of the investment in t goes to zero and stays there.

In Equation (7), α_0 and β_0 are the contiguity or spillin parameters, which are bounded, $0 \leq \alpha_0, \beta_0 \leq 1$. Clearly, if α_0 or β_0 is zero, no spillin occurs, and if α_0 or β_0 is one, then the full impact of research from other states in the region or sub-region is counted as spillin.

In his econometric analysis of state agricultural productivity, Evenson (1980) grouped the 48 contiguous states into three regions: Southern States (commonly referred to as the Appalachian, South East and Delta Regions), Northern States (commonly referred to as the Northeast, Corn Belt and Lake States Regions) and Western States (commonly referred to as the northern plains, southern plains, Mountain and Pacific Regions). He then experimented with different specifications of the empirical productivity equation to obtain the best fitting one. First, he fitted his productivity model separately to each of the three regions using data for all states within the region and then undertook a grid search across sets of timing weights and spillin/continuity weights looking for the pairs that gave the largest partial correlation factor between $\ln(\text{TFP})_t$ and public agricultural research capital, $\ln R_t$. The highest partial correlation occurred for the Northern States at $R(7, 8, 15)$ and $\alpha_0 = 0.5$ ($\beta_0 = 0$), for the Southern States at $R(5, 6, 11)$ and $\alpha_0 = 0.25$ ($\beta_0 = 0$) and for the Western States at $R(7, 8, 15)$ and $\beta_0 = 0.25$ ($\alpha_0 = 0$).[7] He interpreted these results to imply that the trapezoidal timing weighting patterns were similar across the three regions, but spillin effects were more prevalent in the Western States.

The final form of Evenson's (1980) econometric model of state agricultural productivity for the period 1948–71 was:

$$\ln(\text{TFP})_t = \alpha_1 + \alpha_2 \ln(\text{AR})_t + \alpha_3 \ln(\text{AR})_t \times \ln(\text{BR})_t + \alpha_4 \ln(\text{AR})_t \\ \times \ln(\text{EXT})_t + \alpha_5 \ln(\text{ED})_t + \alpha_6 \ln(\text{EXT})_t + \alpha_7 \ln(\text{ED})_t \\ \times \ln(\text{EXT})_t + \alpha_8 \ln(\text{EXT})_t \times \text{PL}_t + \alpha_9 \ln(\text{EXT})_t \times \text{BC}_t + v_t \quad (8)$$

where BR_t is a state's stock of 'basic' public agricultural research. In addition, EXT_t is the state's stock of extension (using exponentially declining weights starting at 0.5 in t), ED_t is an index of years of schooling completed by a state's farmers, PL_t and BC_t are a state's scaling factor (economic slack) and business cycle index, respectively, and v_t is a zero-mean random disturbance term.[8]

When the model as described in Equation (8) was fitted, Evenson obtained relatively good results, except that his estimate of α_3 was not significantly different from zero. Then, in (8), he replaced $\alpha_2 \ln(AR)_t$ with $\alpha_{2S} \ln(AR)_t D_S + \alpha_{2N} \ln(AR)_t D_N + \alpha_{2W} \ln(AR)_t D_W$, where $D_\ell = 1$ if a state in region ℓ = S (Southern region), N (Northern region) or W (Western region) and zero otherwise. The new model gave estimated coefficients that were significantly different from zero at the 5 per cent level (and the estimated coefficients for the applied research stock interacted with the basic research stock and extension stocks were positive). The implied marginal product from investing $1,000 in applied research was $21,200 (= $14,100 intrastate plus $7,100 spillover) in the South, $11,600 (= 5,070 + 6,530) in the North and $12,200 (= 8,270 + 3,930) in the West. He concluded that the implied internal rate of return was 130 per cent in the Southern region, 95 per cent in the Northern region and 55 per cent in the Western region.

Later Measures of Investments in Public Agricultural Research

Over the past 25 years, advances have occurred in state agricultural productivity statistics and measurement of public agricultural research investments. In the mid-1980s, Huffman and Evenson began a two-decade-long research programme to improve agricultural productivity statistics and public agricultural research capital measures for the 48 contiguous US states (Huffman and Evenson, 1993, 2006b). Their efforts to improve productivity statistics were at least partially successful (Huffman and Evenson, 1993), but in 1990, Economics Research Service (ERS) undertook a much larger investment in creating new state-of-the-art agricultural productivity statistics at the state level, using production and index number theory and measurement methods developed by Jorgenson (2002). Eldon Ball led this project, and the ERS data are described in Ball et al. (2002) and Ball et al. (1999). The ERS agricultural productivity statistics are available through 2004 (USDA, 2009).[9] Alston and Pardey (2001) have also worked to create another set of state agricultural productivity statistics. Their methodology is described in Acquaye et al. (2002) and Alston et al. (2010).

Huffman and Evenson (1993) used rich commodity and research problem area (RPA) data from the USDA's Current Research Information System (CRIS) over 1968–86 as the centrepiece for deriving measures of public agricultural research capital over 1950–82. At the start of the twentieth century, the USDA began to locate agricultural research facilities and scientists in the states and regions of the United States (Huffman and Evenson, 1994, pp. 32–54), so the USDA is an important contributor to investments in agricultural research at the state level. Hence, their public agricultural research capital measure aggregated together the investments in research by federal and state agencies in each state. However, they relied heavily upon the use of exponential trends for extending public research investments in agricultural research from 1968 back in time to the 1920s. Moreover, in the late 1990s, they discovered that the above methodology ignored important information that was available over 1948–64 on the commodity focus of research expenditures by state of the SAESs.

In 1998, they initiated a major effort to reconstruct public agricultural research capital measures for the 48 contiguous states over 1960–96 and recently extended these data to 2004. In doing this, they continued to focus on public agricultural research investments by the USDA's research agencies, primarily the ARS and the ERS (or their predecessor organizations), and by SAESs and veterinary medicine colleges and schools.[10] As a result of Evenson's earlier work (Evenson, 1967) and an extensive review of the US public agricultural research system (Huffman and Evenson, 1993, 1994), they continued the practice of aggregating together the public agricultural research expenditures—investments undertaken within state by state and federal agricultural research agencies. For example, little evidence exists that differences exist in the basic versus applied nature of agricultural research in these institutions, given a particular research commodity.

In revising and updating public agricultural research expenditures, the rich data collected by the USDA-CRIS remain the centerpiece.[11] Recall that CRIS collects information from principal investigators or project directors on all research projects underway in the USDA's research institutions—largely the ARS and ERS—and in the state institutions undertaking agricultural research—largely the SAESs and the veterinary schools/colleges of the land-grant universities.

For each new CRIS project, principal investigators or project directors characterize their project by the project investigator's location (for the location of work), his/her choice of one or more research commodities or resources, for example, specific crops (corn, wheat, tomatoes, peaches, cotton, etc.), animals (beef cattle, dairy cattle, swine, sheep, poultry, etc.) and/or resources (land, water, farm structures) on which research is to be undertaken and one or more RPAs (such as soil, plant, water and nutrient relationships; control of weeds and other hazards of field crops and range; improving biological efficiency of field crops) that will be the focus of the research project. He/she also designates the field(s) of science being applied in the project. Over the lifetime of a research project (normally three to five years), salaries and benefits for scientist's time, research assistant's time and other support staff are charged to the projects and reported to CRIS, and the USDA uses this database to prepare annual summary reports, including the annual *Inventory of Agricultural Research*. Hence, CRIS contains a large database describing how resources are allocated to all research projects undertaken by scientists in the USDA and in the agricultural experiment stations and veterinary medicine colleges/schools of the land-grant universities.[12]

Not all public agricultural research expenditures reported to CRIS represent investments in discoveries that will impact US agricultural productivity; exceptions include research on recreation resources (wilderness, parks, campgrounds), clothing and textiles, nonfarm structures and facilities, families and their members, communities and areas (including institutions and organizations), post-harvest marketing and processing firms, including farm cooperatives, and the agricultural economy of foreign countries. Productivity-oriented agricultural research consists of expenditures on research commodities that are farm outputs, inputs or resources and that broadly can be expected to impact farm productivity as reflected in the description of the RPA being pursued.[13] In 1998, CRIS published a revised set of RPAs, and since many of them differ a little and in a few cases they differ a lot from the earlier set of RPA, we had to do some experimenting in order to achieve a smooth transition of agricultural-productivity-oriented research expenditures at the state level over 1996–97 with those over 1998–2001.

How much of a difference does omitting irrelevant research expenditures or investments in public agricultural research make? Using our definitions, 69.5 per cent of the combined research of the USDA, SAESs and veterinary colleges/schools were broadly focused on agricultural productivity in 1970. The share in 1984 was 71 per cent and in 1995 was 69 per cent. Hence, at the national aggregate level, the share of public agricultural research having a broad agricultural productivity focus is significantly below 100 per cent, but also the share has been relatively stable—approximately 70 per cent. However, at the state level, there is more variation in this percentage.

Moreover, failing to exclude irrelevant public agricultural research investments in constructing a measure of public agricultural research capital leads to measurement error in the research capital variable. Why should we care? When ordinary least squares (OLS) are used to estimate the linear ln(TFP) model and research capital is measured with error, the estimated coefficient of research capital will suffer from attenuation bias, that is, the estimated coefficient will be biased toward zero (Greene, 2003, pp. 83–86). Hence, difference measures of public agricultural research capital can be expected to yield different estimates of the impact of research capital on TFP.[14] Moreover, this is not an omitted variable problem, but rather one of obtaining the best proxy for the 'true' measure of public agricultural research capital that explains agricultural productivity.[15]

Starting with the rich data in CRIS over 1970–2002, methods are described for using the lower quality data that are available as we move back in time. Over 1948–65, respectable quality data exist on research expenditures by commodity for each SAES (Huffman and Evenson, 1993, pp. 115–17), but no data exist on the USDA's investments in agricultural research at the state level before 1968. However, data over 1948–65 exist for the SAESs because the USDA required each station to file annual research expenditure reports for 35 subject matter areas, and these unpublished worksheets were available (plus published annual totals by state). Twenty of these subject areas are directly tied to a particular crop or livestock and poultry type, similar to CRIS commodities.

Hence, the challenge for this period is the missing data on the USDA's research activity in each of the states. To construct a measure of agricultural-productivity-oriented research investments over the period from the mid-1920s to 1970, we undertook new work using existing data parsimoniously. We used a new methodology developed by economic historians for interpolating a time series by a related series (Friedman, 1962).[16]

Given that state and national totals exist for SAES research investments in all years, it is useful to undertake some comparisons (see Huffman and Evenson, 1993, 2006a). First, in 1970, the national total for public agricultural productivity research investments (for USDA research agencies and the SAES plus veterinary medicine colleges/schools) is equal to the national total of SAES research investments on all commodities and RPAs. In 1984, the ratio is 1.13, but in 1995 it is 1.02. Hence, over this post-1970 period, the national total for agricultural-productivity-oriented research investments is approximately equal to the total for all SAES research investments. Second, over the pre-1970 period, it is possible to compare the national total of SAES and USDA research investments, but not that of agricultural-productivity-oriented research investments. For example, in 1970, the ratio of the USDA's research investments to that of all SAES research investments is 0.51, in 1960, it is 0.52, and in 1950, it is 0.45. However, a dramatic change in relative and absolute research investments of the USDA and SAES system occurred over 1948–50, and considerable stability in the ratio occurs before 1948 (Huffman and Evenson, 2006b, pp. 104–07). For example, in 1948, this ratio is 1.40, in 1940, it is 1.53 and in 1930, it is 1.98. To circumvent these problems, public agricultural-productivity-oriented research investments (of the USDA, SAES and veterinary medicine schools and colleges) over 1929–69 are measured as an arbitrary re-scaling of state and national totals for all SAES research expenditures. The ratio at the national level is 2.01 over 1929–48 and 1.55 in 1949 and 1.04 in 1950–69, and all states have a similar scaling factor, except that a smaller adjustment was made based on a state's own ratio of agricultural-productivity-oriented research investments undertaken by the USDA, SAES and veterinary medicine colleges/schools to all SAES research in 1970. The national total of these public agricultural investments is 72 per cent of all USDA and SAES agricultural research investments over 1930–48 and 61 per cent in 1950, which compares to 70 per cent in 1970.

Alston et al. (2010, pp. 227–35) present a brief discussion of an alternative measure of public agricultural research expenditures to use in an econometric analysis of state agricultural productivity. They focus on public agricultural research of the SAESs and ARS. They, like us, exclude forestry research from USDA and SAES expenditures. Unlike us, they exclude research expenditures of the state veterinary colleges and schools. Otherwise, they included *all* agricultural research expenditures, even those that could not possibly impact agricultural productivity, including research on recreation resources (wilderness, parks, campgrounds), clothing and textiles, nonfarm structures and facilities, families and their members, communities and areas (including institutions and organizations), post-harvest marketing and processing firms, including farm cooperatives and foreign countries. In addition, they treat all of the USDA's research investments as one large pool of resources at the national level that impacts state agricultural productivity only through a 'spillin' effect, but they ignore the information on the state and regional allocation of these funds that exist in CRIS. The treatment of the USDA's agricultural research investment is especially important over 1910–48 when the USDA's research agencies accounted for two-thirds of the total public agricultural research expenditures and all SAESs for only one-third (Huffman and Evenson, 2006, pp. 105–06).[17] Overall, it seems safe to conclude that Alston et al. (2010) overestimate the size of the investment by federal and state research institutions in agricultural research that contributes to state agricultural productivity. This overestimate is a potential source of measurement error in their productivity analysis.

A Reassessment of Timing and Spatial Weights

After converting state public agricultural-productivity-oriented research expenditures into real terms using a research price index, the next step is to apply temporal and spatial weights to construct empirical research capital measures.[18] Research lag lengths and shapes have been and continue to be debated in the literature. Everyone agrees that current agricultural productivity does not simply depend on the current rate of investment in public agricultural research. In addition, scholars acknowledge that the lags and dynamics are of longer duration and greater importance than for most other types of capital investments. Given these attributes, a free-form representation of the lag structure is unsatisfactory because it asks too much of the available data. Consistent with the historical development of research lags and econometric agricultural productivity analysis, assumptions need to be made not only about the shape and length of the lag pattern, but also about research spillovers (Alston, 2001; Alston et al., 2010; Huffman and Evenson, 1993, 2006).

Outside of agriculture, the standard assumption is that capital either experiences an exponential rate of depreciation over the life of the asset, say 10–40 years, or alternatively, depreciation follows the one-horse-shay pattern where services flow equally over each year of an asset's life and at the end of life, it costlessly disintegrates. Moreover, the industrial R&D literature has assumed an exponential depreciation rate on R&D investments, as in Equation (6), for example, Griliches (1986), Adams (1990) and Nadiri and Prucha (1996). Although there may be good reasons for assuming a constant exponential depreciation rate on industrial R&D capital, this does not seem to fit public agricultural research.

For example, Griliches (1998) argues that the impact of research and development on productivity (or farm output) most likely has a short gestation period (with no impact), then blossoms (a period of rising weights), and eventually becomes obsolete (or depreciates away). Alston et al. (2010) suggest that the lag pattern can be summarized by a gamma distribution spread over roughly 50 years. In their work, the exact shape of the distribution depends on two parameters. However, they conclude that it most likely looks similar to a normal distribution with a mode of 25 years. An alternative approximation is to allow for a short gestation period where the past agricultural research investments have no impact on agricultural productivity and, thereafter, the weights follow a trapezoidal pattern. For example, following Evenson (1980), Huffman and Evenson (1993) chose trapezoidal timing weights but after an initial two-year period with no impact, reflecting start-up cost of major research projects and the time lag from input of effort towards discoveries or significant advances in knowledge are obtained (Huffman and Just, 2000) and then translated into useful farm-level technologies (Huffman and Just, 2006b). Then, positive weights start in year 3 and rise linearly to 0.05128207 in year 9 (seven years of rising weights), the weight remains at a maximum and constant in year 15 (or a total of seven years), and thereafter, weights decline linearly to zero in year 35 (or after 20 years). This trapezoidal pattern can be summarized as $R(1, 7, 7, 20)$ (see Figure 4.2). Hence, the mean lag is 14.5 years, which is three years shorter than the median lag length.[19,20]

Evenson (1980) and Huffman and Evenson (1993) report that public agricultural research spillovers do not follow state boundaries. Moreover, we continue to believe that spillin areas, as defined by the 16 geo-climatic regions and 37 sub-regions (from the 1957 *Agricultural Handbook*), make good econometric sense (Figure 4.1).[21] Given our construction of productivity-oriented agricultural research in each state, these research spillovers include research undertaken by the USDA, SAES and veterinary medicine colleges/schools. Hence, research performed by the USDA (SAES and veterinary medicine college/schools) in a state has a direct impact there and spills over to other states in the same geo-climatic region. Moreover, the spatial weights (see Table 4.1) are based on the share of all agricultural production in a state that is in each of its sub-regions.[22] As indicated above, the methodology underlying these geo-climatic regions is similar to that for the Food and Agriculture Organization (FAO)'s agro-ecological zones.[23,24]

New measures of intra-state public agricultural research capital have been tabulated and plotted by state and year, 1970–2004. These research capital measures are in natural log units, and the patterns or shapes of these plots can be classified into five groups: (a) approximately linear, representing a constant rate of research capital growth, (bi) a somewhat tipped forward and stretched 'S' shape, (c) a roughly 'j' or 'U' shape with a slight counterclockwise rotation, (d) a '∩' shape with a slight counterclockwise rotation and (e) a greatly tipped forward 'S' shape. Only five of the 48 states are included in the first group (Georgia, Idaho, Illinois, Minnesota and Utah), for example, see Figure 4.3 for Idaho; 15 are included in the second group (Alabama, Arizona, Arkansas, Kentucky, Maine, Michigan, Mississippi, Missouri, Nebraska, Nevada, North Carolina, Ohio, Oklahoma, Tennessee and South Dakota), for example, see Figure 4.4 for Alabama; 17 are included in the third group (Colorado, Delaware, Florida, Iowa, Indiana, Kansas, Maryland, Montana, New Mexico, North Dakota, Oklahoma, Pennsylvania, S. Carolina, Texas, W. Virginia, Wisconsin and Wyoming), for example, see Figure 4.5 for S. Carolina; four states are included in the fourth group (California,

Table 4.1:

Specific spillin weight by states across top into states on left (geo-climate sub-region based) (Revised 8-4-10, primarily in northern and southern plains states)

State	MN	IA	WI	IL	IN	OH	MI	PA	ND	SD	NE	KS	MO	KY	TN	WV	SC	NC	GA	VA	MT	AR
MN		0.6	0.9	0.6	0.6	0.5	0.8	0.2	0.9	0.9	0.2	0	0	0	0	0	0	0	0	0	0.2	0
IA	0.6		0.6	0.9	0.9	0.4	0.6	0.2	0.2	0.2	0.4	0.1	0.5	0	0	0	0	0	0	0	0	0
WI	0.9	0.9		0.9	0.7	0.4	0.6	0.2	0.2	0.2	0.4	0.1	0	0	0	0	0	0	0	0	0	0
IL	0.6	0.9	0.8		0.8	0.5	0.6	0.2	0.2	0.2	0.4	0.1	0.4	0	0	0	0	0	0	0	0	0.2
IN	0.6	0.8	0.6	0.8		0.4	0.2	0.1	0.2	0.2	0.4	0.1	0.4	0.1	0.1	0.1	0	0	0	0	0	0.2
OH	0.4	0.6	0.4	0.6	0.6		0.2	0.1	0.2	0.2	0.4	0.1	0.4	0.2	0.2	0.2	0	0.1	0.1	0	0	0.2
MI	0.5	0.5	0.5	0.5	0.5	0.5		0.2	0.2	0.2	0.4	0.1	0.2	0	0	0	0.1	0	0	0	0	0.1
MO	0.4	0.5	0.5	0.5	0.5	0.4	0.4	0	0.1	0.2	0.2	0.5		0.2	0.2	0.2	0.1	0.1	0.2	0.1	0.2	0.2

State	AR	LA	AL	GA	KY	TN	SC	NC	FL	MI	PA	WV	VA	NE	KS	TX	OK	IL	OH	IN	MO	MS
AR		0.8	0.8	0.5	0.1	0.1	0.1	0.1	0	0.1	0.2	0.1	0.1	0.1	0.3	0.3	0.1	0.1	0.1	0.1	0.2	0.8
LA	0.6		0.9	0.8	0	0	0.8	0	0.8	0.6	0.2	0	0	0	0.3	0.3	0.1	0	0	0	0	0.6
AL	0.2	0.8		0.9	0.1	0.1	0.8	0.1	0.9	0.6	0.2	0.1	0	0	0	0	0	0	0	0	0	0.8
GA	0.2	0.8	0.8		0.1	0.1	0.8	0.1	0.9	0.6	0.1	0.1	0	0	0	0	0	0	0	0	0	0.4
KY	0	0	0.1	0.1		0.9	0.7	0.8	0	0.2	0.8	0.7	0.8	0.8	0	0	0	0.1	0.1	0.1	0.2	0
TN	0	0	0.1	0.1	0.1			0.8	0	0	0.8	0	0.8	0.8	0	0	0	0	0	0	0	0.7
SC	0.8	0.7	0.7	0.1	0.2	0.2		0.5	0	0.2	0.7	0.2	0.2	0.2	0	0	0	0	0	0	0	0.5
NC	0.5	0.2	0.2	0.2	0.7	0.7	0.7		0	0.7	0	0.7	0.7	0	0	0	0	0	0	0	0	0.5
FL	0.1	0.9	0.9	0.9	0	0	0.5	0		0.5	0	0	0	0	0	0	0	0	0	0	0	0.3

(Table 4.1 Continued)

(Table 4.1 Continued)

State	AR	LA	AL	GA	KY	TN	SC	NC	FL	VA	WV	TX	OK	IL	OH	IN	MO	MS
VA	0.2	0.2	0.2	0.2	0.2	0.7	0.7	0.7	0	0.8	0	0	0	0	0	0	0	0
WV	0.2	0.2	0.2	0.2	0.2	0.7	0.7	0.7	0.7	0	0	0	0	0	0	0	0	0
MS	0.8	0.6	0.9	0.8	0	0.4	0.7	0.1	0.3	0	0	0.2	0.2	0	0	0	0.1	0

State	MD	DE	NJ	VA	SC	NC	PA
MD		0.8		0.4	0.1	0.3	0.2
DE	0.8		0.8	0.4	0.1	0.3	0.2
NJ	0.8	0.8		0.4	0.4	0.4	0.2

State	WA	OR	CA	NV	ID	MT	WY	UT	CO	AZ	NM	KS	TX	OK	NE	SD	ND	MN	OK
WA		0.9	0.3	0.2	0.2	0.7	0.1	0.2	0.2	0.2	0.2	0	0	0	0	0	0	0	0
OR	0.8		0.4	0.1	0.6	0.2	0.2	0.2	0.1	0.2	0.2	0	0	0	0	0	0	0	0
CA	0.1	0.2		0.2	0.2	0.1	0.1	0.2	0.1	0.4	0.2	0	0.2	0	0	0	0	0	0
NV	0.2	0.2	0.2		0.8	0.2	0.5	0.8	0.6	0.5	0.8	0	0	0	0	0	0	0	0
ID	0.4	0.5	0.3	0.8		0.2	0.6	0.9	0.6	0.5	0.5	0	0.1	0	0	0	0	0	0
MT	0.1	0.1	0	0.2	0.2		0.9	0.2	0.2	0.1	0.1	0	0	0	0.2	0.6	0.8	0.1	0
WY	0.1	0.4	0.2	0.5	0.5	0.2		0.5	0.4	0.4	0.4	0.1	0	0	0	0	0	0	0
UT	0.1	0.4	0.2	0.9	0.9	0.1	0.6		0.6	0.5	0.6	0	0.1	0	0	0	0	0	0
CO	0.1	0.4	0.1	0.7	0.7	0.1	0.4	0.7		0.2	0.2	0.2	0.1	0	0.3	0	0	0	0.1
AZ	0.1	0.3	0.8	0.5	0.5	0.1	0.3	0.5	0.4		0.8	0	0.1	0	0	0	0	0	0
NM	0.1	0.3	0.1	0.9	0.9	0.1	0.5	0.9	0.7	0.4		0	0.5	0	0	0	0	0	0

State	ND	SD	KS	OK	NE	TX	MN	IA	WI	IL	IN	OH	MI	MO	AR	LA	MT	WY	UT	CO	AZ
ND		0.6	0	0	0.2	0	0.2	0	0	0	0	0	0	0	0	0	0.8	0.3	0	0	0
SD	0.8		0.2	0	0.9	0	0.5	0.8	0.5	0.6	0.5	0.3	0.3	0.3	0	0	0.8	0.3	0	0	0
KS	0	0.5		0.3	0.6	0.1	0.4	0.8	0.4	0.8	0.8	0.4	0.4	0.8	0	0	0	0.1	0	0.3	0
OK	0	0	0.8		0.2	0.8	0	0	0	0	0	0	0	0.2	0.4	0.2	0	0.1	0	0.3	0
NE	0.3	0.3	0.2	0.2		0	0.3	0.4	0.4	0.4	0.4	0.4	0.2	0	0	0	0.4	0.1	0.1	0.3	0
TX	0	0	0.2	0.8	0.1		0	0	0	0	0	0	0	0	0.4	0.4	0	0.1	0	0.3	0.2

State	PA	NY	CT	MA	VT	NH	ME	OH	IN	IL	RI
PA		0.9	0.9	0.9	0.9	0.9	0.9	0.1	0.1	0.1	0.8
NY	0.9		0.9	0.9	0.9	0.9	0.9	0	0	0	0.8
CT	0.9	0.9		0.9	0.9	0.9	0.9	0	0	0	0.9
MA	0.9	0.9	0.9		0.9	0.9	0.9	0	0	0	0.8
VT	0.9	0.9	0.9	0.9		0.9	0.9	0	0	0	0.8
NH	0.9	0.9	0.9	0.9	0.9		0.9	0	0	0	0.8
ME	0.9	0.9	0.9	0.9	0.9	0.9		0	0	0	0.8
RI	0.8	0.8	0.9	0.8	0.8	0.8	0.8	0	0	0	

Source: Author's estimates.

Figure 4.3:
Public agricultural research capital, 1970–2004: Idaho

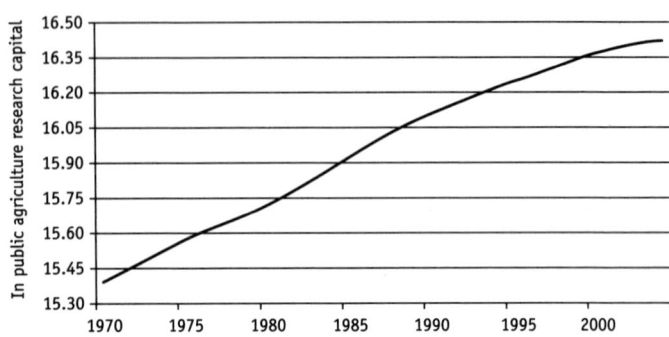

Source: Author.

Figure 4.4:
Public agricultural research capital, 1970–2004: Alabama

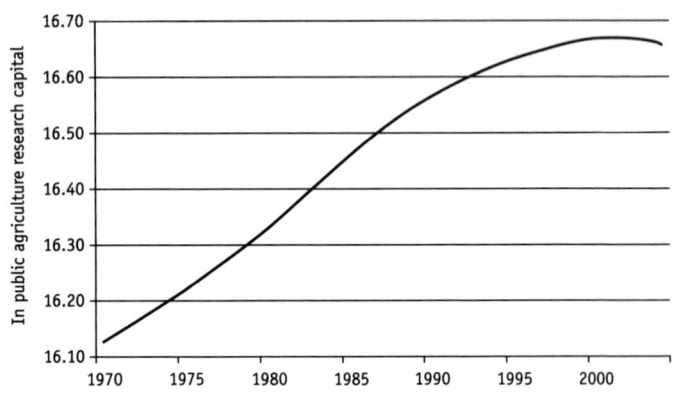

Source: Author.

Louisiana, Virginia and Washington), for example, see Figure 4.6 for Louisiana and five states are included in the fifth group (Connecticut, Massachusetts, New Hampshire, New Jersey and Rhode Island), for example, see Figure 4.7 for Connecticut.[25] It is apparent that the agricultural-productivity-oriented research capital variable is poorly approximated by an exponential trend, but the pay-off to this effect is good performance in explaining state agricultural productivity across states and over time.

Table 4.2 provides a summary of the average growth rate of public agricultural-research-productivity-oriented capital by state units for

Figure 4.5:
Public agricultural research capital, 1970–2004: S. Carolina

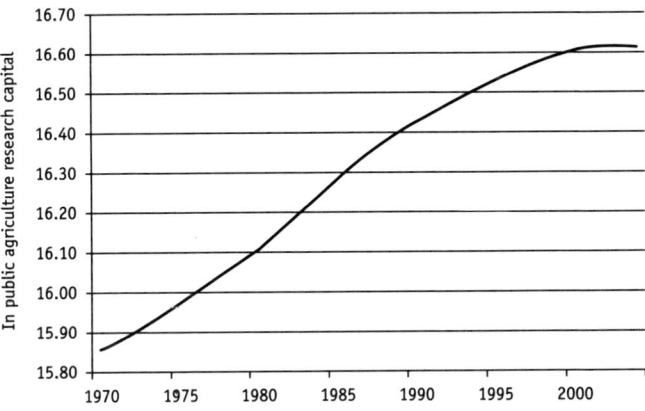

Source: Author.

Figure 4.6:
Public agricultural research capital, 1970–2004: Louisiana

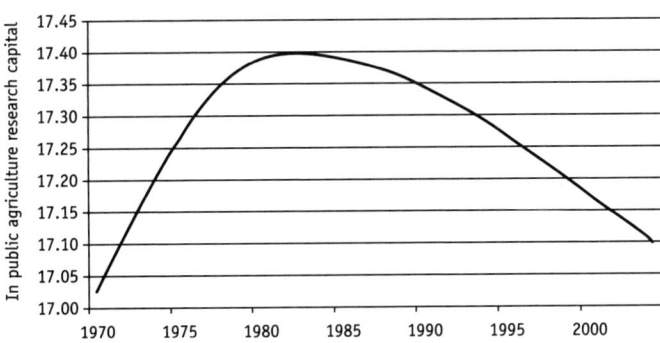

Source: Author.

each of the 48 contiguous states, 1970–2004. To ease the information load, the states are grouped by USDA farm production regions. Individual states that show very rapid growth of public agricultural research capital are Georgia (5.5 per cent), Arizona (4.6 per cent), N. Carolina (4.5 per cent), Nebraska (4.4 per cent) and Nevada and N. Dakota (4.1 per cent). In contrast, states with less than 1 per cent growth of public agricultural research are Massachusetts (0.02 per cent), Connecticut (0.2 per cent), New Hampshire and Ohio (0.8 per cent), Wyoming (0.9 per cent) and

Figure 4.7:
Public agricultural research capital, 1970–2004: Connecticut

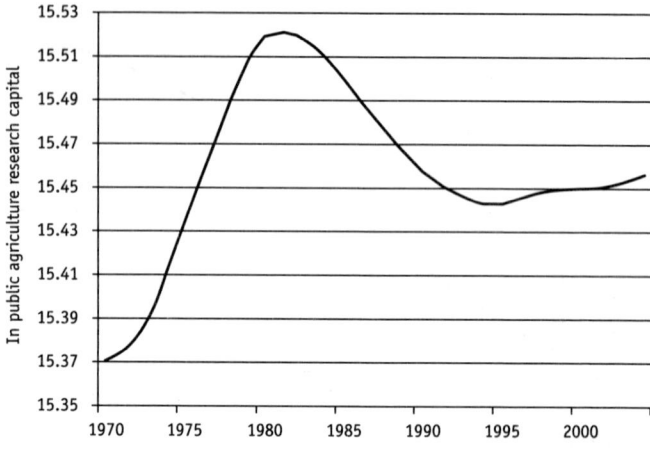

Source: Author.

Table 4.2:
Average annual growth rate for farm output, input, TFP and public agricultural research capital by state grouped by ERS farm production regions, 1970–2004 (rank order is in parentheses)

	Average annual growth rate, 1970–2004 (per cent)				
Region/state	TFP relative level 1996[a]	Total output	Total input	TFP	Public agricultural research capital
New England					
Maine	1.059 (19)	−0.21	−1.87	1.66 (25)	1.27 (41)
N. Hampshire	0.820 (43)	0.25	−1.69	1.94 (11)	1.00 (43)
Vermont	1.006 (27)	0.41	−0.85	1.26 (43)	1.80 (35)
Massachusetts	1.114 (14)	0.10	−1.83	1.94 (12)	0.87 (44)
Connecticut	1.137 (13)	0.96	−1.18	2.14 (5)	0.25 (47)
Rhode Island	1.148 (11)	0.22	−1.89	2.10 (6)	0.44 (46)
Northeast					
New York	1.008 (26)	0.33	−1.08	1.41 (33)	1.79 (36)
New Jersey	1.143 (12)	0.88	−0.90	1.79 (18)	0.80 (45)
Pennsylvania	0.986 (31)	1.58	−0.03	1.61 (30)	2.36 (22)
Delaware	1.293 (6)	2.53	1.17	1.37 (35)	2.05 (31)
Maryland	1.099 (17)	1.41	−0.22	1.64 (27)	2.33 (24)

(Table 4.2 Continued)

(Table 4.2 Continued)

		Average annual growth rate, 1970–2004 (per cent)			
Region/state	TFP relative level 1996[a]	Total output	Total input	TFP	Public agricultural research capital
Lake States					
Michigan	0.941 (34)	1.73	−0.62	2.34 (3)	2.65 (15)
Minnesota	1.030 (23)	1.72	−0.27	1.99 (10)	2.55 (18)
Wisconsin	0.926 (36)	0.80	−0.93	1.74 (19)	2.34 (23)
Corn Belt					
Ohio	0.916 (37)	1.38	−0.67	2.05 (8)	1.32 (39)
Indiana	1.101 (16)	1.86	−0.50	2.35 (2)	2.01 (32)
Illinois	1.193 (10)	1.60	−0.63	2.23 (4)	1.62 (37)
Iowa	1.237 (7)	1.25	−0.80	2.04 (9)	3.19 (9)
Missouri	0.904 (39)	1.29	−0.45	1.81 (17)	2.93 (13)
Northern Plains					
North Dakota	0.979 (32)	2.13	0.08	2.08 (7)	3.34 (7)
South Dakota	1.054 (20)	1.80	−0.26	1.91 (14)	1.91 (33)
Nebraska	1.105 (15)	2.22	0.55	1.67 (24)	3.88 (3)
Kansas	0.990 (30)	1.71	0.60	1.11 (45)	2.97 (12)
Appalachia					
Virginia	0.991 (29)	1.46	−0.02	1.48 (32)	2.58 (17)
West Virginia	0.574 (48)	1.09	−0.20	1.29 (39)	2.63 (16)
Kentucky	1.022 (25)	1.46	0.10	1.37 (36)	2.24 (25)
North Carolina	1.462 (3)	2.10	0.44	1.66 (26)	3.70 (4)
Tennessee	0.837 (41)	0.96	−0.06	1.02 (47)	2.23 (26)
Southeast					
South Carolina	1.088 (18)	1.50	−0.21	1.71 (20)	2.22 (28)
Georgia	1.333 (4)	1.99	0.29	1.70 (21)	4.36 (1)
Florida	1.618 (1)	1.61	0.02	1.59 (31)	3.33 (8)
Alabama	1.000 (28)	1.70	0.08	1.62 (28)	1.56 (38)
Delta States					
Mississippi	1.022 (24)	1.55	−0.14	1.69 (22)	2.23 (27)
Arkansas	1.228 (8)	2.36	0.47	1.89 (15)	2.89 (14)
Louisiana	1.037 (22)	1.01	−0.19	1.27 (41)	0.22 (48)
Southern Plains					
Oklahoma	0.698 (46)	1.48	0.52	1.08 (46)	1.83 (34)
Texas	0.786 (44)	1.77	0.41	1.36 (37)	3.12 (10)

(Table 4.2 Continued)

(Table 4.2 Continued)

		Average annual growth rate, 1970–2004 (per cent)			
Region/state	TFP relative level 1996[a]	Total output	Total input	TFP	Public agricultural research capital
Mountain States					
Montana	0.736 (45)	0.99	−0.28	1.27 (42)	2.18 (29)
Idaho	1.203 (9)	2.32	0.39	1.93 (13)	3.02 (11)
Wyoming	0.576 (47)	0.60	0.03	0.57 (48)	1.28 (40)
Colorado	0.937 (35)	1.13	−0.16	1.29 (40)	3.59 (5)
New Mexico	0.874 (40)	1.98	0.37	1.61 (29)	2.54 (19)
Arizona	1.045 (21)	1.25	−0.57	1.82 (16)	3.93 (2)
Utah	0.829 (42)	1.52	0.17	1.35 (38)	2.43 (20)
Nevada	0.904 (38)	1.35	0.17	1.17 (44)	3.59 (6)
Pacific					
Washington	1.306 (5)	2.47	0.40	1.41 (34)	1.22 (42)
Oregon	0.943 (33)	2.43	0.04	2.39 (1)	2.17 (30)
California	1.549 (2)	2.44	0.75	1.69 (23)	2.41 (21)

Source: Data for input, output and TFP growth rates for the agriculture sector by state are from the USDA-ERS (2009) and the public agricultural research data are my data.
[a] The TFP level is relative to Alabama.

New Jersey (1.0 per cent). All other states experience between 1 and 4 per cent growth rates in public agricultural research capital over 1970–2004. Hence, a substantial amount of variation exists in the rate of growth of public agricultural research capital across the 48 states, and this may help explain variation in agricultural productivity rates across states and over time.

New Evidence of the Contribution of Public Agricultural Research to State Agricultural Productivity

Over 1970–2004, the rate of growth of state agricultural productivity differs considerably across states and regions. The level of TFP in 1996 and rate of growth of TFP for agriculture by state, 1970–2004 (USDA,

2009), are presented in Table 4.2. In the first column, states are ranked relative to Alabama in 1996 (1.00). States with the highest TFP level are Florida (1.62), California (1.55), North Carolina (1.46), Georgia (1.33) and Washington (1.31). On the other end, the states with the lowest level of TFP are West Virginia (0.57), Wyoming (0.58), Oklahoma (0.70), Montana (0.74) and Texas (0.79). The states with the highest average rate of growth of TFP are Oregon (2.39 per cent), Indiana (2.35 per cent), Michigan (2.34 per cent), Illinois (2.23 per cent) and Connecticut (2.14 per cent), and the ones with the lowest rate of growth are Wyoming (0.57 per cent), Tennessee (1.02 per cent), Oklahoma (1.08 per cent), Kansas (1.11 per cent) and Nevada (1.17 per cent). The states with the fastest rate of growth of own-state public agricultural research capital are Georgia (4.36 per cent), Arizona (3.93 per cent), Nebraska (3.88 per cent), North Carolina (3.70 per cent) and Colorado (3.59 per cent), and those with the lowest growth are Louisiana (0.22 per cent), Connecticut (0.25 per cent), Rhode Island (0.44 per cent), New Jersey (0.80 per cent) and Massachusetts (0.87 per cent).

Building on Equation (5), the econometric model of state agricultural productivity for this study is:

$$\ln(TFP)_{ilt} = \beta_1 + \beta_2 \ln(RPUB)_{ilt} + \beta_3 \ln(RPUBSPILL)_{ilt} + \beta_4 \ln(RPUB)_{ilt} \times \ln(RPUBSPILL)_{ilt} + \tau \text{ trend} + \Sigma \, \delta_l \, D_l + u_{ilt}, \quad (9)$$

where TFP_{ilt} is the TFP in state i, region l and year t; $RPUB_{ilt}$ is the public agricultural research capital in state i, region l and year t; $RPUBSPILL_{ilt}$ is the public agricultural research capital spilling in state i, region l and year t and 'trend' is a linear annual time trend.[26] See Table 4.3 for definitions of symbols and summary definitions of variables. The D_ls are dummy variables denoting that a state is located in one of the six arbitrarily chosen regions: Northeast, Southeast, Northern Plains, Southern Plains, Mountain States and Pacific States (see Table 4.4). These dummy variables accommodate research investments that reflect purely regional considerations, for example, as in Hatch Regional Research. The Central region is the excluded region. The random disturbance term u_{ilt} is assumed to have a zero mean and follow a first-order autoregressive process, $u_{ilt} = \rho \, u_{ilt-1} + \xi_{ilt}$, where ξ_{ilt} is assumed to have a zero mean and constant variance. This specification of the autoregressive process imposes the constraint of a single autocorrelation coefficient across all 48 states.

Table 4.3:
Variable names and definitions and summary statistics

Name	Symbol	Mean (SD)	Description
Total factor productivity	TFP	−0.170[a] (0.271)	Total factor productivity for the agricultural sector relative to Alabama in 1996 (Ball et al., 2002)
Public agricultural	RPUB	−0.351[a] (0.891)	Own public-agricultural-productivity-oriented-research capital for an originating state, relative to Alabama in 1996. Capital stock obtained by summing past real research expenditures with 1- through 34-year lag and trapezoidal-shaped timing weights (Figure 4.2)
Public agricultural	RPUBSPILL	1.390[a] (0.499)	The public agricultural research spillin stock for a state, constructed from RPUB research capital spillin and state agricultural sub-region data (Figure 4.1)
Regional indicators	D_1		Dummy variable taking 1 if the state is CT, DE, ME, MD, MA, NH, NJ, NY, PA, RI or VT
	D_2		Dummy variable taking 1 if the state is AL, FL, GA, KY, NC, SC, TN, VA or WV
	D_3		Dummy variable taking 1 if the state is IN, IL, IA, MI, MO, MN, OH or WI
	D_4		Dummy variable taking 1 if the state is KS, NE, ND or SD
	D_5		Dummy variable taking 1 if the state is AR, LA, MS, OK or TX
	D_6		Dummy variable taking 1 if the state is AZ, CO, ID, MT, NV, NM, UT or WY
	D_7		Dummy variable taking 1 if the state is CA, OR or WA
Trend	Trend		Annual time trend

Source: Author's estimates.

[a] Numbers reported in natural logarithms.

Two versions of Equation (9) are fitted to the data for the 48 states over 1970–2004 or 35 years; one version excludes the interaction term between in-state and spillin public agricultural research capital and the other includes this interaction term. The fitting uses the Prais–Winsten estimator in STATA 9.0, which incorporates first-order autocorrelation

Table 4.4:
Econometric estimates of state agricultural productivity equation: Contribution of public agricultural research capital and other factors, 48 US states, 1970–2004 ($N \times T = 48 \times 35 = 1,690$)

	Regression (1)		Regression (2)	
Regressors	Coefficients	z-values[a]	Coefficients	z-values
Intercept	−22.644	5.98	−22.937	6.02
ln(RPUB)	0.125	15.31	0.036	1.31
ln(RPUBSPILL)	0.105	3.95	0.117	4.41
ln(RPUB) × ln(RPUBSPILL)	0.072	3.31		
Regional Indicators				
D_1	0.261	4.88	0.280	5.13
D_2	0.097	2.52	0.108	2.78
D_4	0.045	1.67	0.065	2.30
D_5	−0.039	0.94	−0.028	0.66
D_6	0.040	0.79	0.056	1.07
D_7	0.243	4.51	0.290	5.10
Trend	0.011	5.85	0.011	5.89
R^2	0.404		0.409	

Source: Author's estimates.

Note: The dependent variable is ln(TFP). The central region, D_3 (IN, IL, IA, MI, MO, MN, OH and WI) is the excluded region in the regression equations. Parameters are estimated by Prais–Winsten estimator, where the estimate of the AR(1) parameter r for regression (1) and (2) is 0.719. The estimation was carried out in STATA 9.0 using the panel data routing 'xtpcse' and subroutine 'ar1'. Also, see Beck and Katz (1995).
[a] The absolute size of z-values are reported. They are constructed from standard errors that are corrected for heteroscedasticity across states, that is, clustering, and contemporaneous correlation of disturbances across pairs of states.

and panel-corrected standard errors (PCSE). PCSE adjusts for clustering or unequal variance by state. In both fitted equations (Table 4.4), the estimate of the first-order autocorrelation coefficient (ρ) is 0.72 with a t-value larger than 10, suggesting that only weak dependence exists in the disturbances of the productivity equation and that a unit root is unlikely to be a problem (Greene, 2003; Enders, 2010).

The empirical results for regression (1) are impressive. The estimated coefficient for a state's own public agricultural research capital is relatively large, 0.125, and significantly different from zero at better than the 0.1 per cent level and has a tight 95 per cent confidence interval of (0.109, 0.141). In addition, the estimate of the coefficient for public agricultural research spillin capital is large at 0.105, and it is significantly different from zero at better than the 0.1 per cent level and has a 95 per cent confidence interval of (0.053, 0.157). The large size and strong statistical significance of these estimates cast doubt on any measurement error issues in our research capital measures. The coefficient of trend is 0.011 which implies that trend-dominated factors on net contribute 1.1 per cent per year to state TFP growth over the study period. Only three of the six regional dummy variables are significantly different from zero. The R^2 for this equation is 0.404.

When the interaction term between own-state public agricultural research capital and spillin research capital is included in the fitted model, the contributions of research capital are rearranged. The coefficient for own-public agricultural research capital is reduced to 0.036 (significantly different from zero at the 20 per cent level) and for spillin research capital is 0.072 (significantly different from zero at 0.1 per cent level), but then the estimated coefficient of the interaction term is now 0.072 (significantly different from zero at the 0.1 per cent level). The positive sign of the estimated coefficient of the interaction term implies that own-state agricultural research capital and spillin research capital are a type of complement, that is, in order for a state to benefit fully from spillin effects it must also undertake agricultural research within its boundaries. In this second regression, the estimated coefficient of trend is exactly the same as in the first regression, and the estimated coefficients for the regional dummy variables are similar. The size of the R^2 for this equation is only slightly higher than for the first equation, 0.409 versus 0.404.

Table 4.5 presents the estimates of the elasticity of state agricultural TFP with respect to the own-state and spillin agricultural research capital. It shows that the impact elasticity for own-state public agricultural research capital is slightly higher in regression (2) than (1), 0.136 versus 0.125, and that the impact elasticity of spillin research capital is a little smaller in regression (2) than in regression (1), 0.092 versus 0.105. The own-state impact elasticity is very similar to that of Huffman and Evenson (2006b), but the impact elasticity for research spillin is 2.5 times larger than in that paper.[27] Overall, the two different TFP regressions provide similar

Table 4.5:
The marginal impacts of agricultural research and extension on TFP: State data[a]

Marginal impact	Regression (1)	Regression (2)
$\partial \ln(TFP)/\partial \ln(RPUB)$	0.125	0.136
$\partial \ln(TFP)/\partial \ln(RPUBSPILL)$	0.105	0.092

Source: Author's estimates.
[a] Estimated regression coefficients are taken from Table 4.4. The marginal effect for the spillin variable in regression (2) is evaluated at the sample mean of ln(RPUB).

information about the impact elasticity of public agricultural research—own-state and spillin—on TFP, and the impact elasticity of increasing own-state and spillin public agricultural-research-oriented research capital by one percentage point is 0.23, which is very large compared to earlier estimates. Hence, the new empirical results reported in Tables 4.4 and 4.5 paint a simple but powerful story about the contributions of public agricultural research to state agricultural productivity over 1970–2004. These results are even more impressive, given that they surface when a linear trend is included in the regressions. These excluded trend factors include public agricultural extension capital and private agricultural research capital and other things.

Discussion

Alston et al. (2010) provide an alternative set of data and econometric analysis of state agricultural TFP over 1949–2002, which shows a large investment of effort on their part. They use their constructed state TFP data series rather than the official USDA state TFP date, which we use. They define public agricultural research investments differently and use their constructed data set to derive measures of public agricultural research and extension capital. Moreover, in the base model (Alston et al., 2010, p. 317), public agricultural research and extension expenditures are aggregated together to create a single variable.[28] The combination of public agricultural research and extension into one variable is unusual at this time. Although Griliches (1964) combined research and extension expenditures together in his early analysis of US agricultural production, starting with Evenson (1967) and continued, for example, in Huffman and Evenson (1993, 1994, 2006b), public agricultural extension has been treated as a separate variable. The main reason is that agricultural extension primarily delivers information to farmers, and agricultural research is a discovery or innovative process. Inputs of public agricultural extension largely speed up the adoption of new and profitable technologies by helping farmers understand how to use the new technologies effectively and form good expectations about future farm input and output prices. In addition, Huffman (1974, 1981) and Wozniak (1984, 1993) provide evidence of an immediate impact of public agricultural extension on farmers' decision-making.

It is equally surprising that Alston et al. (2010, p. 286) chosen not to use the rich data available in CRIS on the allocation of the USDA's research investments among the states, but instead treat the USDA's own research effort as occurring in Washington, DC.[29] For many years, USDA scientists have been located frequently at or near major land-grant universities so that they can easily interact with SAES scientists (Huffman and Evenson, 1993, pp. 40–41). In addition, Huffman and Evenson (1993, 2006b) found that combining all public agricultural research activity undertaken within a state—that by the SAES, veterinary medicine colleges/schools and USDA research agencies—into one variable to be a useful simplification for econometric modelling of state agricultural TFP.[30]

An unusual statistical property of the basic econometric model of TFP in Alston et al. (2010, p. 315) is that it does not include a time trend. A time-series econometric model of ln(TFP) without a time-trend regressor almost certainly does not have the important property of being covariance-stationary. This, then, results in the estimated regression coefficients of these models being biased and otherwise having poor statistical properties.[31] For example, if trend is omitted from regression (1), the estimated coefficient of ln(RPUB) becomes 0.195, which is 1.56 times or 56 per cent larger than reported in Table 4.4. The estimated coefficient of ln(RPUBSPILL) becomes 0.294, which is 2.8 times or 180 per cent larger than reported in Table 4.4. Put another way, the sum of coefficients on the two research capital variables increases from 0.230 to 0.489, which is 2.12 times the size of the sum in regression (1), or 112 per cent larger. Hence, these supplemental results confirm that the econometric results from fitting an ln(TFP) regression may be quite sensitive to omitting trend.

Given the use of different data, different definitions, different econometric models and a large volume of results from many different specification of their TFP model, it is difficult to make valid comparisons between the results of Alston et al. (2010) and our results. For example, it is very difficult to determine exactly how their models were fitted, including the exact set of regressors and t- or z-values associated with estimated coefficients, and this affects the interpretation.

Conclusion

In this chapter, a new model of state agricultural productivity has been developed. Second, a methodology for constructing a measure of intangible public agricultural research capital has been described and applied to generate revised and updated measures of public agricultural research capital at the state level for the contiguous 48 states during 1970–2004. This required constructing estimates of public agricultural research expenditures extending back to 1935, which, given the nature of the available early data, makes this a difficult task and requires much ingenuity. In our work, we exclude public agricultural research expenditures that are not focused on agricultural commodities or resources and research problem areas that do not have an agricultural productivity component. These exclusions include research on recreation resources (wilderness, parks, campgrounds), clothing and textiles, nonfarm structures and facilities, families and their members, communities and areas (including institutions and organizations), post-harvest marketing and processing firms, including farm cooperatives, and the agricultural economy of the United States and foreign countries. After constructing the measure of within-state public agricultural research capital, spillin public agricultural research capital was constructed using these data and a geo-climatic region map. This map contains geo-climatic regions that are not bound to geo-political units. Finally, the new model and data performed well in explaining state agricultural productivity over 1970–2004, and the results suggest an impact elasticity on TFP for a 1 per cent increase in public agricultural-productivity-oriented research capital of 0.23, which is very large.

The rate of US agricultural TFP growth over 1990–2004 averaged 1.82 per cent per year. However, real investments in public agricultural-productivity-oriented research peaked in 2004 and have declined a bit over 2004–2009, and the 2010 CRIS data may show a further decline. Our econometric model has provided new evidence on the linkage of investments in public agricultural research and TFP of US agriculture. Moreover, the slowdown in the growth of recent and current public investments in agricultural research will reduce the potential for growth of the US agricultural productivity over the next three decades (from what they otherwise would have been). Although our model shows a trend rate of TFP growth of 1.1 per cent per year, a flattening of public investments in productivity-oriented agricultural research will translate into a 0.7 per cent per year reduction relative to the 14-year period 1990–2004. This has the potential to erode existing international competitive positions of US grains, beans and livestock products. Moreover, given that the research lags are long, another decade of no growth in US investments in public agricultural-productivity-oriented research could mean that it will be very difficult to attain future productivity rates that are above the 1.1 per cent trend rate.

Almost a half-century ago, Griliches initiated the first attempt to econometric estimation, the contribution of public agricultural research (and extension) capital to agricultural productivity. Bob Evenson and others have followed his example and have improved upon the data, measures and models of productivity. The research on productivity decomposition in agriculture continues to lead the research on the measurement of intangible research capital and its contribution to productivity.

Notes

1. He estimated an unrestricted Cobb–Douglas production function. There were five input categories plus education per farm worker and public expenditures on research and extension (dissemination of research results) capital per farm. He fitted the model to data for 39 of the largest 48 states for 1949, 1954 and 1959.
2. Evenson (1967) also provided evidence that after 1940, public agricultural research expenditures were allocated primarily to pay the salaries of scientists, research assistants, laboratory technicians and other support staff. Less than 6 per cent were allocated to investment in equipment and structures.

3. Evenson (1967) used annual time-series data on aggregate US agriculture, 1938–63, to explore the impact of public agricultural research and extension on national agricultural productivity.
4. Evenson (1980, pp. 216–21) also developed the first set of state agricultural productivity statistics, covering 48 states over 1948–71.
5. The philosophy that underlies the USDA's land resource regions (USDA, 2006) and FAO's agro-ecological zones (FAO, 1978–81) is similar to the USDA's designation of geo-climate regions and sub-regions (USDA, 1957).
6. For example, factors contributing to the failure of the Flavr-Savr tomato in the early 1990s include that the gene for extended shelf life was inserted into a tomato variety developed for processing and for growing in the low humidity, hot summers of California, but these tomatoes when transferred to Florida to be grown as winter season fresh tomatoes. The winter Florida growing season is cool and damp relative to summers of California, and this led to serious new pest problems (Soil Association, 2005).
7. For each region, there were 12 different pairs of trapezoidal weights tested: $R(3, 4, 7)$, $R(3, 4, 11)$, $R(5, 6, 11)$, $R(5, 6, 15)$, $R(7, 8, 15)$, $R(7, 8, 19)$, $R(7, 8, 25)$, $R(11, 12, 25)$, and $R(15, 20, 25)$. Nine sets of contiguity weights tested were: $(\alpha_0=0, \beta_0=0)$, $(\alpha_0=0.25, \beta_0=0)$, $(\alpha_0=0.5, \beta_0=0)$, $(\alpha_0=0.75, \beta_0=0)$, $(\alpha_0=1.0, \beta_0=0)$, $(\alpha_0=0, \beta_0=0.25)$, $(\alpha_0=0, \beta_0=0.5)$, $(\alpha_0=0, \beta_0=0.75)$ and $(\alpha_0=0, \beta_0=1.0)$.
8. Equation (6) also included regional dummy variables.
9. Alston and Pardey (2001) also engaged in a separate effort to generate a new set of state productivity statistics (Acquaye et al., 2002) and a measure of public agricultural research capital and Alston et al. (1998).
10. State veterinary medicine colleges and schools undertake important research on livestock diseases and other pests that impact livestock productivity.
11. In the transition to the new CRIS data system, the data were noisy in 1968 and 1969, and they do not match the quality of later data.
12. Although there may occasionally be an aberration in data submitted by the states to CRIS, many intra-institution reporting errors cancel out due to the law of large numbers (Greene, 2003).
13. In our initial attempt to sort through RPAs, we received assistance from a knowledgeable staff member of the USDA, Cooperative States Research Service (CSRS).
14. If the TFP series are different, this is another source of different regression results.
15. Omitted variable bias can be positive, negative or zero, depending on the correlation between the omitted and included variables and the true sign of the omitted variable in the relationship under investigation (Wooldridge, 2002, pp. 61–63).
16. The fact that the USDA's annual data on agricultural research expenditures over 1925–69 are sometimes noisy weighed into the choice of our estimating procedure (Huffman and Evenson, 2006a, pp. 145–46).
17. After 1950, the proportions reversed.
18. We use the Huffman and Evenson (2006b, pp. 105–06) research price index to deflate nominal agricultural research expenditures. Over 1970–2005, the Huffman and Evenson research price index performs very similarly to the Pardey et al. (1989) research price index; the difference in the average rate of price change is only 0.06 per cent per year. See ERS website for recent values of the Pardey et al. (1989) index over 1970–2005.
19. A few researchers have included free-form lags of public agricultural research expenditures without much structure in aggregate productivity analysis (Alston et al., 1998), but this particular set of weights has the undesirable outcome that signs on estimated

coefficients of successive lag terms tending to oscillate. The trapezoidal and gamma timing weights can also be viewed as similar to the Bayesian's parameter smoothing priors (Kitagawa and Gersch, 1996; Geweke and Kean, 2005).

20. A major problem with long lags is that one must go far back in time to start a research capital series, and the quality of the data declines and the nature of research changes as we move back in time. Although the Hatch Act was passed in 1887, US public agricultural research did not get organized for research and graduate education until about 1920. For example, in 1910, 51 per cent of the SAES scientists had only a BS degree, and it was not until the 1950s that 50 per cent or more of the SAES scientists held a PhD degree (Huffman and Evenson, 2006, pp. 77–82).

21. The 16 geo-climatic regions in Figure 4.1 are very similar to the USDA's 20 land resource regions for the contiguous USA (USDA, 2006), and both have similarities to FAO's agro-ecological zones.

22. In addition, a public agricultural research spillin variable constructed from geo-climatic-based spatial weights performed significantly better (t-values and R^2) than spatial weights derived by grouping states into the 10 ERS farm production regions (see Table 4.2 for these state groups). Earlier, we found no support for the hypothesis that public agricultural research undertaken in one state spills over to all other states.

23. The spillin matrix was revised recently for five states because it was obvious that some impacts had been omitted earlier.

24. Alston et al. (2010, p. 286) use a spillin-weighting scheme designed around a matrix of particular crop and livestock products produced by states. This methodology was designed for industrial research, where all research is private and intellectual property rights retard spillovers. In contrast, the SAES system was established in 1887 because geo-climatic conditions differ across states in a way that affects the types of agricultural research discoveries that benefit local farmers, especially for plant-based agriculture. Moreover, farmers have sometimes complained that SAES research is not useful to them unless it is tested on agricultural research substations near them (Huffman and Evenson, 1993, pp. 38–39). Of course, these concerns are greater in very large states, such as Texas, California, Florida, than in small ones, such as Vermont, New Hampshire, Rhode Island and New Jersey.

25. No spillin public agricultural research capital is included in these measures.

26. Public extension expenditures and activities are allocated to one of four major programme areas: agriculture and natural resource, home economics and human nutrition, 4-H youth and community and rural development programmes. Over 1973–92, agricultural and natural resource extension accounted for roughly 42 per cent of the total (Huffman and Evenson, 2006, p. 137). After 1992, the extension service discontinued collecting and publishing data on the breakdown by state and year of expenditures across these four programmes (Ahearn et al., 2002, ERS). Hence, data unavailability for the later years is the reason for excluding public agricultural extension from the TFP model.

27. The main reason for the different impact of spillin research is that there were errors in the spillin matrix for five states in the Southern and Northern Plains States. These errors have been fixed.

28. It is unclear from the discussion in Alston et al. (2010, pp. 235–36, 297–98) how public agricultural extension is measured. Although public agricultural extension consists of four major programmes, the agricultural and resource programme is less than 50 per cent of the total.

29. The USDA's research facilities were limited to the immediate Washington, DC, area up to1910, but by 1931, the USDA had 51 field states in 24 states (Huffman and Evenson, 1994, pp. 31–32). Recent administrators of ARS also provide evidence of the strong tie of ARS research to the states.
30. The USDA established a major research centre at Beltsville, MD, starting in 1910 and the size of this facility grew at least up to the early 1970s. This makes expenditures on public agricultural research in the state of Maryland stand out as being very large relative to its size of the agricultural sector and we made some downward adjustments.
31. The issue is that the mean of $\ln(\text{TFP})_{it}$ and most of the regressors is not a finite constant over all t, including as t goes to infinity, for a given state i (see Enders, 2010, pp. 54).

References

AAUP (American Association of University Professors). *Academe: Bulletin of the AAUP*, various years.
Acquaye, A.K., J.M. Alston and P.G. Pardey. 2002. 'A disaggregated perspective on postwar productivity growth in U.S. agriculture: Isn't that spatial?', in V.E. Ball and G.E. Norton (eds), *Agricultural Productivity: Measurement and Sources of Growth*. Norwell, MA: Kluwer Academic Publishers, pp. 37–84.
Adams, J.D. 1990. 'Fundamental stocks of knowledge and productivity growth', *Journal of Political Economy*, 98: 673–702.
Ahearn, M., J. Lee and J. Bottom. 2002. 'Regional Trends in Extension Resources', Paper presented at Southern Agricultural Economics Association Meeting, Orlando, FL.
Alston, J.M. 2002. 'Spillovers', *Australian Journal of Agricultural and Resource Economics*, 46(3): 315–46.
Alston, J.M., M.A. Anderson, J.S. James and P.G. Pardey. 2010. *Persistence Pays: U.S. Agricultural Productivity Growth and the Benefits from Public R&D Spending*. New York, NY: Springer.
Alston, J.M., B. Craig and P.G. Pardey. 1998. 'Dynamics in the Creation and Depreciation of Knowledge, and the Returns to Research', International Food Policy Research Institute EPTD Discussion Paper No. 35.
Alston, J.M. and P.G. Pardey. 2001. 'Attribution and other problems in assessing the returns to agricultural R&D', *Agricultural Economics* 25: 141–52.
Ball, V.E., J. Butault and R.F. Nehring. 2002. 'United States Agriculture, 1960–96: A Multilateral Comparison of Total Factor Productivity', in Ball, V.E. and G.E. Norton (eds), *Agricultural Productivity: Measurement and Sources of Growth*. Norwell, MA: Kluwer Academic Publishers, pp. 11–36.
Ball, V.E., F. Gollop, A. Kelly-Hawke and G. Swinand. 1999. 'Patterns of productivity growth in the U.S. farm sector: Linking state and aggregate model', *American Journal of Agricultural Economics*, 81: 164–79.
Beck, N. and J.N. Katz. 1995. 'What to do (and not to do) with time-series cross-section data', *American Political Science Review*, 89: 634–47.
Cornes, R. and T. Sandler. 1996. *The Theory of Externalities, Public Goods and Club Goods*. New York, NY: Cambridge University Press.

Enders, W. 2010. *Applied Econometric Time Series*. 3rd ed. Hoboken, NJ: John Wiley & Sons, Inc.
ERS. 'R&D deflator', Available at: http://www.ers.usda.gov/Data/AgResearchFunding/
ERS. 'Trends in Extension Staffing', Available at: http://www.ers.usda.gov/Data/extension/
Evenson, R.E. 1967. 'The contribution of agricultural research to production', *Journal of Farm Economics*, 49: 1415–25.
Evenson, R.E. 1968. 'The Contribution of Agricultural Research and Extension to Agricultural Production', PhD completed, The University of Chicago, Chicago, IL.
―――― 1980. An evaluation of methods for examining the quality of agricultural research (with B. Wright), in *An Assessment of the United States Food and Agricultural Research System, Vol. II*. Office of Technology Assessment, Washington, D.C.: U.S. Government Printing Office.
―――― 1996. 'Two blades of grass: Research for U.S. agriculture', in Sumner, D. and J. Antle (eds), *Papers in Honor of D. Gale Johnson, The Economics of Agriculture, Vol 2*. Chicago, IL: The University of Chicago Press.
―――― 2001. 'Economic impacts of agricultural research and extension', in Gardner, B.L. and G. Rausser (eds), *Handbook of Agricultural Economics, Vol. 1A*. New York, NY: North-Holland, pp. 574–628.
FAO. 1978–81. 'Report on the Agro-Ecological Zone Project'. World Soil Resources Report 48, FAO, Rome, Italy.
Friedman, M. 1962. 'The interpolation of time series by related series', *Journal of the American Statistical Association*, 57: 729–57.
Geweke, J. and M. Keane. 2005. 'Bayesian cross-sectional analysis of the conditional distribution of earnings of men in the United States, 1967–1996', in Upadhyay, S.K., U. Singh and D.K. Dey (eds), *Bayesian Statistics and Its Applications*. New Delhi: Anamaya Publishers.
Greene, W.H. 2003. *Econometric Analysis*. 5th Edition. New York, NY: Prentice Hall.
Griliches, Z. 1960. 'Hybrid corn and the economics of innovation', *Science* 132: 275–80.
―――― 1964. 'Research expenditures, education, and the aggregate agricultural production function', *American Economic Review* 54: 961–74.
―――― 1986. 'Productivity, R&D, and basic research at the firm level in the 1970s', *American Economic Review*, 76: 141–54.
Griliches, Z. 1998. *R&D and Productivity: The Econometric Evidence*. Chicago, IL: The University of Chicago Press.
Huffman, W.E. 1974. 'Decision making: The role of education', *American Journal of Agricultural Economics*, 56: 85–97.
―――― 1981. 'Black-white human capital differences: Impact on agricultural productivity in the U.S. South', *American Economic Review*, 71: 94–107.
―――― April 2006. 'Economics of Intellectual Property Rights in Plant Materials', Iowa State University, Department of Economics Working Paper #06016.
Huffman, W.E. and R.E. Evenson. 1993. *Science for Agriculture: A Long-Term Perspective*. Ames, IA: Iowa State University Press.
―――― 1994. *The Development of U.S. Agricultural Research and Education: An Economic Perspective*. Ames, IA: Iowa State University, Department of Economics. (Available at the Iowa State University Library, National Agricultural Library and libraries of 8 major land-grant universities.)
Huffman, W.E. and R.E. Evenson. 2006a. 'Do formula or competitive grant funds have greater impacts on state agricultural productivity?', *American Journal of Agricultural Economics*, 88: 783–98.

Huffman, W.E. and R.E. Evenson. 2006b. *Science for Agriculture: A Long-Term Perspective.* Ames, IA: Blackwell Publishing.

Huffman, W.E. and R.E. Just. 2000. 'Setting efficient incentives for agricultural research: Lessons from principal-agent theory', *American Journal of Agricultural Economics*, 82: 828–41.

Jorgenson, D.W. 2002. 'Foreword', in Ball, V.E. and G.E. Norton (eds), *Agricultural Productivity: Measurement and Sources of Growth.* Norwell, MA: Kluwer Academic Publishers, pp. i–iii.

Jorgenson, D.W., S. Mun and K.J. Stiroh. 2005. *Information Technology and the American Growth Resurgence.* Cambridge, MA: The MIT Press.

Khanna, J., W.E. Huffman and T. Sandler. May 1994. 'Agricultural research expenditures in the United States: A public goods perspective', *Review of Economics and Statistics*, 76: 267–77.

Kitagawa, G. and W. Gersh. 1996. *Smoothness Priors Analysis of Time Series.* Springer Lecture Notes Vol. 116. New York, NY: Springer-Verlag.

Nadiri, M. and I.R. Prucha. 1996. 'Estimation of the depreciation rate on physical and R&D capital in the U.S. total manufacturing sector', *Economic Inquiry*, 34: 43–56.

Pardey, P.G. and B. Craig. 1989. 'Causal relationships between public sector agricultural expenditures and output', *American Journal of Agricultural Economics*, 71: 9–19.

Pardey, P.G., B.J. Craig and M.L. Hallway. 1989. 'U.S. agricutual research deflators: 1890–1985', *Research Policy*, 18: 289–296.

Schultz, T.W. 1953. *The Economic Organization of Agriculture.* New York, NY: McGraw-Hill Book Co. Inc.

Soil Association. 2005. 'Flavr-Savr Tomato & GM Tomato Puree: The Failure of the First GM Foods', Briefing Paper, 29 November 2005. Available at: http://www.soilassociation.org/web/sa/sweb.nsf/.

STATA Corp. 2005. *Stata Statistical Software* 9.0. College Station, TX: Stata Corporation.

U.S. Department of Agriculture. 2009. 'Agricultural Productivity in the United States: Data Files'. Available at: http://www.ers.usda.gov/data/agproductivity.

―――― 1903–1942. *Annual Reports, Office of Experiment Stations.* Washington, DC: U.S. Government Printing Office.

―――― 1944–1954. *Report on the Agricultural Experiment Stations.* Washington, DC: U.S. Government Printing Office.

―――― 1955–1961. *Report on State Agricultural Experiment Stations.* Washington, DC: U.S. Government Printing Office.

―――― 1962–1969. *Funds for Research at State Agricultural Experiment Stations.* Washington, DC: U.S. Government Printing Office.

―――― 1957. *Yearbook of Agriculture, 1957: Soils.* Washington, DC: U.S. Government Printing Office.

U.S. Department of Agriculture, CSRS, CRIS. February 1993. *Manual of Classification of Agricultural and Forestry Research.* Revision V. Beltsville, MD: USDA-CSRS.

U.S. Department of Agriculture, CSREES, CRIS. December 1998. *Manual of Classification of Agricultural and Forestry Research.* Revision VI. Beltsville, MD: USDA-CSREES.

U.S. Department of Agriculture, CSRS, CRIS. 1968–2000. *Inventory of Agricultural Research, 1967–1999.* Washington, DC: U.S. Dept of Agri., Cooperatives States Research Service (later named the Cooperative States Research, Education and Extension Service).

U.S. Department of Agriculture, Natural Resources Conservation Service. 2006. *Land Resource Regions and Major Land Resource Areas of the United States, the Caribbean and the Pacific Rim*. U.S. Department of Agriculture Handbook 296.

U.S. President. 2008. *Economic Report of the President, 2008*. Washington, DC: U.S. Government Printing Office.

Wooldridge, J.M. 2002. *Econometric Analysis of Cross Section and Panel Data*. Cambridge, MA: The MIT Press.

Wozniak, G.D. 1984. 'The adoption of interrelated innovations: A human-capital approach', *Review of Economics and Statistics*, 66: 70–79.

——— 1993. 'Joint information acquisition and new technology adoption: Later versus early adoption', *Review of Economics and Statistics*, 75: 438–45.

5

Access to Markets and Farm Efficiency: A Study of Bicol Rice Farms over Two Decades

Sanjaya DeSilva

Introduction[1]

Rice farming communities in South and Southeast Asia have undergone a tremendous change particularly since the introduction of Green Revolution technologies in the 1960s (Evenson and Gollin, 2003). Perhaps the most dramatic and lasting difference between peasant villages then and now is the pervasiveness of markets and the consequent integration of previously isolated and subsistence-oriented farmers with regional, national and even global markets for commodities and factor inputs.

This chapter presents an empirical investigation of the relationship between the spatial and temporal spread of market institutions and improvements in the productivity and efficiency of rice farms over a period of two decades. The analysis synthesizes two lines of research on developing agricultural economies that have benefitted greatly from the contributions of Professor Robert Evenson over four decades: (a) the study of transaction costs in agricultural markets (DeSilva et al., 2006; Evenson and Roumasset, 1986; Lanzona and Evenson, 1997; Naseer et al., 2007) and (b) the analysis of the determinants of farm productivity and efficiency (Bravo-Ureta and Evenson, 1994; Evenson and Mwabu, 2001; Rosegrant and Evenson, 1992).

Our primary goal is to estimate how differences in access to markets affect yields, unit costs and efficiency at the farm level, and how this relationship has changed over time. Using results of the econometric models, we draw policy inferences regarding the value of improved institutional conditions that facilitate the functioning of agricultural commodity and factor markets to farmers.

Data

This study utilizes household-level data from Camarines Sur, one of the six provinces that form the Bicol Region in the Philippines. Although rainfall is relatively abundant and water is generally plentiful in the Bicol River basin, Bicol has long been the poorest region in the country (Lanzona and Evenson, 1997). Among the reasons for the economic backwardness of Bicol is its location in the relatively isolated south-eastern end of the Luzon Island and the terrain that is mostly mountainous. The data used in this study were collected as part of the Bicol Multipurpose Survey, a rich multi-year household and *barangay* (village) survey carried out in 1978, 1983, 1994 and 2003 through a series of collaborations between the Bicol River Basin Development Program and the Economic Growth Center of Yale University (Bicol River Basin Development Program, 1998; Lanzona, 1997; Naseer et al., 2007).[2] The sample used here contains an unbalanced panel from 413 households in 1983 and 196 households in 2003. The households come from 59 villages (barangays), 40 of which are located in rural areas, 9 are in towns (poblacions) and 10 are in cities. The availability of an unbalanced panel spanning over two decades is unusual for a micro data set from a rural agricultural region of a developing country and provides us with a sufficiently long time frame to study changes in institutional conditions.

Theoretical and Empirical Background

Spatial differences in the productivity of farms have long attracted the attention of economists. Schultz (1953) predicted in his 'urban-industrial impact hypothesis' that the 'locational matrix' of economic development has a centre that is 'primarily industrial-urban in composition' and that 'those parts of agriculture which are situated favorably in relation to such a center' will benefit from well-functioning economic organizations (p. 147). The urban-industrial centre, in Schultz's view, functions as a source of technological innovation and contains relatively efficient factor and product markets. More generally, the advantage of the urban-industrial centre can be encapsulated in its ability to minimize transaction costs associated with 'information, search, negotiation, screening, monitoring, coordination and enforcement' (Sadoulet and de Janvry, 1995, p. 254). As Evenson and Roumasset (1986) describe, 'In highly developed market economies, transactions are low cost. The public sector provides goods and standards that facilitate transactions. Communication is low cost (p. 141).'

The 'costs of engaging in market transactions vary a great deal over the development process' (Evenson and Roumasset, 1986 p. 141) and are particularly 'significant in rural economies where communications and transportation facilities are poor, markets are segmented, and access for market participation is restricted' (Lanzona and Evenson, 1997, p. 1). For example, spot markets for labour are subjected to high transaction costs in rural labour markets where 'institutions such as labor and contract law and formal employment assistance mechanism are not in place' (DeSilva et al., 2006, p. 851). In an analysis of the same Bicol villages that are the subjects of the present study, DeSilva et al. (2006) found that farmers engage in the costly activity of directly supervising hired labour more intensively in villages that are less urbanized and located far from market towns.

In the peripheral areas of Schultz's locational matrix, formal markets are often weak or, in some cases, absent. For example, Bicol villages with high transaction costs have a low incidence of wage labour market participation and low earnings (Lanzona and Evenson, 1997). In place of formal markets, the peripheral villages are typically served by 'a pattern of market organizations with heavy reliance on traditional institutions' (Evenson and Roumasset, 1986, p. 141). According to a vast literature in the tradition of new institutional economics, informal institutions such as the family farm, sharecropping and social networks respond to and help overcome the information and enforcement problems that arise from missing or incomplete markets (Hoff et al., 1993; Lanjouw, 1999; Otsuka et al., 1992). In the Philippines, it has been documented that high transaction costs encourage households with large farms to have larger families (Evenson and Roumasett, 1986). There is also some evidence that community-based social organizations help to alleviate disadvantages faced by farmers in remote villages (Naseer et al., 2007).

Although there is widespread acknowledgement of the crucial role played by informal institutions in developing agrarian economies, there is a consensus that these institutions are second best efficient (Hoff et al., 1993). For example, the reduction in the technical efficiency of rice farmers in Bicol due to adverse institutional conditions is only partly offset by the direct supervision of workers (Evenson et al., 2000). Focusing on markets for insurance on price and weather risks, Larson and Plessmann (2009) found that 'community-based informal arrangements are subject to failure when adverse events are extreme or occur with unusual frequency' (p. 30). As documented in two studies of Bicol (Lanzona and Evenson, 1997; Larson and Plessmann, 2009), weak or absent markets for labour, credit and insurance have adverse implications not only for efficiency but also for income distribution when market imperfections impose disproportionate costs on the poor and the landless. With the development of market infrastructure, informal institutions lose their advantage vis-à-vis formal markets. Their erosion is accelerated by the weakening of traditional methods of enforcement that typically accompanies the process of development (Hoff et al., 1993). Larson and Plessman (2009) found that the technical efficiency of Bicol rice farms is higher in villages with favourable market conditions, proxied in their study by the barangay-level price of rice.

Whether a rural farmer is able to benefit from the transaction cost advantages of well-developed markets depends on his or her proximity and connectivity to the core (Benziger, 1996; Jacoby, 2000). In the context of rural Filipino villages, the core does not need to be urban or industrial as suggested by Schultz; rural towns (e.g. poblacions in the Philippines) can provide access to well-functioning markets for commodities and factors of production and serve as sources of technical knowledge and skills. With infrastructural improvements such as new roads and rural electrification (Evenson, 1986; Jacoby, 2000) and with the development of agricultural extension services (Flores-Moya et al., 1978; see Birkhaeuser et al., 1991 and Evenson, 2001 for a review) that bring ideas and technologies from the centre to the periphery, locational disadvantages of remote villages vis-à-vis market towns can be overcome. Development of well-functioning market institutions in the peripheral villages can be promoted through government action (e.g. improvements in legal systems and infrastructure) and be aided by population growth that often accompanies economic development. With good roads, communication networks, extension services, well-developed market supporting institutions and high-population densities, the physical distance from the centre matters less for the peripheral villages.

Based on insights from the studies discussed, the primary hypothesis of this study is that greater access to markets promotes growth in agricultural productivity and efficiency in rice farms. In the next section, we present a descriptive portrait of the correlation between market access and farm outcomes across time and space. In the following section, results of a regression analysis that explores this correlation in greater detail are discussed.

Descriptive Portrait

The empirical analysis relies on two proxy variables to measure access to markets at the village level: (a) the distance from the village (barangay) to the market town (poblacion) measures the location of the peripheral rural village in relation to the market core; and (b) the population density of the village (barangay) is a measure of the level of market development in the barangay itself. To verify whether these two variables adequately capture barangay-level transaction cost conditions, we examine their correlation with the ratio of buying and selling prices of rice at the barangay level, a variant of the 'price wedge' measure of transaction costs developed for the Bicol Region by Lanzona and Evenson (1997).[3] As expected, distance to the poblacion is positively correlated ($\rho = 0.2818$) and population density is negatively correlated ($\rho = -0.2453$) with the 'price wedge' measure of transaction costs.[4] Insofar as localized price wedges or differences between the buying and selling prices of commodities reflect localized transaction costs (Lanzona and Evenson, 1997), this correlation provides evidence that the two measures of market access we have chosen are reasonable, albeit imperfect, proxies for transaction cost conditions at the barangay level.

Yields and Unit Costs

If the hypothesized advantages of market access exist, we would expect villages with greater access to markets to have higher average farm yields (output per hectare) and lower unit costs (cost per unit of output). This was indeed the case for yields in 1983, with the average yield progressively lower in barangays that were located away from the poblacion (Figure 5.1 left panel) and in barangays that were less densely populated (Figure 5.2 left panel).[5]

Over the two decades, the associations between the yield and the two measures of market access had become markedly weaker. While yields had improved across the board by 2003, the improvement was relatively larger in the more distant (Figure 5.1 right panel) and more sparsely populated (Figure 5.2 right panel) barangays. The spatial convergence may be a result of improvements in roads and communications infrastructure or in the institutional environment, both of which enhance access to the poblacions and promote the development of market institutions in the peripheral barangays. With better roads, electricity and telephone lines and with better legal systems and contract enforcement, physical distance to the market matters less. There has also been population growth in the peripheral villages (shown in Figures 5.1 and 5.2 by the relative size of the bubbles) resulting

Figure 5.1:
Yield and distance to poblacion

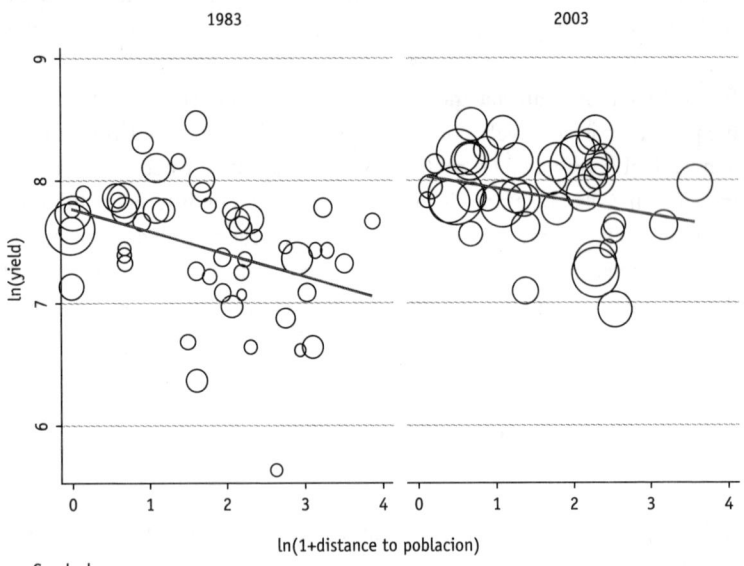

Graphs by year

Figure 5.2:
Yield and population density

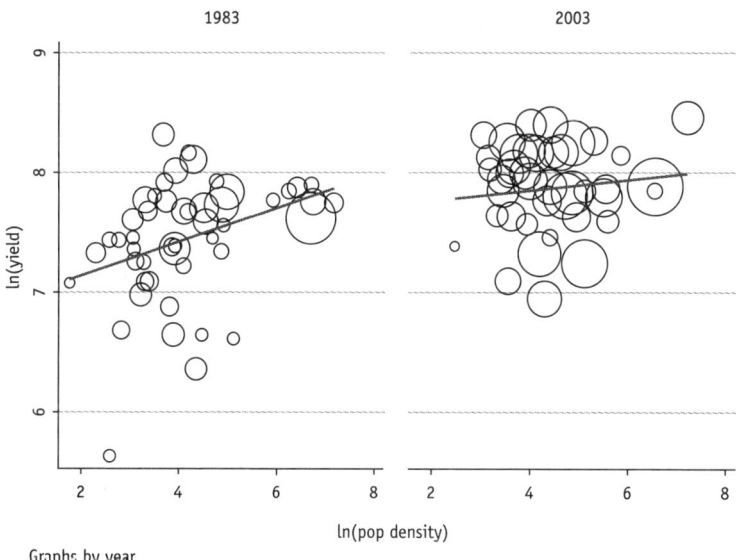

Graphs by year

in a noticeable decline in the variation in population densities (Figure 5.2 right panel). As peripheral villages become more densely populated, an environment conducive to development of efficient markets is created, lowering transaction costs.

Changes in unit costs follow a moderately different pattern. In 1983, unit costs were not strongly correlated with distance to poblacion (Figure 5.3 left panel) but decreased sharply with barangay population density (Figure 5.4 left panel). Between 1983 and 2003, the decrease in unit cost (measured in 1994 prices) was most pronounced in barangays that were close to the poblacion and those with low population densities (Figures 5.3 and 5.4, right panels).

The relation between distance to poblacion and unit cost was noticeably steeper in 2003 than in 1983 whereas the relation between population density and unit cost was flatter. Conforming with our expectations, unit costs vary with population density much similar to the way yields vary with population density. However, the pattern with respect to the distance to poblacion is different; the relatively flat relationship in 1983

Figure 5.3:
Unit cost and distance to poblacion

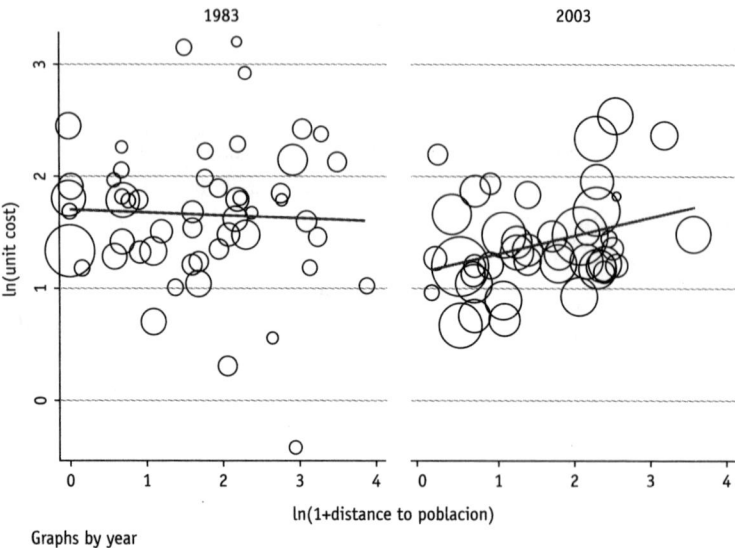

Figure 5.4:
Unit cost and population density

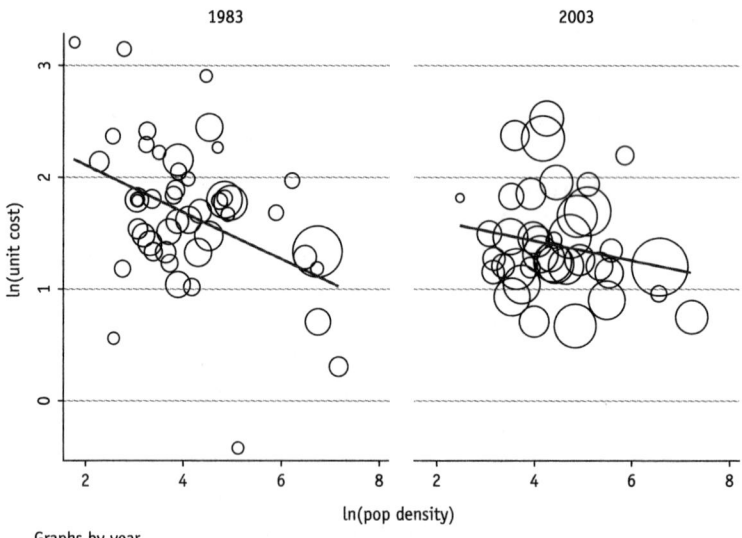

arose possibly because lower input costs (land rents and wages) in remote areas compensated for lower yields. The inter-temporal pattern possibly reflects factor prices that had evolved unfavourably for the distant barangays. While greater competition may lower prices for inputs such as fertilizer and tractors in the proximate barangays, the advantage which the remote barangays had in terms of lower labour and land costs may have eroded with greater labour mobility and improved market infrastructure. Improvements in the markets for factors that are relatively abundant in the remote villages do not necessarily lower production costs.

Table 5.1 presents the patterns illustrated graphically in the form of simple regression coefficients estimated at the farm level. For farms located in the poblacion itself, i.e. at zero distance from the poblacion, the average yield increased by 25 per cent in the two decades. In addition, the elasticity of the yield with respect to distance to the poblacion decreased substantially in magnitude but remained significantly negative. Unit cost decreased by about 25 per cent, but the association between distance to poblacion and unit cost was not statistically significant in either year even though the elasticity increased in magnitude (as seen in Figure 5.3). In 1983, population density was positively associated with yields and negatively associated with unit costs, whereas both relationships were smaller and statistically insignificant in 2003.

Table 5.1:
Simple regression results

	ln Yield 1983	ln Yield 2003	ln Unit cost 1983	ln Unit cost 2003
ln dist pob	−0.160***	−0.078**	0.016	0.051
	(0.036)	(0.037)	(0.037)	(0.034)
_cons	7.624***	7.879***	1.453***	1.194***
	(0.073)	(0.070)	(0.074)	(0.064)
N	387	195	387	195
ln pop density	0.132***	0.065	−0.087**	−0.050
	(0.041)	(0.045)	(0.042)	(0.041)
_cons	6.845***	7.473***	1.825***	1.479***
	(0.160)	(0.195)	(0.163)	(0.178)
N	398	187	398	187

Note: * $p < 0.1$, ** $p < 0.05$, *** $p < 0.01$; standard errors in parentheses.

Technical and Cost Efficiency

High yields (partial factor productivity) and low unit costs do not necessarily make a farm efficient (Bravo-Ureta and Evenson, 1994; Rosegrant and Evenson, 1992). Yields and unit costs are influenced by best-practice technologies, i.e. production and cost frontiers, and the ability of farmers to utilize them. A farm is technically efficient when it maximizes output conditional on input levels and the technology, operating on the production possibilities frontier (or the equivalent iso-quant); a farm is cost efficient when it minimizes costs conditional on the technology, level of output and factor prices, operating on the cost frontier (or the equivalent iso-cost curve). Stochastic frontier analysis (Aigner et al., 1977; Meeusen and van der Broeck, 1977; see Kumbhakar and Lovell, 2000, for overview) provides us with a method with which farm-level data on inputs, outputs and prices can be utilized to quantify levels of farm efficiency. Two examples of the application of this method are a study of cotton and cassava farmers in eastern Paraguay by Bravo-Ureta and Evenson (1994) and a study of Bicol rice farmers by Evenson et al. (2000). We investigate both channels: technical (or production) efficiency and cost efficiency, keeping in mind that technically efficient farm may still be cost inefficient if it is unable to achieve allocative efficiency and the utilization of inputs such that marginal returns equate relative factor prices (Coelli, 1996; Farrell, 1957).

The model and estimate procedure are outlined in the first section of the technical appendix. We found that the average technical efficiency in our sample increased from 0.642 to 0.846 in the 20 years; that is, conditional on technology and inputs, the average farm's production has increased from 64.2 per cent to 84.6 per cent of the maximum attainable output. Analogously, cost inefficiency has decreased from 1.997 to 1.129; the production costs of the average farm decreased from 99.7 per cent to 12.9 per cent above the minimum attainable costs. Both results indicate substantial improvements in efficiency.

Our hypothesis is that access to markets enhances both technical and cost efficiency of farms. With high transaction costs and incomplete markets, the input mix used by farmers may not equate marginal returns with factor prices and the farming methods used may not reflect the best-practice technologies. With greater access to markets, farmers in peripheral villages benefit from technological change, shifting the production possibilities frontier outward and the cost frontier inwards, and experiencing gains in the technical and allocative efficiency. Figures 5.5 and 5.6 correlate the estimated distance from the production frontier

Figure 5.5:
Production efficiency and distance to poblacion

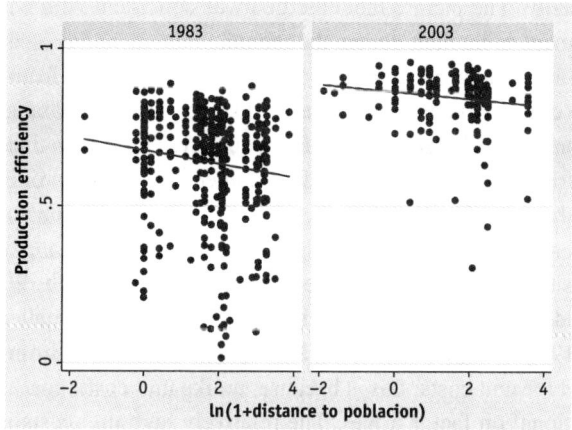

Figure 5.6:
Production efficiency and population density

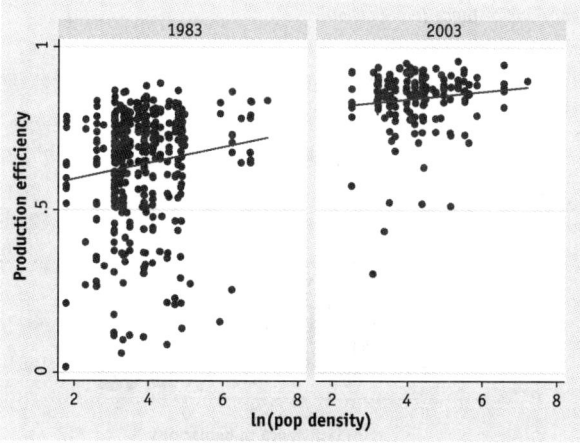

(technical efficiency) of farms against the distance from the barangay to the poblacion and the barangay population density.

The variation in technical efficiency across barangays mirrors that of the yield. In 1983, technical efficiency was higher, on an average, in barangays that were close to the poblacion and in barangays that were relatively densely populated. In 2003, both relationships were much weaker even though there was a general increase in technical efficiency across all barangays and the variation in technical efficiency had decreased substantially. The picture that emerged was consistent with widespread development of markets across the region.

Figures 5.7 and 5.8 correlate the estimated distance from the cost frontier (cost inefficiency) against the distance from the barangay to the poblacion and the barangay population density. The pattern that emerges supports our hypothesis of market development further; the strong negative relationship between cost efficiency and distance to the poblacion and the strong positive relationship between cost efficiency and barangay population density observed in 1983 were weakened considerably by 2003. Both mean and variance of cost inefficiency were also much smaller in 2003 than in 1983. The pattern of cost efficiency was different from what we observed for unit costs; this is because, unlike unit costs, cost efficiency is conditional on factor prices. The relatively high unit costs in remote barangays in 2003 appears to be caused by higher factor prices rather than

Figure 5.7:
Cost efficiency and distance to poblacion

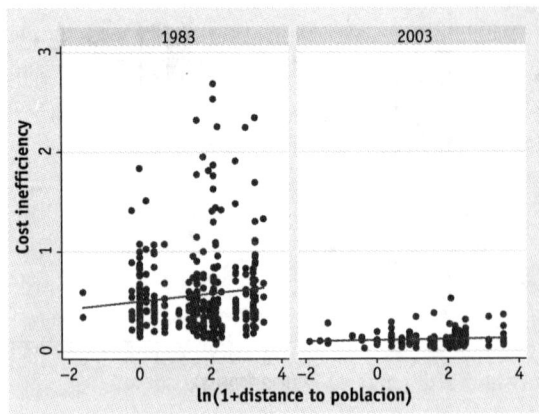

Figure 5.8:
Cost efficiency and population density

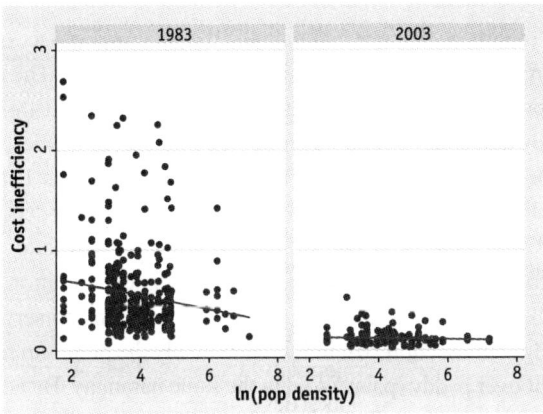

the cost inefficiency, i.e. the ability to use the optimal input mix for given factor prices of farmers.

Determinants of Yields, Unit Costs and Efficiency

This section investigates the underlying causes of the differences in farm productivity and efficiency that were observed over time and across barangays in the previous section, paying particular attention to the role played by the two market access variables. In the cross-sectional estimates, we are interested in examining how differential access to markets across barangays manifest in the differences in farm productivity and efficiency. In the inter-temporal comparison, we distinguish changes in productivity and efficiency that is attributable to inter-temporal changes in the market access variables from the changes in productivity and efficiency that arise from inter-temporal changes in the marginal influence of these variables. For example, a distant and small barangay may have experienced population growth over the two decades; at the same time, the construction of new roads may have alleviated the disadvantages that arise from its small size and remoteness.[6] Both these developments have implications for the productivity and efficiency of farmers in this barangay.

Yield and Unit Cost Estimates

Table 5.2 reports coefficient estimates of the two market access variables obtained in four sequential regression model specifications. The first model contains only the two market access variables. The next three models sequentially add control variables for household characteristics, barangay-level factor prices and barangay-level institutional conditions. Estimates are carried out separately for each year and for the two dependent variables, yield and unit cost. For institutional conditions, we add a dummy variable for whether the barangay has roads in good condition, a dummy variable for whether the barangay has access to extension services and a 'price wedge' transaction cost index, specifically the premium received by rice bought over paddy (palay) sold in the same barangay. By adding these sets of variables sequentially, we are able to ascertain how each set mediates the relation between the two market access variables on the one hand and the farm yields and unit costs on the other hand.

Table 5.2:
Step-by-step regression estimates of the market access effect

	(1)	(2)	(3)	(4)
	Market access	(1) + Household attributes	(2) + Factor prices	(3) + Barangay conditions
ln Yield 1983				
ln dist pob	−0.131***	−0.079**	−0.031	−0.046
	(0.038)	(0.039)	(0.055)	(0.053)
ln pop density	0.055	0.036	0.016	−0.010
	(0.057)	(0.059)	(0.065)	(0.063)
Joint significance of added variables				
F-statistic		13.04	9.39	5.37	2.96
p-value		0.00	0.00	0.00	0.03
N	373	372	329	329
ln Yield 2003				
ln dist pob	−0.082**	−0.101**	−0.116***	−0.108**
	(0.040)	(0.043)	(0.041)	(0.042)
ln pop density	0.002	−0.029	−0.040	−0.044
	(0.057)	(0.069)	(0.070)	(0.075)

(Table 5.2 Continued)

(Table 5.2 Continued)

	(1)	(2)	(3)	(4)
	Market access	(1) + Household attributes	(2) + Factor prices	(3) + Barangay conditions
Joint significance of added variables				
F-statistic	3.20	2.83	3.21	0.53
p-value	0.00	0.01	0.01	0.66
N	187	164	164	153
ln Unit Cost 1983				
ln dist pob	−0.046	−0.044	−0.004	0.023
	(0.044)	(0.044)	(0.059)	(0.059)
ln pop density	−0.142**	−0.130*	−0.124*	−0.098
	(0.065)	(0.069)	(0.074)	(0.074)
Joint significance of added variables				
F-statistic	2.45	0.77	4.87	3.40
p-value	0.09	0.63	0.00	0.02
N	373	372	329	329
ln Unit cost 2003				
ln dist pob	0.039	0.027	0.030	0.035
	(0.050)	(0.049)	(0.049)	(0.051)
ln pop density	−0.021	−0.013	0.010	0.028
	(0.049)	(0.060)	(0.063)	(0.067)
Joint significance of added variables				
F-statistic	0.82	1.02	1.87	1.44
p-value	0.44	0.42	0.12	0.24
N	187	164	164	153

Note: Standard errors in parentheses; * $p < 0.1$, ** $p < 0.05$, *** $p < 0.01$.

In the model that contains only the two market access variables (column 1), a 1 per cent increase in the distance to the poblacion lowered the yield, on an average, by 13.1 per cent in 1983 and 8.2 per cent in 2003, whereas controlling for distance to the poblacion, population density has a statistically insignificant effect of 5.5 per cent and 0.2 per cent, respectively. For every 1 per cent increase in the population density, unit costs decreased by 14.2 per cent in 1983. Neither the effect of population density on unit costs in 2003 nor the effect of distance to poblacion on unit costs in both years was statistically significant. In both years, the market access variables were jointly significant at the 5 per cent level in the determination of yields but not unit costs. The results of this simple benchmark model suggest that lack of market access created two problems for farmers with distance to poblacion negatively influencing yields and low population density negatively influencing unit costs in 1983. Both these disadvantage were reduced by 2003.

When we add household-level control variables, the qualitative results are not affected. The disadvantage of remote farms in 1983 is in part attributable to the less favourable attributes (e.g. lower levels of education) of the farmers themselves. However, the distance to poblacion had a larger effect on yields in 2003 than in 1983. When we introduce controls for barangay-level factor prices, the same pattern is reinforced; now the effect of distance to poblacion on yield is not statistically significant. When controls are introduced for barangay-level institutional conditions—road quality, transaction costs and availability of extension services—the effect of population density on unit costs also becomes smaller and statistically insignificant. This exercise reveals that, in 1983, distance to the poblacion influenced yields because the distant barangays had unfavourable household attributes and factor prices, and that population density influenced unit costs because sparsely populated barangays had insufficient roads, inadequate access to extension services and, more generally, high transaction costs. For both years, market access effects, if they had existed, are explained by the control variables. The one exception, however, was the effect of distance to poblacion on yields in 2003; this effect remains robust and statistically significant in all four specifications, suggesting that remoteness influences yields in ways that are not accounted for by the included variables.

In Table 5.3, we report five variants of the fully specified model for the determination of yields. The ordinary least squares (OLS) (first column) and random effects (fourth column) specifications for the pooled sample yield similar results.[7] These models identify several important determinants of high yields: (a) the availability of gravity and pump irrigation improves yields by as much as 35.5 per cent and 32 per cent, respectively; (b) elementary schooling (of the household head) improves yields by 32.1 per cent and secondary schooling brings an additional gain of 15.3 per cent. Tertiary schooling, on the other hand, has no effect. Such diminishing returns to schooling, with especially high gains at the elementary level, have been observed widely in developing country farms; (c) farm size has

Table 5.3:
Yield equation estimates

	(1)	(2)	(3)	(4)	(5)
	Pooled OLS	1983 OLS	2003 OLS	Random effects	Fixed effects
Production Environment					
ln area	−0.149***	−0.170***	−0.103*	−0.149***	−0.161
	(0.037)	(0.048)	(0.060)	(0.036)	(0.137)
Irrigation Dummies (Reference = Lowland Rain-fed)					
Upland rain-fed	−0.301*	−0.224	−0.621**	−0.301*	−0.600
	(0.180)	(0.220)	(0.305)	(0.155)	(0.642)
Gravity irr	0.355***	0.402***	0.145	0.355***	0.724**
	(0.079)	(0.101)	(0.134)	(0.081)	(0.345)
Pump irr	0.320***	0.456***	0.108	0.320***	−0.010
	(0.084)	(0.112)	(0.130)	(0.093)	(0.371)
Household Head Characteristics					
Age	−0.001	−0.001	0.000	−0.001	−0.026
	(0.002)	(0.003)	(0.003)	(0.002)	(0.023)
Education Dummies (Reference = No Schooling)					
Primary	0.321**	0.400***	0.002	0.321**	0.221
	(0.126)	(0.142)	(0.241)	(0.130)	(0.758)
Secondary	0.474***	0.590***	0.160	0.474***	0.796
	(0.145)	(0.173)	(0.258)	(0.151)	(0.915)
Tertiary	0.297	0.182	0.250	0.297	0.000
	(0.219)	(0.304)	(0.293)	(0.210)	(.)
Barangay Conditions					
ln dist pob	−0.071**	−0.047	−0.111**	−0.071*	−0.028
	(0.036)	(0.054)	(0.042)	(0.037)	(0.278)
ln pop density	−0.004	−0.010	−0.035	−0.004	0.335
	(0.048)	(0.064)	(0.078)	(0.045)	(0.303)
Good roads (dummy)	0.103	0.100	−0.064	0.103	0.685*
	(0.103)	(0.117)	(0.190)	(0.093)	(0.374)
Trans cost index	0.005	−0.029	−0.344	0.005	−0.098
	(0.051)	(0.056)	(0.272)	(0.055)	(0.172)
Extension (dummy)	0.292*	0.366**	−0.262	0.292**	0.615
	(0.151)	(0.160)	(0.692)	(0.118)	(0.519)
Barangay Level Prices					
ln male wage	−0.035	−0.019	−0.501	−0.035	0.219
	(0.205)	(0.238)	(0.979)	(0.203)	(0.599)

(Table 5.3 Continued)

(Table 5.3 Continued)

	(1) Pooled OLS	(2) 1983 OLS	(3) 2003 OLS	(4) Random effects	(5) Fixed effects
ln pr paddy	0.075 (0.240)	−0.035 (0.294)	−0.636 (0.714)	0.075 (0.262)	−0.781 (0.885)
ln pr fertilizer	−0.205*** (0.075)	−0.270*** (0.085)	−0.433 (0.592)	−0.205*** (0.070)	−0.277 (0.287)
ln pr seed	0.572** (0.240)	−0.154 (0.371)	1.326 (1.044)	0.572** (0.246)	0.192 (1.511)
ln pr tractor	−0.093** (0.043)	−0.135*** (0.050)	−0.053 (0.484)	−0.093** (0.041)	0.045 (0.130)
ln pr animal	0.021 (0.060)	0.032 (0.068)	−0.091 (0.230)	0.021 (0.057)	−0.355 (0.262)
Year 2003 (dummy)	−0.001 (0.007)			−0.001 (0.007)	
Intercept	9.388 (14.471)	8.415*** (1.302)	10.392*** (3.082)	9.388 (14.681)	7.606** (3.247)
N	482	329	153	482	482
R-sq	0.267	0.283	0.196		0.470

Note: Standard errors in parentheses; * $p < 0.1$, ** $p < 0.05$, *** $p < 0.01$.

a negative effect on yields, confirming another well-established finding in the literature that there are diseconomies of scale in rice farming; (d) low prices for fertilizer and tractors—the two factors most associated with the modernization of agriculture—are associated with high yields with elasticities of −0.21 and −0.09, respectively. The seed price has the opposite effect, possibly because seeds are heterogeneous and higher seed prices are associated with better high-yielding varieties. Neither wages nor the price of animals—two traditional inputs—has a significant effect on yields; and (e) at the barangay level, distance to the poblacion decreases yields with an elasticity of −0.071 and the availability of extension services boosts yields by 29.2 per cent. Neither the condition of roads nor the transaction cost index has a significant effect.

The estimates by year (reported in the second and third columns) show that the variables included predict yields quite well in 1983 but not in 2003. Two notable exceptions are the dummy variables for upland rain-fed farming and distance to poblacion. Although the advantage of gravity and pump irrigation over lowland rain-fed farming had decreased in 2003, upland farms had lower yields compared with all other farms. Distance to poblacion had a strong negative effect on yields even after controlling for other barangay level variables, none of which had a significant effect on yields in 2003.

The fixed effects estimates (reported in the fifth column) rely exclusively on within-household covariance of the yields with the independent variables, allowing us to establish the sources of the growth in yields over the two decades. The results indicate that the two primary determinants of within-household inter-temporal changes in yields were the increased availability of gravity irrigation at the household level and the improvement in road conditions at the barangay level. The fixed-effects results suggest that weakening of the relationship between market access and yields over the two decades was caused by infrastructural improvements in the peripheral villages.

Table 5.4 reports results of a similar set of regressions estimated for unit costs. In the pooled sample (OLS in first column and random effects in fourth column) estimates, only the factor prices have a significant effect on unit costs; wages and fertilizer prices increase unit costs, whereas seed prices have a negative effect. The latter result is consistent with what we found in the yield regression, which may be a consequence of the positive association between price and quality of seeds. When the models are estimated for each year, there is little predictive power in 2003 except for a negative coefficient for farm size, suggesting the presence of economies of scale. This contrasts with the previous finding, in the yield regression, that there were diseconomies of scale in 1983. Barangay-level institutional conditions do not explain any cross-sectional differences in unit costs. Fixed-effects estimates reveal, however, that increases in the population density and improvements in road conditions are associated with reductions in unit costs. In addition, inter-temporal changes in farmer age and fertilizer prices are positively correlated with inter-temporal changes in unit costs.

The results reported in Tables 5.3 and 5.4 provide several insights on how barangay-level accessibility to markets influences farm outcomes: (a) remote barangays had lower yields in 2003; (b) increases in the population density had a positive effect on reductions in unit costs from 1983 to 2003; (c) improvements in road conditions had a positive effect on improvement in yields and reduction of unit costs from 1983 to 2003; (d) farms in barangays with access to extension services had higher yields

Table 5.4:
Unit cost equation estimates

	(1) Pooled OLS	(2) 1983 OLS	(3) 2003 OLS	(4) Random effects	(5) Fixed effects
Production Environment					
ln area	−0.054	−0.017	−0.145**	−0.053	0.073
	(0.040)	(0.054)	(0.065)	(0.038)	(0.135)
Irrigation Dummies (Reference = Lowland Rain-fed)					
Upland rain-fed	−0.010	−0.068	0.354	−0.007	0.372
	(0.208)	(0.261)	(0.321)	(0.162)	(0.631)
Gravity irr	−0.027	0.000	0.024	−0.026	−0.162
	(0.088)	(0.121)	(0.115)	(0.084)	(0.339)
Pump irr	0.029	0.007	0.052	0.030	0.002
	(0.088)	(0.121)	(0.135)	(0.097)	(0.365)
Household Head Characteristics					
Age	0.002	0.001	0.003	0.002	0.043*
	(0.002)	(0.003)	(0.004)	(0.003)	(0.023)
Education Dummies (Reference = No Schooling)					
Primary	−0.088	−0.111	0.165	−0.088	0.045
	(0.134)	(0.150)	(0.250)	(0.136)	(0.745)
Secondary	−0.152	−0.252	0.127	−0.153	−1.003
	(0.146)	(0.170)	(0.257)	(0.159)	(0.900)
Tertiary	−0.051	0.038	0.014	−0.050	0.000
	(0.235)	(0.325)	(0.307)	(0.220)	(.)
Barangay Conditions					
ln dist pob	0.017	0.018	0.036	0.017	−0.032
	(0.041)	(0.059)	(0.051)	(0.039)	(0.273)
ln pop density	−0.052	−0.094	0.025	−0.052	−0.543*
	(0.052)	(0.075)	(0.070)	(0.047)	(0.298)
Good roads (dummy)	−0.045	−0.083	0.123	−0.047	−0.761**
	(0.112)	(0.132)	(0.228)	(0.097)	(0.368)
Trans cost index	0.048	0.067	0.295	0.046	0.020
	(0.048)	(0.055)	(0.243)	(0.057)	(0.169)
Extension (dummy)	−0.224	−0.269	−0.846	−0.223*	−0.096
	(0.171)	(0.182)	(0.554)	(0.124)	(0.510)
Barangay Level Prices					
ln male wage	0.208	0.392	0.138	0.208	0.193
	(0.220)	(0.265)	(0.783)	(0.212)	(0.589)

(Table 5.4 Continued)

(Table 5.4 Continued)

	(1) Pooled OLS	(2) 1983 OLS	(3) 2003 OLS	(4) Random effects	(5) Fixed effects
ln pr paddy	−0.394 (0.255)	−0.206 (0.339)	0.203 (0.581)	−0.399 (0.273)	0.334 (0.870)
ln pr fertilizer	0.196** (0.077)	0.300*** (0.087)	−0.252 (0.467)	0.196*** (0.074)	0.545* (0.282)
ln pr seed	−0.293 (0.265)	0.280 (0.397)	0.612 (0.899)	−0.297 (0.257)	−1.435 (1.486)
ln pr tractor	0.000 (0.047)	0.039 (0.057)	−0.398 (0.428)	0.001 (0.043)	0.153 (0.128)
ln pr animal	−0.041 (0.064)	−0.048 (0.073)	0.182 (0.207)	−0.041 (0.059)	0.071 (0.257)
Year 2003 (dummy)	−0.009 (0.008)			−0.009 (0.008)	
Intercept	19.222 (15.843)	−0.549 (1.507)	0.621 (2.864)	19.257 (15.341)	1.326 (3.193)
N	482	329	153	482	482
R-sq	0.097	0.097	0.145	0.394	

Note: Standard errors in parentheses; * $p < 0.1$, ** $p < 0.05$, *** $p < 0.01$.

in 1983; and (e) the transaction cost index in the rice market had no effect on either levels or changes in yields and unit costs.

Technical and Cost Efficiency Estimates

Table 5.5 presents the results of the joint maximum likelihood estimates of the production frontier and the associated mean efficiency function for the pooled sample and the two years separately (see section 2 of technical appendix for details of the model). For each sample, estimates are carried out with and without the three barangay-level institutional variables—road conditions, availability of extension services and the transaction cost index—to determine whether the efficiency effects of distance to poblacion and population density are mediated by these three variables. In the pooled sample, technical efficiency is negatively affected by the age of the farmer and distance to poblacion and positively affected by education and population density. When the three institutional variables are added, the population density effect becomes insignificant but the distance to poblacion remains significant. We also see that good roads and access to extension service have a positive effect on technical efficiency while transaction costs have a negative effect. The year dummy tells us that there has been an increase in technical efficiency over the 20 years. The results were broadly similar in the subsample of 1983. In 2003, however, none of the included variables significantly predicted technical efficiency. In the frontier itself, the results conform to our expectations: land, labour, seeds, fertilizer and tractors are significant contributors to farm output; gravity and pump irrigation improves output, whereas upland irrigation decreases it; there had been increasing returns to scale and there had been no significant shift of the production frontier from 1983 to 2003.

Table 5.6 presents the analogous estimates for a cost frontier and the associated efficiency equation. In the pooled sample, only the secondary education of the household head has a positive effect on cost efficiency among the household-level variables. At the barangay level, distance to poblacion increases and population density decreases cost inefficiency as hypothesized, and these effects are robust to the inclusion of the three other institutional variables. Among the institutional variables, transaction costs increase cost inefficiency and availability of extension services decreases it. Road conditions do not affect cost efficiency. The results for 1983 were qualitatively similar, whereas there were no significant predictors of cost efficiency in 2003. In the frontier itself, the results mostly conform to that of the previous models; there is evidence of economies of scale in both years; wages and fertilizer prices increase costs in the pooled sample in 1983, whereas higher seed prices and tractor prices were associated with lower costs possibly due to the heterogeneous quality of these inputs. In 2003, the results for prices were different, with fertilizer and tractor prices negatively correlated and seed prices positively correlated with costs.

The frontier analysis reported in Tables 5.5 and 5.6 presents several additional insights on how barangay-level accessibility to markets influenced farm outcomes in 1983: (a) distance to market centres and village-level transaction costs increase both technical and cost inefficiency; (b) availability of extension services increases both technical and cost efficiency; (c) population density has a positive effect on cost efficiency but not technical efficiency; and (d) road conditions improve technical efficiency but not cost efficiency. None of the barangay-level variables

Table 5.5:
Production frontier and efficiency estimates

	Pooled		1983		2003	
Production Frontier						
Intercept	5.134***	5.059***	4.957***	4.929***	5.391***	5.006***
	0.242	0.247	0.278	0.296	0.517	0.470
ln land	0.966***	0.946***	0.867***	0.869***	1.123***	0.984***
	0.088	0.088	0.104	0.108	0.174	0.158
ln labour	0.181***	0.197***	0.177***	0.183***	0.207**	0.311***
	0.042	0.042	0.050	0.050	0.085	0.084
ln seed	0.100**	0.105**	0.164***	0.166***	0.015	0.035
	0.044	0.043	0.051	0.055	0.072	0.073
ln fertilizer	0.063***	0.063***	0.069***	0.067***	0.059	0.024
	0.011	0.011	0.012	0.012	0.037	0.036
ln animal	0.008	0.002	−0.001	−0.006	0.034	0.028
	0.013	0.013	0.014	0.015	0.025	0.024
ln tractor	0.066***	0.059***	0.066***	0.062***	0.047	0.028
	0.018	0.017	0.017	0.020	0.042	0.041
Upland rain-fed (d)	−0.302***	−0.267***	−0.248**	−0.193*	−0.426**	−0.392*
	0.093	0.098	0.106	0.116	0.212	0.205
Gravity irr (d)	0.253***	0.238***	0.240***	0.242***	0.273***	0.214**
	0.055	0.055	0.071	0.070	0.089	0.091
Irr Pump (d)	0.172***	0.140**	0.145*	0.132	0.193**	0.105
	0.062	0.061	0.086	0.084	0.097	0.091
2003 (d)	0.046	0.045				
	0.063	0.060				
Distance from Frontier						
Intercept	−2.806	−4.827	−1.364	−6.080	−3.491	−8.333
	1.975	3.465	2.789	4.856	8.256	10.585
Age	0.017*	0.020*	0.005	0.026	0.012	0.013
	0.010	0.012	0.012	0.022	0.024	0.028
Primary ed (d)	−1.656**	−1.318	−2.768**	−2.178**	1.203	1.053
	0.723	0.827	1.352	1.051	2.597	2.005
Secondary ed (d)	−3.157**	−3.121*	−5.693**	−4.922**	1.092	0.911
	1.374	1.736	2.896	2.292	2.570	2.141
Tertiary ed (d)	−1.670*	−1.436	−2.073	−0.662	0.041	−2.714
	1.004	1.209	1.415	0.992	1.900	5.920
ln dist pob	0.721**	0.886**	0.679	0.969**	0.325	0.395
	0.300	0.410	0.437	0.493	0.554	0.521

(Table 5.5 Continued)

(Table 5.5 Continued)

	Pooled		1983		2003	
ln pop density	−1.248**	−0.682	−1.677**	−0.702	−0.065	0.054
	0.600	0.429	0.804	0.494	0.296	0.353
Good roads (d)		−1.701*		−3.037*		3.292
		0.889		1.653		4.283
Trans cost index		0.518**		0.585**		1.752
		0.256		0.274		1.679
Extension (d)		−3.627**		−4.652*		−2.359
		1.731		2.458		2.942
2003 (d)	−3.119**	−1.862*				
	1.522	1.076				
	4.588**	5.380*	5.547*	6.798*	1.093	1.323
	1.925	2.930	2.898	3.959	1.650	1.367
	0.978***	0.982***	0.983***	0.986***	0.918***	0.932
	0.010	0.010	0.010	0.009	0.103	0.073
Log Likelihood	−432.0	−416.2	−317.0	−310.7	−110.6	−96.4
LR Test	140.14	152.66	110.33	122.92	14.13	23.62
Restrictions	9	*	8	*	8	*
No. households	461	456	372	372	164	153
No. years	2	2	1	1	1	1
No. obs	536	525	372	372	164	153

Note: $* \, p < 0.1$, $** \, p < 0.05$, $*** \, p < 0.01$.

Table 5.6:
Cost frontier and efficiency estimates

	Pooled		1983		2003	
Cost Frontier						
Intercept	1.454***	1.703***	−0.503	−0.895	4.561***	1.928
	0.556	0.618	0.940	0.967	1.301	1.592
ln output	0.759***	0.778***	0.760***	0.800***	0.741***	0.750***
	0.026	0.022	0.034	0.036	0.035	0.036
ln male wage	0.595***	0.420***	0.820***	0.662***	−0.194	0.467
	0.138	0.151	0.160	0.161	0.416	0.498
ln pr fertilizer	0.096*	0.116**	0.168**	0.205***	−0.206	−0.551*
	0.055	0.054	0.067	0.066	0.265	0.311
ln pr seed	−0.285*	−0.147	0.174	0.402	0.790	1.210**
	0.153	0.155	0.285	0.272	0.543	0.583

(Table 5.6 Continued)

(Table 5.6 Continued)

	Pooled		1983		2003	
ln pr Tractor	−0.119***	−0.128***	−0.094**	−0.074**	−0.592**	−0.648**
	0.032	0.032	0.043	0.038	0.270	0.279
ln pr animal	−0.032	−0.041	−0.040	−0.039	0.182	0.089
	0.046	0.046	0.052	0.045	0.144	0.157
2003 (d)	−0.110	−0.077				
	0.101	0.102				
Distance from Frontier						
Intercept	1.996***	−0.098	3.478***	0.385	−20.824	−9.107
	0.876	1.260	1.247	1.796	20.163	8.419
Age	0.008	0.011*	0.002	0.011	0.087	0.033
	0.006	0.007	0.008	0.016	0.089	0.023
Primary ed (d)	−0.274	−0.423	−0.457	−0.773	5.741	4.460
	0.343	0.350	0.489	0.584	3.963	4.189
Secondary ed (d)	−0.546	−0.858*	−1.211	−2.353*	7.215	5.153
	0.416	0.452	0.769	1.356	5.720	4.465
Tertiary ed (d)	0.140	0.115	0.198	0.740	3.351*	2.582
	0.624	0.659	0.764	1.040	1.722	2.956
ln dist pob	0.216***	0.387***	0.265	0.633**	0.238	0.130
	0.104	0.132	0.188	0.313	0.445	0.193
ln pop density	−1.020***	−0.870***	−1.542***	−1.789**	0.160	0.222
	0.106	0.161	0.526	0.823	0.577	0.256
Good roads (d)		0.164		−0.231		1.720
		0.215		0.446		2.191
Trans cost index		0.640***		0.838*		2.185
		0.193		0.450		1.466
Extension (d)		−1.079**		−1.907*		−6.311
		0.464		1.188		4.667
2003 (d)	−3.066***	−3.412***				
	0.375	0.364				
	1.297***	1.842***	1.288***	3.126*	3.050	1.110*
	0.142	0.586	0.257	1.952	2.345	0.665
	0.858***	0.909***	0.813***	0.935***	0.970***	0.917***
	0.029	0.033	0.032	0.048	0.021	0.051
Log Likelihood	−435.4	−313.6	−310.0	−104.8	−93.3	
LR Test	68.497	39.624	46.841	21.576	33.804	
Restrictions	9		8	*	8	*
No. households	437		330	330	185	174
No. years	2		1	1	1	1
No. obs	515		330	330	185	174

Note: * $p < 0.1$, ** $p < 0.05$, *** $p < 0.01$.

predicted technical or cost efficiency in 2003. It appears that the relationship between barangay-level institutional conditions and farm efficiency has weakened dramatically during the period of the two decades.

Conclusions

Four broad themes emerge from the analysis of Bicol rice farms presented in this chapter. First, as predicted by the urban-industrial hypothesis of Schultz (1953), there was a significant inverse relationship between distance from the market and farm productivity and efficiency in 1983. Second, in the two decades from 1983 to 2003, rice farms in Bicol have experienced substantial (about 25–35 per cent) gains in terms of yields, unit costs and efficiency. Third, gains in productivity and efficiency were larger in the more remote and sparsely populated villages, weakening the relationship between institutional conditions and farm outcomes. Fourth, development of markets in the peripheral villages and the improved connectivity between the peripheral villages and market centres was facilitated by population growth, infrastructural investments (specifically, irrigation and roads) and the availability of agricultural extension programmes.

Compared with 20 years ago, the present-day farmers in the more remote villages of Bicol face fewer disadvantages in obtaining access to technologies and factors of production relative to those in and around market towns. This convergence of peripheral villages with the poblacions is a reflection of the spread of markets that is in part attributable to rapid population growth; costs of market transactions are lower in densely populated areas. However, our analysis suggests that public investments—in irrigation, roads and extension services—have played a vital role in developing market institutions and helping the more remote villages overcome their locational disadvantages. With better roads and extension services, farmers in the previously isolated corners in the periphery are now able to receive information on market prices, learn new farming technologies and obtain modern inputs such as tractors and fertilizer at competitive prices. With greater connectivity between the villages and the cities and market towns, the physical isolation of a rice farm is less of a handicap, and the Schultz hypothesis may no longer be relevant even in the poorest region of the Philippines.

Notes

1. This study builds on an unfinished book-length collaboration of the author with Professor Robert Evenson on 'Transaction Costs and Household Behavior in the Philippines'. The author gratefully acknowledges the insights and contributions of Professor Evenson that were pivotal to the analysis presented here. The author is also thankful to the Economic Growth Center at Yale University for research support and to Boriana Handjyiska for able research assistance.
2. Professor Evenson played an instrumental role in the design and implementation of all four waves of the Bicol Multipurpose Survey. His long and fruitful engagement with the Philippines began with the 3 years he spent at the University of the Philippines at Los Banos from 1974 to 1977 and has resulted in a series of studies including several that are cited in this chapter.
3. Their measure was constructed by 'measuring the effect of village dummies on the observed prices wedges' between the buying and selling prices of rice (p. 2).
4. In Lanzona and Evenson (1997), distance to poblacion is the variable most strongly correlated with the barangay-level differences in the price 'wedge' and the buying and selling prices of rice.
5. The figures are drawn in a log scale. Also note that one is added to distance to poblacion so that farms in the poblacion (zero distance from the poblacion) have a defined log value. The bubbles are proportional to the population of each barangay.
6. Unlike population density, the distance to the poblacion is relatively time invariant.
7. This is not surprising because our unbalanced panel contains very few households for which data are available for both years.

Technical Appendix

1. Technical and Cost Efficiency Computations Using Stochastic Frontier Models

We compute the technical and cost efficiency of farm households by estimating the following production and cost frontier equations.

The Cobb–Douglas production frontier estimated is the version proposed for unbalanced panel data by Battese and Coeli (1992):

$$y_{it} = \beta_0 + \beta_1 x_{it} + u_{it} - v_{it}$$

where y_{it} is the logarithm of the output of farm i at time $t = 1, ..., T$, x_{it} is a vector of the logarithms of inputs and dummy variables for type of irrigation and year, u_{it} is an error term that is distributed i.i.d. $N \sim (0, \sigma_u^2)$,

$v_{it} = v_i \exp(-\vartheta(t - T))$, v_i follows an i.i.d. distribution that is truncated at zero of $N \sim (\mu, \sigma_v^2)$. The non-negative error term v_{it} is assumed to be independently distributed from the random error u_{it} and represents the technical inefficiency, or distance from the production frontier, for each farm i at time t.

The dual cost frontier estimate has the following analogous form:

$$c_{it} = \beta_0 + \beta_1 y_{it} + \beta_2 p_{it} + u_{it} + w_{it}$$

Here, c_{it} is the logarithm of the total cost of farm i at time t, p_{it} is a vector of factor prices and w_{it} is an analogous truncated non-negative error term that represents cost inefficiency or the distance from the cost frontier. Note that cost inefficient farms are located above the cost frontier, whereas technically inefficient farms are located below the production frontier.

Both production and cost frontier models are estimated in two specifications: the first assumes that the non-negative error term is drawn from a distribution that is truncated at zero, whereas the second estimates the truncation point as a parameter (μ). The second model is favoured because the estimates of the more flexible model reveal that the truncation point is significantly different from zero.

At the farm level, the technical efficiency can be estimated by computing the ratio of observed to maximum feasible output where the latter is determined by the stochastic production frontier (Lovell, 1993). The technical efficiency can be interpreted as mean distance below the production frontier and is defined (for a Cobb–Douglas frontier) as follows:

$$TE_{it} = \frac{\exp(\beta_0 + \beta_1 x_{it} + u_{it} - v_{it})}{\exp(\beta_0 + \beta_1 x_{it} + u_{it})} = \exp(-v_{it})$$

Analogously, the cost efficiency of a farm can be interpreted as the mean distance above the cost frontiers and is defined as follows:

$$CE_{it} = \frac{\exp(\beta_0 + \beta_1 y_{it} + \beta_2 p_{it} + u_{it} + w_{it})}{\exp(\beta_0 + \beta_1 y_{it} + \beta_2 p_{it} + u_{it})} = \exp(w_{it})$$

Note that technical efficiency varies between 0 (inefficient) and 1 (efficient), whereas cost efficiency varies from 1 (efficient) to infinity (inefficient).

2. Estimation of the Determinants of Technical and Cost Efficiency

Linear regressions are not suited for the estimation of the determinants of technical and cost efficiency because the dependent variable, by construction, follows a truncated normal distribution (Kumbhakar et al., 1991; Reifschneider and Stevenson, 1991). We utilize the panel data variant of a two-equation stochastic frontier model (Battese and Coelli, 1995) where each of the frontier is estimated using a maximum likelihood estimator jointly with a corresponding linear equation for the determination of the one-sided error term:

$$v_{it} = \delta_0 + \delta_1 z_{it} + \epsilon_{it}$$
$$w_{it} = \gamma_0 + \gamma_1 z_{it} + \varepsilon_{it}$$

The vector z_{it} represents a vector of determinants of farm efficiency. These two equations modify the simple production frontier by asserting that the non-negative error terms, v_{it} and w_{it}, come from truncated (at zero) distributions of the $N(\delta_0 + \delta_1 z_{it}, \sigma_\epsilon^2)$ and $N(\gamma_0 + \gamma_1 z_{it}, \sigma_\varepsilon^2)$, respectively. In our model, the predictors of farm efficiency are the same household-level and barangay-level attributes that were included in the yield and unit cost regressions.

References

Aigner, D.J., C.A.K. Lovell and P. Schmidt. 1977. 'Formulation and estimation of stochastic frontier production function models', *Journal of Econometrics*, 6: 21–37.

Battese, G.E. and T.J. Coelli. 1992. 'Frontier production functions, technical efficiency and panel data: With application to paddy farmers in India', *Journal of Productivity Analysis*, 3: 153–69.

——— 1995. 'A model for technical inefficiency effects in a stochastic frontier production function for panel data', *Empirical Economics*, 20: 325–32.

Benziger, V. 1996. 'Urban access and rural productivity growth in post-Mao China', *Economic Development and Cultural Change*, 44(3): 539–70.

Bicol River Basin Development Program. 1998. Bicol Multipurpose Survey (BMS), 1983: Philippines [Computer File]. ICPSR Version. Camarines ur, Philippines: Bicol River Basin Development Program [producer], 1983. Ann Arbor, MI: Inter-university Consortium for Political and Social Research [distributor].

Birkhaeuser, D., R.E. Evenson and G. Feder. 1991. 'The economic impact of agricultural extension: A review', *Economic Development and Cultural Change*, 39(3): 607–50.

Bravo-Ureta, B.E. and R.E. Evenson. 1994. 'Efficiency in agricultural production: The case of peasant farmers in eastern Paraguay', *Agricultural Economics*, 10(1): 27–37.

Coelli, T.J. 1996. 'A Guide to FRONTIER Version 4.1: A Computer Program for Stochastic Frontier Production and Cost Function Estimation', CEPA Working Paper 96/7, Department of Econometrics, University of New England, Armidale, NSW, Australia.

DeSilva, S., R.E. Evenson and A. Kimhi. 2006. 'Labor supervision and institutional conditions: Evidence from Bicol Rice Farms', *American Journal of Agricultural Economics*, 88(4): 851–65.

Evenson, R.E. 1986. 'Infrastructure, output supply and input demand in Philippine agriculture: Provisional estimates', *Journal of Philippine Development*, XIII(1–2): 62–76.

────── 2001. 'Economic impacts of agricultural research and extension', in Pingali, P. and R.E. Evenson (eds.) *Handbook of Agricultural Economics*. Vol. 1 Part 1, pp. 573–628.

Evenson, R.E. and D. Gollin. 2003. 'Assessing the impact of the Green Revolution, 1960–2000', *Science*, 300(2 May): 758–62.

Evenson, R.E., A. Kimhi and S. DeSilva. 2000. 'Supervision and Transaction Costs: Evidence from Rice Farms in Bicol, the Philippines', Center Discussion Paper No. 814, Economic Growth Center, Yale University, New Haven, Connecticut.

Evenson, R.E. and G. Mwabu. 2001. 'The effect of agricultural extension on farm yields in Kenya', *African Development Review*, 13: 1–23.

Evenson, R.E. and J.A. Roumasset. 1986. 'Markets, institutions and family size in the rural Philippines', *Journal of Philippine Development*, 23(18): 141–62.

Farrell, M.J. 1957. 'The measurement of productive efficiency', *The Journal of the Royal Statistical Society*, 120(3): 253–90.

Flores-Moya, P., R.E. Evenson and Y. Hayami. 1978. 'Social returns to rice research in the Philippines: Domestic benefits and foreign spillover', *Economic Development and Cultural Change*, 26(3): 591–607.

Hoff, K., A. Braverman and J.E. Stiglitz. 1993. *The Economics of Rural Organization: Theory, Practice, and Policy*. Oxford: Oxford University Press.

Jacoby, H.C. 2000. 'Access to markets and the benefits of rural roads', *Economic Journal*, 110(465): 713–37.

Kumbhakar, S.C., S. Ghosh and T.J. McGuckin. 1991. 'A generalized production frontier approach for estimating the determinants of inefficiency in U.S. dairy farms', *Journal of Business and Economic Statistics*, 9: 279–86.

Kumbhakar, S.C. and C.A.K. Lovell. 2000. *Stochastic Frontier Modeling*. Cambridge, UK: Cambridge University Press.

Lanzona, L.A. 1997. Bicol Multipurpose Survey (BMS), 1994: Philippines [ComputerFile]. ICPSR Version. New Haven, CT: Yale University [producer], 1994. Ann Arbor, MI: Inter-university Consortium for Political and Social Research [distributor].

Lanzona, L.A. and R.E. Evenson. 1997. 'The Effects of Transaction Costs on Labor Market Participation and Earnings: Evidence from Rural Philippine Markets', Center Discussion Paper No. 790, Economic Growth Center, Yale University, New Haven, Connecticut.

Larson, D.F. and F. Plessmann. 2009. 'Do farmers choose to be inefficient? Evidence from Bicol', *Journal of Development Economics*, 90: 24–32.

Lovell, C.A.K. 1993. 'Production Frontiers and Productive Efficiency', in Fried, H.O., Lovell, C.A.K. and Schmidt, S.S. (eds), *The Measurement of Productive Efficiency*. New York: Oxford University Press, pp. 68–119.

Meeusen, W. and J. van den Broeck. 1977. 'Efficiency estimation from Cobb-Douglas production functions with composed error', *International Economic Review*, 18: 435–44.

Naseer, F., R.E. Evenson and S. DeSilva. 2007. 'Social capital, efficiency and transaction costs in the Philippines', Unpublished Manuscript.

Otsuka, K., H. Chuma and Y. Hayami. 1992. 'Land and labor contracts in agrarian economies: Theories and facts', *Journal of Economic Literature*, 30: 1965–2018.

Reifschneider, D. and R. Stevenson. 1991. 'Systematic departures from the frontier: A framework for the analysis of firm inefficiency', *International Economic Review*, 1991, 32: 715–23.

Rosegrant, M. and R.E. Evenson. 1992. 'Agricultural productivity and sources of growth in South Asia', *American Journal of Agricultural Economics*, 74(3): 757–61.

Sadoulet, E. and A. de Janvry. 1995. *Quantitative Development Policy Analysis*. Baltimore: The Johns Hopkins University Press.

Schultz, T.W. 1953. *The Economic Organization of Agriculture*. New York: McGraw-Hill.

SECTION III

Technology Transfer, National Innovation Systems and Industrial Development

6

Global Innovation Networks and Industry-University Interaction: A Study of India's ICT Sector

*K.J. Joseph and Vinoj Abraham**

Introduction

UNCTAD (2005) reported that by 2004 China had emerged as the most attractive location for R&D affiliates after the United States and the United Kingdom followed by India (sixth position). The study also projected that the pace of R&D internationalization is likely to accelerate with greater participation of developing countries. In a similar vein, Cantwell (1995) observed that global innovation networks (hereafter GINs) have expanded in the traditional high-tech regions in the United States, the European Union (EU) and Japan along with new locations in Asia, especially in India and China. Ernst (2009) argued that the emergence of GINs is real, and it is not merely something that can be expected to occur in the future. Similar view has also been held by scholars like Athreye and Prevezer (2008) and Athukorala and Kohpaiboon (2010). Globalization of R&D and resultant formation of GINs with increasing participation by developing countries need to be seen as one of the competitive strategies of the multinational corporations (MNCs) under heightened competition resulting from globalization. However, our understanding on the implications of such strategies on the national innovation systems in developing countries in general and the knowledge generation and diffusion in particular remains rudimentary.

The most significant outcome of GINs on knowledge production and diffusion is through their influence on universities and research

---

* This chapter is based on the INGINEUS project sponsored by the EU. The financial support from the EU and the feedback from the research team are gratefully acknowledged.

institutions (hereafter RIs). Scholars have highlighted the critical role of innovation capacity for reaping the potential advantages of GINs. Studies on innovation using the national system of innovation perspective have assigned key role for the interaction of universities and public research institutions (PRIs) with the industry in building up innovation capabilities. The literature on GINs, as Britto et al. (2013) argues, definitely displays awareness of the significant relationship between GINs and universities and RIs, but this is typically implicit and largely unexplored. Britto et al. (2013) based on a survey of existing literature on GINs have highlighted the following main types of interactions:

1. Local firms interacting with local and/or foreign universities,
2. Multinational enterprises (MNEs) interacting only with their local home-based universities,
3. MNEs interacting both with their local (home-based) universities and foreign (host country) universities and
4. International consortia between firms and network of universities.

It is not that all these forms of interactions will be present in all the GINs. The type of interaction that is found to exist is likely to be conditioned by a host of factors. This includes the characteristics of both the universities and sectors/firms involved along with the institutional arrangements and policy environment, or the system of innovation in the host country and receiving country at the sectoral, regional and national level.

The present study focusing on the industry–university interaction under GINs by taking the case of India's information and communications technology (ICT) sector assumes importance for more than one reason. To begin with, industry–university interaction in India remains almost an unexplored area. While there are a few studies, with the possible exception of Joseph and Abraham (2009), all of them are related to either a specific institution (Chandra, 2007 on IIT and Mashelkar, 1996 on NCL), a specific region (Basant and Chandra, 2007) or a specific industry (D'Costa 2006 on software). More importantly, none of the studies have explored the interaction between universities and firms in the context of GINs while India is known to have emerged as an active location for foreign R&D centres (Basant and Mani, 2012). In such a context, any study on industry–university interaction, neglecting the involvement of firms and universities in GIN, is likely to present only a partial picture. The selection of ICT sector is guided by the fact that India's presence in GINs has been most notable in the sphere of information technology (IT) that accounted for almost 40 per cent of foreign direct investment in R&D in India (TIFAC, 2006).

The remainder of this chapter is organized as follows. The section 'The Analytical Context' presents the analytical context of the study by situating the issue at hand in the broader context of studies on GINs. Database made use of in the study is introduced in the section 'On Database'. The section 'International Knowledge Flows in the ICT Sector in India' presents the empirical evidence on the interaction between the ICT firms within the country and abroad with universities and RIs within and abroad and highlights the underlying factors behind the observed pattern of interaction. The section 'Interaction of MNCs with Universities and Public Research Institutes: Two Case Studies' presents the insights from two case studies, and the final section concludes the chapter.

The Analytical Context

The emergence and spread of GINs according to Ernst (2009) poses new challenges and opportunities for the policymakers as well as the academia. In a sense, GINs could be viewed as a double-edged sword for both the MNCs and the host countries. As far as the MNCs are concerned, in the current context of globalization and heightened competitive pressures, there are many gains from GINs, especially an enhancement of the rate of return on R&D investment. At the same time, there are also pains associated with GINs like loss of control over the technology. The situation is summed up by Kaiser and Grimpe (2010); R&D outsourcing is certainly beneficial to innovation performance and that it may result in increased efficiency, reduced cost or foster innovation by getting access to valuable resources not available internally. At the same time, it might lead to dilution of firms' resource base, deterioration of integrative capabilities as well as rise in cost of coordination. From the perspective of developing countries, there are a number of instances wherein the participation in GINs has been instrumental in domestic capability building. At the same time, in the absence of appropriate policy measures and institutional interventions, GINs could also turn out to be global innovation traps, or 'poisoned chalice' to use the term used by Ernst (2009), *inter alia*, on account of brain drain and focus on low end of the value chain.

It is often held that rising R&D costs, increasing risk and complexity associated with technology development activities along with shorter product life cycle and intense competition in domestic and global markets have compelled firms to locate their R&D activities outside the borders of their home countries (Stembridge, 2007). Drawing on the ownership–location–internalization (OLI) framework by John Dunning, it has been argued that the internationalization of R&D is the result of a complex interaction between the ownership advantages of MNCs and the location advantages of regions. Cantwell (1995), for example, predicts that in a globalized world, MNCs will relocate the R&D activities to exploit regions' differential advantage in production and R&D. Such gains can arise through lowering the costs of routine R&D by rationalization of human capital-intensive activities and the opportunity to source new types of skills and knowledge by harnessing the science base in emerging regions. Another set of scholars, drawing insights from the resource-based view of the firm, analyzed the implication of R&D globalization and argued that these activities could be important instruments to acquire external technological knowledge that is subsequently integrated into firm's own knowledge base. Scholars have also approached the problem from transaction cost economics and transaction value perspective that combines both resource-based theory and transaction cost economics as a reasonably parsimonious theoretical approach to the idea of outsourcing as an alliance strategy (Mudambi and Tallman, 2010).

Earlier studies have identified a host of 'centripetal forces' that induce the firms to centralize the R&D activities in the headquarters and the 'centrifugal forces' that work towards the dispersion of R&D activities across different locations. The centripetal forces included the need to protect firm-specific technology as relocation of R&D that could lead to unwanted R&D leakage (Rugman, 1981). Patel and Pavitt (1991) argued that despite the increasing globalization of business, technological activities of large firms tended to stay in their countries of origin. This has to be viewed against certain key features of major innovations like the tacit nature of technological knowledge, need for closer coordination in decision-making in the face of uncertainty of innovation and the person-embeddedness of multidisciplinary scientific research. Further, the firm-specific technical advantages tend to evolve from and mirror home market conditions, and the strengthening of such advantages necessitates continued close contact with the domestic market. This turns out to be a vital factor leading to centralization of R&D. Another line of argument centred around the significance of scale economies in R&D and the difficulties in achieving minimum scale in case of decentralized laboratories. Some of the pioneering studies also found the high cost of coordination and control (Vernon, 1974) as a centripetal force.

The centrifugal forces include the need to be nearer to the export market. Very often, the establishment of R&D units in the host countries is preceded by exports and later the setting up of sales subsidiary and finally the manufacturing facilities. However, the transfer of manufacturing technologies from the parent firm calls for a substantial adaptation to suit the local market conditions. This induces the firms to establish local R&D units. Needless to say, such R&D units oriented towards adaptive R&D are found to be in countries with large domestic market. There could also be cases where in government regulations reinforce the inducement to setting up of such R&D units. When it comes to supply-side factors operating as centrifugal forces, most important one appears to be the access to scientific and technological skill, including scientific infrastructure that are available at more advantageous terms than in the domestic market. For instance, it has been shown that by 2010 China would have more science and engineering doctorates than the United States (Freeman, 2005; National Science Board, 2008).

The unprecedented increase in the pace of GINs formation tends to suggest that in the real world, the centrifugal forces prevail over the centripetal forces. There are two important factors for this rebalancing and resultant increase in the mobility of knowledge as argued by Albuquerque et al. (2012). The first one relates to the improvement in the information communication infrastructure and its extensions around the world, and second one relates to the liberalization of trade and investment policies that helped firms to exploit the benefit of technological change.

Though there are indications to suggest that disadvantages associated with centralization of R&D are often counterbalanced by the advantages arising from centrifugal forces, GINs cannot be viewed as purely a benign phenomenon without any associated challenges. Exploiting the opportunities while addressing the challenges, however, calls for a deeper understanding of the complexities involved in this network, *inter alia*, in terms of the inducement mechanism and their outcomes. This is the central issue addressed by the present study by taking the case of India's ICT sector.

On Database

The chapter is largely based on the survey of India's IT firms. The survey was designed to be implemented to all IT firms in India. The firms were chosen from the National Association of Software and Services Companies (NASSCOM) Directory of ICT firms. The 2009–10 Directory provided the information of 1,287 firms in IT industry that accounted for about 95 per cent of ICT production in the country (Table 6.1).

Initially, a web-based survey was implemented, but the yield was very low. Hence, direct interviews were planned. For the face-to-face interviews, time and resource constraint did not permit us to cover the entire

Table 6.1:
Survey design

Cities chosen for survey	Number of firms as per NASSCOM survey (2009–10)	Number of firms surveyed manually	Percentage of firms surveyed manually
Bangalore	281	50	17.79
Delhi/Noida/Gurgaon	256	75	29.30
Mumbai	185	68	36.76
Pune	72	20	27.78
Chennai	147	39	26.53
Trivandrum	184	20	10.87
Hyderabad	107	25	23.36
Kochi	55	10	18.18
Total	1,287	307	23.85
Online total		18	
All total		325	

Source: Primary survey (2009–10).

country. Instead, we chose cities/IT clusters that together represented nearly 93 per cent of all firms according to the NASSCOM directory. A survey team was developed and face-to-face interviews were implemented in cities/IT clusters selected for the survey. The survey was implemented in eight cities during the period March–April 2010, ending up with a sample of 325 completed questionnaires representing a favourable response rate of 24 per cent.

The primary survey was supplemented by the case study of two MNCs, one with headquarters in India and the other with headquarters in the United States. The case study interviews were conducted with high-ranking representatives of the firms in India.

International Knowledge Flows in the ICT Sector in India

Ernst (2009) argued that absorptive capacity is critical for upgrading innovative capacity and the firms must increase R&D investment and other innovative activities to avoid diminishing returns on network integration. Thus, the new geography of knowledge cannot be left to market forces alone. In such a context, to explore the possibilities for exploiting various potential advantages of GINs, we need to have an understanding of the present state of innovative capacity as evident from the R&D investment by firms in the ICT sector.

The survey showed that more than 60 per cent of the firms in ICT sector are engaged in R&D activities. Moreover, it was the MNC headquarters in India that had the highest share of firms that undertook R&D. While more than 78 per cent of the MNC headquarters are having in-house R&D units, it was 75 per cent for the MNC subsidiaries. In case of stand-alone firms, only 46 per cent of them were found engaged in R&D activities (see Table 6.2). The high share of MNCs, both headquarters in India and subsidiaries, in R&D activity suggests that the firm-specific characteristics in terms of organizational status do matter in R&D, and being MNCs induces them to engage in R&D activity, while not so much for the stand-alone firms.

The evidence further tends to suggest that the low prevalence of R&D activities among the stand-alone firms might set limits to their ability to

Table 6.2:
Existence of R&D activity in firms (per cent)

	Stand alone	Subsidiary of MNCs	Headquarter of MNCs	Total
R&D no	53.89	25.47	22	39.63
R&D yes	46.11	74.53	78	60.37
Total	100	100	100	100

Source: Primary survey (2009–10).

take advantage of the potential benefits of GINs as compared to their foreign and large local counterparts who appears to be well positioned (given the high R&D orientation) in the 'global innovation race'.

We also find significant inter-industry variation in the nature of R&D activities. In an earlier study (Joseph and Abraham, 2009) that covered six groups of industries, it revealed that out of the firms that invested in R&D, 43.5 per cent undertook R&D in a regular and centralized manner, whereas for 48.5 per cent of firms, R&D has been reported as occasional (see Table 6.3). Coming to inter-industry variation, while 81 per cent of the firms in chemicals (including pharma and biotech) reported R&D as a regular

Table 6.3:
Nature of R&D activities in firms

	Regular and centralized	Regular and decentralized	Regular	Occasional and centralized	Occasional and decentralized	Occasional	Total
Pharma, chemical and biotech	67.3	13.5	80.8	7.7	11.5	19.2	100
IT and electronics	48.1	16.5	64.6	7.6	27.9	35.4	100
Automobile	48.6	8.6	57.1	20.0	22.9	42.9	100
Textile and garments	10.0	0.0	10.0	5.0	85.0	90.0	100
Machine tools	39.7	4.8	44.4	30.2	25.4	55.6	100
Others	41.9	4.3	46.2	17.1	36.8	53.8	100
Total	43.5	8.0	51.6	15.0	33.4	48.5	100

Source: Joseph and Abraham (2009).

activity, only 10 per cent of the firms in textile and garments reported R&D as regular activity. In IT and electronics also, a substantial share of firms (65 per cent) were found engaged in regular R&D. In machine tools, and other industries, R&D was an occasional activity for majority of the firms. Thus, it appears that firms operating in technologically more dynamic industries are likely to undertake R&D as a regular activity.

Scale and Pattern of Interaction with Universities

Of the 324 ICT firms that were surveyed, majority of them (53 per cent) reported that they had some form of interaction with universities (local or foreign) or public-funded RIs in India. Among the firms, about 47 per cent stated that they had interacted with local universities or RIs. About 31 per cent of the firms interacted with foreign universities or RIs, whereas 24 per cent of the firms had interacted with both foreign and local universities. This represents a moderately high level of interaction between universities and firms in India's ICT sector.

The above aggregate picture conceals more than what is revealed, as there are significant differences across different organizational categories in terms of their interaction with universities and RIs. What is striking is the higher level of interaction observed in case of MNC headquarters as compared to the MNC subsidiaries and stand-alone companies regardless of the type of interaction. While 84 per cent of the MNC headquarters are found interacting with any university or RIs, among their foreign counterparts, only 55 per cent are found interacting, and the corresponding percentage in case of stand-alone firms is found lower (42 per cent). The difference becomes all the more striking when it comes to interaction with local universities wherein the observed percentages are 74 per cent, 46 per cent and 39 per cent, respectively, in case of MNC headquarters, MNC subsidiaries and stand-alone firms. The same pattern is observed with respect to their interaction with foreign universities or with both local and foreign universities wherein the stand-alone companies record much lower level of interaction (see Table 6.4).

Given that the interactions were for the purpose of innovation, as is evident from the survey, Indian MNCs and foreign MNCs are utilizing the resources from public sources such as universities and RIs much more

Table 6.4:
Firms that reported any form of university–industry interaction during the last three years that was important for an innovation for them

Firm type	Any university/RI	Any local university/RI	Any foreign university/RI	Both local and foreign university/RI
Stand-alone firms	42.51	39.52	19.76	16.77
MNC subsidiaries	55.66	46.23	36.79	27.36
Indian MNCs	84	74	54	44
Total	53.25	47.06	30.65	24.46
Total number	172	152	99	79

Source: Primary survey (2009–10).

intensively as compared to the stand-alone firms, indicating the bearing of firm characteristics in university–industry interaction.

Sectoral and Regional Patterns

Apart from firm characteristics, the sector specificities also appear to influence the interaction between universities and industry. From Table 6.5, it is evident that the occurrence of interaction/collaboration with universities varies significantly across industries. While the firms operating in the modern knowledge-intensive industries such as pharmaceuticals, biotechnology, chemicals and IT reported a relatively higher level of

Table 6.5:
Occurrence of interaction (collaboration) with universities/PRIs

Sector	Collaboration	No collaboration	Total
Chemical, pharma and biotech	23.44	76.56	100
IT and electronics	20.22	79.78	100
Automobile and parts	5.36	94.64	100
Textile and garments	0	100	100
Machine tools	11.76	88.24	100
Others	5.13	94.87	100
Total	11.33	88.67	100

Source: Joseph and Abraham (2009).

interaction, their counterparts in low technology industries like textiles reported hardly any interaction (Table 6.5).

The survey also indicated a substantial regional variation in the incidence of university–industry interaction, indicating that regional innovation system does matter. While in Bangalore, 94 per cent of the firms reported having interaction with universities, in Delhi, it was 77 per cent and in Mumbai, it was as low as 13 per cent. In other regions like Pune, Trivandrum and Hyderabad, it was moderate. The incidence of interaction with foreign universities was relatively less compared to local universities in most regions. In Bangalore, while the share of firms that interacted with the local universities/RIs was 90 per cent, only 40 per cent of the firms interacted with foreign firms (Table 6.6). Similarly, in Delhi, while 71 per cent of the

Table 6.6:
Regional variation in the number of firms that reported any form of interaction with university and RIs during the last three years that was important for an innovation

City		Any interaction	Local interaction	Foreign interaction	Total no. of firms
Bangalore	Number	46	44	20	49
	Percentage	93.88	89.8	40.82	100
Chennai	Number	22	22	11	41
	Percentage	53.66	53.66	26.83	100
Cochin	Number	8	8	8	10
	Percentage	80	80	80	100
Hyderabad	Number	12	8	10	27
	Percentage	44.44	29.63	37.04	100
Mumbai	Number	9	9	0	71
	Percentage	12.68	12.68	0	100
Delhi/Noida/Gurgaon	Number	59	54	37	76
	Percentage	77.63	71.05	48.68	100
Pune	Number	4	2	2	20
	Percentage	20	10	10	100
Trivandrum	Number	7	0	7	20
	Percentage	35	0	35	100
Total	Number	167	147	95	314
	Percentage	53.18	46.82	30.25	100

Source: Primary survey (2009–10).

firms interacted with local universities and RIs, only 49 per cent of the firms interacted with foreign universities. In Mumbai, where the overall interaction levels were very weak, there were no instances of interaction with foreign universities. As noted by earlier studies, the centres like Bangalore and Delhi are characterized by relatively more vibrant regional innovation system with the presence a number of pubic-funded RIs, public sector industrial units and universities, which in turn acted as an inducement factor for the clustering of ICT firms (Kumar and Joseph, 2005). Here it needs to be noted that the observed pattern with respect to Mumbai is an aberration and against the commonly observed trend.

However, in relatively smaller cities such as Trivandrum, Pune and Cochin, even though their overall interaction levels are relatively less, their foreign interaction seems to be as high as compared to interaction with local universities. This difference in interaction patterns based on city size probably indicates the lack of availability of knowledge and information base at the local-level universities and RIs compared to the ones in the larger cities and the weak regional innovation system.

Interaction with Local versus Foreign Universities

The interaction of the firms with the local universities was more with universities in the region than with universities away from the region (Table 6.7). Moreover, the interaction with universities in the local region as well as with universities at the national level was much higher for

Table 6.7:
Firm interaction with local universities

Firm type	With universities in the local region			With universities in the country other than regional		
	No interaction	*Any interaction*	Total	*No interaction*	*Any interaction*	Total
Stand alone	75.45	24.55	100	82.63	17.37	100
MNC subsidiaries	73.58	26.42	100	81.13	18.87	100
MNC headquarters	52	48	100	70	30	100
Total	71.21	28.79	100	80.19	19.81	100

Source: Primary survey (2009–10).

the MNC headquartered in India compared to other types of the firms. Interestingly, there is not much difference in the share of stand-alone or MNC subsidiaries that interact with regional or national universities.

In terms of institutional linkages across countries, study found highest collaboration of Indian firms with the North American universities and RIs. This needs to be seen in terms of the fact that North America accounted for highest share of IT export from India. Of the 314 firms that reported their location, 38 firms (12.1 per cent) had interaction with North American universities (see Table 6.8). The next largest occurrence of interaction was with South American firms. This could perhaps be seen in the context of various initiatives towards greater integration between India and Mercusul (especially Brazil) after the New Delhi declaration at the instance of the president of Brazil in 2004. West European universities and RIs came to only a distant third in interaction with Indian firms in almost all regions across the country. This in turn underscores the relevance of recent initiatives being undertaken to promote greater integration between EU and India. In fact, while Bangalore and Chennai have their university interaction largely originating from the North America, in case of Hyderabad and Delhi, the main university interactions came from South America. Finally, interaction of Indian firms with Asian economies was found only marginal, underlying the need for institutional interventions and policy measures to promote such interaction, especially in a context wherein India's foreign policy provides for south–south cooperation and looks for east strategy in particular.

The observed nature of interactions is such that most interactions with foreign universities were formal in nature. Of the 99 firms that reported any foreign university linkages, 62 of them had formal linkages. Only nine firms reported explicitly that they had informal linkages with the universities in foreign countries. The rest of the firms did not respond on their status of formality of their relationship. The degree of formal interaction with foreign universities is suggestive of rigid, inflexible and structured forms of interaction that is typical of developing country patterns of interaction with universities. The literature talks of such interactions of formal type not being fruitful to the extent of informal interactions that lead to flexible and multiple forms of benefits.

The observed pattern of university interaction also corresponds largely to the export markets of these firms. The largest number of firms interacts

Table 6.8:
Number of firms that reported any form of foreign university–industry interaction during the last three years by region and country

		State	Rest of country	North America	South America	West Europe	C&E Europe	Africa	Jap, Aus, Asia	Rest of Asia	Total no. of firms
Bangalore	No.	0	0	12	2	3	3	0	2	3	49
	%	0	0	24.49	4.08	6.12	6.12	0	4.08	6.12	100
Chennai	No.	0	0	7	6	2	1	0	0	3	41
	%	0	0	17.07	14.63	4.88	2.44	0	0	7.32	100
Cochin	No.	0	0	1	2	6	4	0	0	0	10
	%	0	0	10	20	60	40	0	0	0	100
Hyderabad	No.	0	1	2	7	0	2	0	0	0	27
	%	0	3.7	7.41	25.93	0	7.41	0	0	0	100
Mumbai	No.	0	0	0	0	0	0	0	0	0	71
	%	0	0	0	0	0	0	0	0	0	100
Delhi/Noida/ Gurgaon	No.	0	1	12	13	5	1	2	8	2	76
	%	0	1.32	15.79	17.11	6.58	1.32	2.63	10.53	2.63	100
Pune	No.	0	0	1	0	1	0	0	0	0	20
	%	0	0	5	0	5	0	0	0	0	100
Trivandrum	No.	0	0	3	1	2	4	1	0	0	20
	%	0	0	15	5	10	20	5	0	0	100
Total	No.	0	2	38	31	19	15	3	10	8	314
	%	0	0.64	12.1	9.87	6.05	4.78	0.96	3.18	2.55	100

Source: Primary survey (2009–10).

with North American universities, and they also have their largest export market in North America. The second largest market seems to be the South America, and correspondingly, university interaction in South America also seems to be relatively high compared to other regions. Similar is the case with Western Europe and other regions as well. This suggests that interactions with universities are marked with local market for their products. However, we do not know what purpose these interactions serve. Theory talks about interactions for local product adaptation, understanding the local market and also getting skilled workforce from their respective markets.

Own R&D versus University–Industry Interaction

While the traditional literature on innovation highlighted the role of R&D, studies on innovation systems (Lundvall, 1992; Nelson, 1993), conceptualizing innovation as an evolutionary process involving interaction between different agents involved in the generation and diffusion of knowledge, consider interaction between universities and firms as an important factor in innovation. In the literature, there is a line of argument that the university research is a substitute for firm R&D and that one might expect firms with higher absorptive capacity to rely less on universities (Thursby and Thursby, 2002). Here, research by universities and public laboratories are considered as a substitute for in-house R&D by firms.

However, evidence presented in Table 6.9 tends to suggest that there is a high degree of association between university–industry interaction and the firm's R&D activity. Both the activities seem to occur together in a large number of firms. Nearly 70 per cent of the firms stated that they undertake R&D activity and have some form of interaction with the universities/PRIs. While this was the average among all firms, there were some differences across organizational types. Among the stand-alone firms and MNC subsidiaries, more than 60 per cent of the firms that engaged in in-house R&D activity were also interacting with universities (Table 6.9). But for the MNC headquarters, more than 95 per cent of the firms undertaking R&D activity were found interacting with the universities. These patterns suggest that university research in general

Table 6.9:
Incidence of R&D activity and university–industry interactions among firms (per cent)

	No interaction	Any interaction	Total
Stand alone			
Firms without R&D	73.33	26.67	100
Firms with R&D	38.96	61.04	100
MNC subsidiaries			
Firms without R&D	74.07	25.93	100
Firms with R&D	34.18	65.82	100
MNC headquarters			
Firms without R&D	54.55	45.45	100
Firms with R&D	5.13	94.87	100
All firms			
Firms without R&D	72.09	27.91	100
Firms with R&D	30.26	69.74	100

Source: Survey.

acts as a complement to the industry's own research efforts, rather than being a substitute to own R&D efforts. Further, we have already seen that the prevalence of R&D is high among the MNC headquarters and MNC subsidiaries, and they interact more intensively with the university system. This tends to support the view that MNCs search for such locations that are capable of offering the complementary capabilities (Ernst, 2000; Ernst and Lundvall, 2000). But the observed trend with respect to the stand-alone firms regarding their interaction with universities is not very encouraging, especially when viewed along with the evidence from Table 6.1 to the effect that prevalence of R&D activity among the stand-alone firms is low as compared to their counterparts. The preliminary evidence, therefore, further reinforces our argument that the innovative behaviour of large segment of firms operating in India's ICT sector are not attuned to reap the potential benefits of GINs in the context of global innovation race. At the same time, we also observe that interaction with universities is not an agenda of a large proportion of firms. To be more specific, over 72 per cent of the firms are not engaged in R&D and about 30 per cent of those undertaking R&D are not having any interaction. As already noted, the interaction is much lower with foreign universities. To the extent that various studies have highlighted the role of such interaction in innovation, we now explore the underlying factors.

Behind the Low Level of University–Industry Interaction

Lower level of interaction that firms have with the universities and research centres, especially with those from abroad, reflect the firms' internationalization strategies. Most firms did not outsource the functions to other firms. Rather the delegation of functions seems to be among the subsidiaries of the same firm. The functions of the firm are still centralized, though there was some delegation with regard to technology and process development to subsidiaries in the developed countries. More than 73 per cent of the firms stated that technology and process development was internal to the firm (Table 6.10).

Only 18 per cent had subsidiaries of the firm in developed country and 11 per cent of the firms had process development being delegated to subsidiaries in developing countries. Even then most technology development was firm specific and internally generated. Thus viewed, the old model of global production networks still characterizes the GINs, wherein firms' functions are distributed mostly among their own subsidiaries and the partnerships with other agents either in the developing or in the developed countries are remaining weak. While this could be seen as an indication of the premature nature of innovation systems in developing countries, it is in tune with the findings of Pavitt and Patel (1991) that innovative activities among the largest firms in the world were among the least internationalized of their functions.

Why global knowledge collaboration has not taken root among firms in India? Majority of the firms (55 per cent) perceived that finding relevant knowledge across the globe was a serious or moderate barrier for such collaborations to develop, of which 23 per cent reported this as a serious barrier (Table 6.11). As seen earlier, stand-alone firms do not make attempts to internationalize their knowledge sources; hence, they may not find the issue as a serious barrier. Among the MNC subsidiaries in India, 24 per cent reported finding knowledge of relevance as a serious barrier.

Table 6.10:
Performance of various functions of the firm and internationalization

	Firm type	By your unit in your location	subsidiaries of firm in developed country	subsidiaries of firm in developing country	Outsourced to partner in your country	Outsourced to a partner outside your country in a developed location	Outsourced to a partner outside your country in a developing location
Product development	SA	0.0	12.6	4.2	0.0	1.2	0.6
	MNCSUB	0.0	20.8	21.7	2.8	2.8	1.9
	MNCHQ	0.0	20.0	42.0	4.0	2.0	0.0
	Total	0.0	16.4	15.8	1.6	1.9	0.9
Technology and process development	SA	82.0	10.2	3.6	0.0	0.0	0.0
	MNCSUB	69.8	22.6	17.0	0.9	0.9	0.0
	MNCHQ	54.0	36.0	24.0	2.0	2.0	0.0
	Total	73.7	18.3	11.2	0.6	0.6	0.0

Source: Primary survey 2009–10.

Table 6.11:
Factors that represent a challenge or a barrier to international innovation collaboration

	Firm type	Extreme/ serious barrier	Extreme barrier	Serious barrier	Moderate barrier	Small barrier	Not a barrier at all	Total
Finding relevant new knowledge across the globe	SA	16.8	7.7	9.0	34.8	26.5	21.9	100
	MNCSUB	23.6	5.7	17.9	26.4	33.9	16.0	100
	MNCHQ	42.0	6.0	36.0	34.0	12.0	12.0	100
	Total	23.2	6.8	16.4	31.8	26.7	18.3	100

Source: Primary survey (2009–10).

It was the MNCs headquartered in India that found operational knowledge collaboration for innovation very difficult. More than 42 per cent of the firms felt that it was an extreme or a serious barrier. The MNCs emerging from developing country locations, such as India, seem to find partners either among other firms or from universities/RIs.

With respect to the institutional arrangements to promote internationalization of innovation, nearly 70 per cent of the firms felt that public-funded centres of innovation carried a negative attitude towards internationalization efforts of the firm (Table 6.12). While this is true in case of all types of firms, a much larger share of the MNCs from India (82 per cent) felt that public support for innovation was very negative. This was true in case of public support and incentives, international exposure to universities and training of labour force for internationalization. In all of these factors of internationalization, the firms largely felt that the role of these institutions were negative and did not encourage internationalization. Thus, the state policies and institutions are yet to fully appreciate the potential benefits and initiate proactive measures to support internationalization of firm's innovative activities. Yet another factor came out from the survey related to the availability of labour force required for internationalization. This finding assumes added importance in a context wherein it is generally believed that there exists an abundant supply of skilled manpower that facilitates India's participation in GINs, whereas it appears that with low level of enrolment in higher education, India needs to travel a long distance to accomplish this perception into reality, and the present situation is one wherein there is intense competition for the available labour force between different players in GINs.

Table 6.12:
Factors influencing the internationalization of innovation activities

		Highly negative	Moderately negative	Moderately positive	Highly positive	Factor not experience	Total
Practical support from centres for the internationalization of innovation	SA	22.08	40.26	3.25	4.55	29.87	100
	MNCSUB	37.86	33.98	12.62	2.91	12.62	100
	MNCHQ	34	48	2	4	12	100
	Total	29.32	39.41	6.19	3.91	21.17	100
Public incentives and economic support available for internationalizing	SA	28.1	30.72	9.8	1.96	29.41	100
	MNCSUB	36.89	35.92	8.74	3.88	14.56	100
	MNCHQ	46	34	6	2	12	100
	Total	33.99	33.01	8.82	2.61	21.57	100
The international contacts and international exposure of universities, university and public research	SA	23.68	38.82	6.58	3.29	27.63	100
	MNCSUB	39.81	33.01	8.74	3.88	14.56	100
	MNCHQ	36	42	4	2	16	100
	Total	31.15	37.38	6.89	3.28	21.31	100
Labour force training specific to the needs of the internationalization	SA	31.58	37.5	7.89	1.97	21.05	100
	MNCSUB	45.63	29.13	8.74	2.91	13.59	100
	MNCHQ	50	28	10	2	10	100
	Total	39.34	33.11	8.52	2.3	16.72	100

Source: Primary survey (2009–10).

Interaction of MNCs with Universities and Public Research Institutes: Two Case Studies

The case studies of two MNCs, one headquartered in India and other with headquarters in the United States, are presented in this section to throw further insights on the issue subjected to discussion in the previous section.

Case 1: Integration with Universities as a Conduit for Ensuring Skill Supply

Firm A, an MNC with headquarters in India, is one of the largest ICT firms, with export intensity (export as a proportion of sales) higher than 90 per cent. The firm began with customized services during the early stages and gradually moved up towards consultancy and product development, and the traditional demand for its services and products comes from the US market. However, the market is increasingly getting diversified to Europe. In the recent years, especially after the financial crisis, there is a trend towards increased domestic market orientation. The firm focused mainly on customized services in its early stages, but now gradually moving to products as well.

This firm considers its clients and suppliers as the main sources of knowledge for innovation. The firm follows a much closed model for knowledge seeking. The firm does not seek interactions with other agencies in R&D, be it competitor companies or universities. The firm has a history of acquiring foreign firms for expanding its knowledge of foreign markets and getting access to soft skills that are required to operate in foreign markets. The firm has two R&D centres, one in India and the other in another developing country, and there is a close constant interaction between employees of the firm and their R&D centres. These interactions, which are essentially internal to the firm, are spurred usually by the specificities of the customer's demand. Thus, it may be stated that this firm's innovation is demand driven. To illustrate, on the sources of knowledge for innovation, the interviewee stated that 'We basically start looking at the problems that our customers are facing, issues with services and those reports and our in-house knowledge tries to solve the problems'.

However, this is not to say that there is no interaction with the universities or research centres. The most common form of interaction with the academic world was through student recruitment by the firm. This firm has an elaborate network of student placements every year in the firm, recruiting annually a few thousand employees from all over the country. The interviewee stated that from repeated placement programmes for many years, they recognized that there was a vast gap between the students' knowledge pool and the industry's skill requirements. This had led to the establishment of an educational intervention programme by the firm in select colleges across the country, wherein the senior members of the firm would join hands with the locally available faculty in training students to be industry ready. In some cases, the firm had been successful in bringing about changes in the university/college syllabus such that the regular taught programmes reflect the industry skill requirements. Moreover, the firm has also invested heavily in training by establishing a massive training facility with a capacity of 20,000 per annum, much more than any non-affiliating university in India.

To conclude, it appears that this firm represents the case typical of university–industry interaction in an immature innovation system. The firm does not see universities as partners in research or innovation rather its internal sources are the core sources of innovation along with the feedback from customers and suppliers. While the firm interacts with the universities, the basic objective is to influence the teaching and training in such a way that the products from the university system is industry ready to enhance its international competitiveness by reducing the cost involved in training. Such cost reduction appears especially important given the high export orientation of the firm.

Case 2: Interaction with Universities as a Means of Market Creation

Firm B is the Indian subsidiary of an MNC with headquarters in the United States and a global market leader in networking and telecommunication. This firm was founded in 1984 by a small group of computer scientists and produced the first router in 1986. Over the years, the firm emerged as the worldwide leader in networking. In 2010, the company had its presence in 165 countries with 550 offices and around 30 manufacturing sites and employed 72,600 with a sales turnover of US$44 billion.

Being a company operating in a high-tech industry with relatively shorter product cycle under highly competitive conditions, it has been providing top priority for innovation. The R&D expenditure as a proportion of sales by the firm has been maintained consistently at a two-digit level even during the period of economic crisis indicates the high priority to research and development attached by the company. Innovation by firm is perceived not as technological innovation alone but also as business innovation (capturing the current transitions at societal level and create cooperation with partners and competitors to grow). The firm, for example, is involved in the creation of smart cities (especially in emerging markets like China, India and Mexico). The role of innovation as is evident from statements in various annual reports. To illustrate:

> In our opinion, the key to long-term success in the high-technology industry is ongoing strategic investment and innovation, and we intend to continue to take good business risks. Our innovation strategy requires a unique combination of internal development, partnerships, and acquisitions. In our opinion, for companies to lead in the technology industry they must be able to do all three. (Annual Report, 2005)

The company is a typical case of open innovation within a global network. Apart from being highly R&D intensive, the firms' innovation strategy is reflected in the strong internal network between the headquarters and 550 offices, and 30 manufacturing sites spread across 165 countries. Apart from alliances, the firm also actively pursued, since late 1980s, a strategy of acquiring new companies with competence in niche areas from both developed and developing countries to strengthen its technological competence. So far the company has acquired 140 companies of which 14 of them were since 2008.

The firm first came to India in 1995 to establish a representative office to provide data communications solutions to various customers. In 1999, the firm established a Global Engineering Development Center in Bangalore with an investment of US$20 million to develop and test a particular network management software, application-specific integrated circuits (ASICs), and other technologies like ATM (asynchronous transmission mode) and VoIP (Voice over Internet Protocol). For this firm, India, besides from being a market, is a source of 'engineering and talent' to develop its networking products.

The Globalisation Centre in Bengaluru, established in December 2006, serves as a 'mirror site' to many headquarters functions, including R&D, IT, sales and customer support and finance, and is termed as the Eastern Headquarters. The 12.5-hour time difference between San Jose and India allowed the firm to be run 12 hours out of India and 12 hours in India. It provides every other function as part of a global team, participating in global work. In the words of the chief technology officer, 'The Globalization Center in Bangalore is not a center for India [the] market, it is a center for global markets'.

Being a truly innovative, global and networked firm, the strategy of building and sustaining technological competence by the firm involves networking not only within the firm and other firms, suppliers and customers but also with other knowledge-generating organizations like universities around the world. It collaborates with universities worldwide both for educational purpose (e.g. special graduated programmes) and for innovation purpose. The main collaborations are in US and in India.

A major programme is Global Talent Acceleration Programme (GTAP) established in different countries. Networking Academy aims to provide a consistently enriching learning experience by partnering with public and private institutions such as schools, universities, businesses, nonprofits and government organizations to develop and deliver innovative ICT courses, improve the effectiveness and accessibility of the programme, increase access to education and career opportunities and help ensure that students and instructors have the resources they need to accomplish their goals. These initiatives are equally helpful in creating new markets and expanding the existing ones.

In India, though there are many instances of interaction with academic institutions like the Indian Institute of Science, Bangalore, a full-fledged and formal linkage with institutions is yet to take place. The firm interacts with other institutions as well, but mostly informal in nature. The initiative for such interactions comes from both the firm and the institutes. Hiring of course continues to be the major form of interaction with the academic institutions. There is a small focus now growing in India towards research interactions as well, but mostly informal in nature. The subsidiary in

India has some highly knowledgeable persons who have joined from the academia, and they are the ones who usually established linkage with the academic circles. Also its personnel participate in research seminars, workshops, conferences in the various academic institutions. In addition, interactions also take place in the form of internship for students, which is approximately 150–200 or even more in some years.

The case study of this highly innovative and globally integrated firm indicates that it considers innovation as an interactive process wherein collaboration with users, suppliers and other knowledge-generating entities like universities (both within an outside the countries) is as important as in-house R&D activities. The firm under study explicitly states three key elements in the innovation strategy; they are internal development, partnership and acquisitions. The firm has also been establishing a global network for promoting skilled manpower, especially in the core area of its concern. In addition, the firm has been developing strategic alliances, even with competing firms and local governments, apart from their conscious effort to acquire firms with skills set in niche areas. The key question is to what extent the policy environment and institutional arrangements within developing countries like India and China are capable of harnessing the presence of such firms for making their innovation system more vibrant. The answer to the above issue, going by the evidence from India, appears to be not very encouraging. While the firm believes in interactive learning and collaboration with universities, no formal arrangements exist until now in India. At the same time, there are various initiatives that are oriented more towards market creation and expansion as compared to promoting innovation.

Concluding Observations

The most pertinent characteristic of GINs, viewed from the perspective promoting innovation both in the developing and developed countries, is the process of interaction between universities and RIs that it brings about. While the literature on GINs acknowledged the significance of the interface between universities and industries in GIN formation, our understanding of the underlying process and its implications at best remains rudimentary. In this context, the present study explored the interaction of

ICT firms within the country and abroad with universities and RIs within and abroad and highlighted the underlying factors.

The study found that the interaction with the universities is highest in case of MNCs headquartered in India as compared to the subsidiaries of MNCs and stand-alone companies, indicating that the Indian MNCs and foreign MNCs are utilizing the resources of universities and RIs much more intensively as compared to the stand-alone firms. Collaboration was found highest with North American universities and RIs. This needs to be seen in of the light of the fact that North America accounted for the highest share of IT export from India. While there are indications of growing interaction with universities in the South America and Indian MNCs, that with the European firms is found much lower calling for the immediate attention of policymakers.

Apart from firm characteristics, the sector specificities also appear to influence the interaction between universities and industry. While the firms operating in the modern knowledge-intensive industries such as pharmaceuticals, biotechnology, chemicals and IT reported a relatively higher level of interaction, their counterparts in low-technology industries like textiles reported hardly any interaction at all. There seems to be a substantial regional variation in the incidence of university–industry interaction, indicating that regional innovation system does matter.

The study also observed the low prevalence of R&D activities among the stand-alone firms as compared to their MNC counterparts. To the extent that there is high association between firms' in-house R&D and interaction with universities, the observed behaviour of stand-alone companies tends to suggest that they are poorly positioned to take advantage of the potential benefits of GINs as compared to their foreign and local counterparts who appears to be well positioned (given the high R&D orientation) in the 'global innovation race'.

The study provides some insights on why global knowledge collaboration has not taken root among firms in India. Majority of the firms perceived that finding relevant knowledge across the globe was a serious or moderate barrier for such collaborations to develop. It was also transpired that the state policies and institutions are yet to fully appreciate the potential benefits and initiate proactive measures to support internationalization of firm's innovative activities. Yet another factor that came out from the study related to the availability of labour force required for internationalization. This finding assumes added importance in a context wherein it is generally believed that India provides an abundant supply of skilled manpower.

The case study of two MNCs, one based in India and other based in abroad, tends to suggest that while interaction with universities does exist, the major focus is not on promoting innovation. The nature of university interaction by the MNC based in India tends to suggest that the underlying objective at present is to ensure that graduates from the universities are industry ready such that the cost of in-house training is reduced. In case of the MNC based in abroad, interaction with the university has the major objective of generating new markets for their products. Going by the findings of this study, the low level of R&D activities and low interaction with the universities observed in case of relatively small stand-alone firms in India's ICT sector have to be a major point of concern. Similarly, in case of both domestic and foreign MNCs, while interaction with the universities does exist, its ultimate objective is not to build up innovation capabilities.

References

Athukorala, P. and A. Kohpaiboon. 2010. 'Globalization of R&D by US-based multinational enterprises', *Research Policy*, 39: 1335–47.
Athreye, S. and M. Prevezer. 2008. 'R&D offshoring and the domestic science base in India and China', Centre for Globalisation Research School of Business and Management, Working Paper 26, University of London.
Basant, R. and P. Chandra. 2007. 'Role of educational and R&D institutions in city clusters: An exploratory study of Bangalore and Pune regions', *World Development*, 35(6): 1037–55.
Basant, R. and S. Mani. 2012. 'Foreign R&D centres in India: An analysis of their size, structure and implications', Working Paper No. W.P. No. 2012-01-06, Indian Institute of Management, Ahmadabad.
Britto, G., O. Camargo, G. Kruss and E. Albuquerque. 2013. 'Global interactions between firms and universities', *Innovation and Development*, 3(1): 71–89.
Cantwell, J.A. 1995. 'The globalization of technology: What remains of the product cycle model?', *Cambridge Journal of Economics*, 19: 155–74.
D'Costa, A.P. 2006. 'Exports, University-Industry Linkages, and Innovation Challenges in Bangalore, India', Policy Research Working Paper 3887, World Bank, Washington, DC.
Ernst, D. 2000. 'Global production networks and the changing geography of innovation systems: Implications for developing countries', *Economics of Innovation and New Technology*, 11(6): 497–523.

Ernst, D. 2009. 'A new geography of knowledge in the electronics industry: Asia's role in global innovation networks', Policy Studies 54, East-West Centre, Hawaii.
Ernst, D. and B.A. Lundvall. 2000. 'Information technology in the learning economy: challenges for developing countries', East-West Centre Working Paper 8, East-West Center, Honolulu.
Freeman, R.B. 2005. 'Does globalisation of the scientific/engineering workforce threaten U.S. economic leadership', NBER Working Paper 11457, National Bureau of Economic Research, Cambridge, MA.
Joseph K.J. and V. Abraham. 2009. 'University–industry interactions and innovation in India: Patterns, determinants, and effects in select industries', *Seoul Journal of Economics*, 22(4): Winter.
Kaiser, U. and C. Grimpe. 2010. 'Balancing internal and external knowledge acquisition: The gains and pains from R&D outsourcing', *Journal of Management Studies*, ZEW Centre for European Economic Research, University of Zurich and Institute for Strategy and Business Economics.
Kumar, N. and K.J. Joseph. 2005. 'Export of software and business process outsourcing from developing countries: Lessons from India', *Asia Pacific Trade and Investment Review*, 1(1): 91–108.
Patel, P. and K. Pavitt. 1991. 'Large firms in the production of the world's technology: An important case of "non-globalisation"', *Journal of International Business Studies*, 22: 1–21.
Rugman, A.M. 1981. 'Research and development by multinational and domestic firms in Canada', *Canadian Public Policy*, 7(4): 604–16.
Stembridge, B. 2007. Eastward Ho! The Geographic Drift of Global R&D Knowledge Link newsletter from Thomson Scientific.
Tallman S. and M.M. Susan. 2010. 'Make, buy or ally? Theoretical perspectives on knowledge process outsourcing through alliances', *Journal of Management Studies*, Temple University: University of Richmound.
Thursby, J. and M. Thursby. 2002. 'Who is selling the ivory tower? Sources of growth in university licensing', *Management Science*, 498(1): 90–104.
TIFAC. 2006. *FDI in the R&D Sector: Study for the Pattern in 1998–2003*. New Delhi, TIFAC: TIFAC and Academy of Business School.
UNCTAD (United Nation Conference of Trade and Development). 2005. 'World Investment Report: Transnational Corporations and the Internationalization of R&D', Geneva.
Vernon, R. 1974. 'The location of economic activity', In Dunning, J. (ed), *Economic Analysis and the Multinational Enterprise*. London: George Allen & Unwin, pp. 89–114.

7

Globalization of Industrial R&D in Developing Countries: A Sociological Perspective

Binay Kumar Pattnaik

Introduction

Globalization is a comprehensive process of socio-economic change that has emerged from Reaganomics and Thatcherism of the early 1980s followed through (Uruguay Round) Trade-Related Aspects of Intellectual Property Rights (TRIPS) and Trade-Related Investment Measures (TRIMs) of the 1990s via General Agreement on Tariffs and Trade (GATT) to present-day World Trade Organization (WTO). Although essentially it is an economic phenomenon, it is backed by a long process of political, legal and administrative reforms that otherwise have come to be known as liberalization. If the essence of globalization as an economic phenomenon is the optimization of profits through international trade by making the global market as one, the key to this is located in the drive for efficiency and competitiveness. It refers to the ultimate form of market economics (so far). However, this economic process backed by a political, legal and administrative reform process has other repercussions that are social and cultural by nature. The other allied institutions like education and science and technology (S&T), which are linked to the institutions of economy, are not left untouched, have rather consequentially undergone extensive changes. Hence, globalization, the process, through two decades, now is a multifaceted reality. The multifacetedness is not confined to various institutional aspects alone, as it has cultural aspects too. It is of course beyond the scope of the present exercise to deal with globalization, the phenomenon in its entirety. To be more specific to the context, I shall be articulating globalization of the institution of S&T from the viewpoint of developing countries only. It is worth being specific here as globalization is not the same for developed and for developing countries. Mostly, these two refer to the two ends of the same process of globalization, as the two worlds are related by a continuum. Although elsewhere (Pattnaik, 2005) I have spelt out what globalization of S&T regimes means in more general terms, it would be highly pertinent to spell out very concisely what it means for developed countries. It becomes more essential to do so as subsequently I have to discern it from globalization of S&T regimes in developing countries. Certain primary features of globalization of the S&T regimes in developed countries could be identified as follows:

1. Finding out upcoming and growing large markets (e.g. in China and India) for their industrial technologies, capital goods and high-quality products.
2. Making more profit through international trade and imposition of their intellectual property right (IPR) regime particularly in that section of the developing countries that was not under the control of international IPR regimes and the countries that could not have developed IPR regimes for themselves.
3. Tapping more resources likes cheap technical manpower for their technical/research and development (R&D) activities and tapping newer traditional knowledge bases (indigenous) for the development of new designs/processes (maybe through mergers, acquisitions, collaborations and direct entry through foreign direct investment [FDI]).
4. Making use of the cheap technical manpower in developing countries for services in developed countries, that is, outsourcing services through business process outsourcing (BPO)/call centres (for e-commerce, banking data management and medical transcription), etc.
5. Reaching out to the previously inaccessible investment markets (e.g. those in former socialistic countries and in semi-socialistic countries like India where although the industrial sector is partly open, sectors like finance/insurance, higher education and print and electronic media are still not fully open) through the strategy of international joint ventures/collaborations at the early stage and through the strategy of deep penetration (through subsidiaries or direct merchandise of industrial semi-finished products, intermediary goods, etc.) at a later stage.

6. Making of international scientific and technological consortiums for addressing common research questions/interests (in both the government and corporate sectors).
7. Expansion of the university–industry linkages and emergence of entrepreneurial universities (the ultimate goal of course is attaining an information-based society and knowledge-based economy in home countries).

Effects of globalization in developing countries cannot be perceived as an independent phenomenon as globalization entails a qualitatively changed relationship between the centre and the periphery countries although it is inclusive of (globalization-induced) interactional benefits among the peripheral countries. Of course, globalization of the centre refers predominantly to the interactional benefits among the countries at the centre and significantly the interactional benefits between the countries at the centre and those at the periphery. Hence, the effects of globalization at the periphery/developing countries are greatly reflections of the globalization of the centre and (globalization-induced) interactional benefits among peripheral countries. Needless to say, that the centre is always a self-starter and auto-centric because of the inherent dynamism in industrial capitalism. In addition, the periphery being dependent on the centre mostly experiences growth as a reflection of the dynamism at the centre. This argument unfolds itself in my articulation of globalization effects in the developing countries, as the bulk of the features identified in the course of discussion in this chapter are in fact consequences of initiatives triggered by the countries at the centre. In their ongoing patron and client relationship, which of course is a matter of reciprocity, countries at both the centre and periphery have gained from each other. However, it is noteworthy that some countries at the periphery have immensely benefited under this globalized regime and have successfully transformed themselves into what Wallerstein calls the 'semi-periphery' status by displaying some amount of dynamism of their own. The mark of distinction of these greatly globalized semi-peripheral countries is that there too the business driven by scientific and technological innovations is local as well as global. Further, most notable is that these new global drivers of innovations are non-European, non-American and non-white, notes Friedman (2006: 11).

The limited space here does not allow us to make a comprehensive presentation of all the major features of the 'globalization' of S&T in developing countries which could be articulated under the following four headings: (a) internationalization of higher education, (b) emergence of Entrepreneurial universities, (c) internationalization and globalization of R&D and the (d) globalizing social impacts of the information and communications technology (ICT) revolution. Hence, we shall confine our attempt here to the analysis of internationalization and globalization of industrial R&D in developing countries. While doing so, the theoretical footing of our analysis remains to be anchored on the new-dependence school in general and on Emanuel Wallerstein's 'world system perspective' in particular. Because the analysis entails a qualitatively changed relationship between the centre and some selected peripheral countries, which implies that those developing countries have successfully transferred themselves from peripheral to semi-peripheral stage, the analysis is based on secondary data and already observed case studies.

Multinationalization and Globalization of Industrial R&D

The three concepts of internationalization, multinationalization and globalization are elusive concepts particularly in this context of R&D, although the three are being used often interchangeably, as if they are describing the same phenomenon. The globalization of technology and economy is not a new phenomenon, if it refers to the internationalization of flows of technical knowledge, new materials, semi-finished and finished products as well as services within the main industrial sectors. Similarly, multinationalization of national enterprises through the gradual extension of multinational corporation (MNC) activities, through the setting up of direct subsidiaries, through acquisitions or through various types of cooperative agreements (e.g. commercial, financial, pure technological or any combination of the three) has characterized industrial development for nearly a century now. However, globalization is a nascent economic phenomenon that refers to: (a) creation, distribution and consumption of goods and services, using structures organized on a global basis for economic exploitation of material and non-material means of production, for example, by means of patents databases and highly advanced training of human talents; b) the development of world markets or markets that eventually are becoming so to be regulated by universal standards; and (c) organizations acting on a global basis, whose organizational culture is in conformity with a global strategy. Globalization does not necessarily signify standardized products for homogeneous world markets (e.g. even for consumer goods). On the contrary, it signifies adaptation to a variety of local markets. Above all, it is the circulation of products, production methods, organizational structures, decision-making and supervisory processes and entrepreneurial strategies that are global (Petrella, 1992). Being part of the same process of transformation of industrial financial capitalism, internationalization, multinationalization and globalization are obviously interconnected. This overlapped process is further demarcated by the disappearance of a system/systemic features referred to as 'national' (e.g. national bank, national education, national economy, national culture, etc.). Of course, this process of disappearance has already been ushered in by the process of internationalization and multinationalization through the relentless interactions among states, economies and cultures. Probably globalization signifies the end of 'national' as a point of strategic relevance, but 'national' continues to be one of the levels of significant relevance in the process of technological innovation. Hence, it would be a mistake to analyze the globalization of R&D activities in isolation from the processes of national R&D strategies and economic development, internationalization and multinationalization of R&D. Internationalization of R&D occurs when two or more industrial firms, research bodies or universities from different countries carry out joint R&D programmes. It can be bilateral or multilateral. Similarly, R&D multinationalization occurs when an industrial firm establishes one or more R&D activities centres in one or more countries other than its country of origin. Not so differently, globalization of R&D occurs when an industrial firm has developed a global strategy and vision of R&D activities at both the internal level (i.e. in-house R&D activities, through internationalization and multinationalization) and external level (i.e. alliances of R&D with other firms, mergers and acquisition, agreements with universities, national research bodies, or governments from different countries, participation in international scientific and technical cooperative programmes or mere outsourcing of support services like database management, etc., from firms of different countries particularly from developing countries with cheap labour) in all the areas of R&D. Thus, globalization has emerged as a more inclusive concept that thematically subsumes the process of internationalization and multinationalization. Petrella (1992) has mapped out the indicators of globalization process in R&D organizations, like industrial firms, universities and public research organizations. However, the context of his articulation was that of the developed countries, whereas we are here using the same indicators (may even be adapted) in the context of developing countries, of course, keeping in mind that the same phenomena could be looked at from the vantage of developing countries as well.

With respect to industrial firms in the developing countries, the indicators of globalization could be the following:

(i) Increasing investment for R&D activities and facilities located abroad (particularly through subsidiaries/affiliates in developing countries and in former socialistic countries that have liberalized their policies),
(ii) Increasing intensity of technological cooperation between the different foreign affiliates of an MNC,
(iii) Mergers and acquisition of firms/foreign affiliates with their own R&D centres,
(iv) Joint R&D activities with foreign affiliates,
(v) Ad hoc, short-term contracts with foreign universities and/or public research organizations (in both developed and developing countries).

With respect to universities in developing countries, the indicators of globalization could be the following:

(i) Being part of multivariant, multinational research networks,
(ii) Evolving joint research projects and programmes with foreign universities,
(iii) Research contracts with industrial firms both within and outside the country,
(iv) Being part of international/global research activities/programmes/ projects.

With respect to public research organizations in developing countries, the indicators of globalization could be the following:

(i) Liberalizing the normative framework/procedures for facilitating participation of foreign firms/R&D organization in the technical projects/programmes within the country and on the contrary facilitating participation of domestic S&T personnel to be part of similar projects/programmes abroad,
(ii) Opening up overseas branches/subsidiaries with technical and financial participation from overseas partners,
(iii) Availability of facilities for such research organizations to allow foreign investment as well as technical collaborations (particularly in hi-tech sectors),
(iv) Becoming part of an increasing number of intergovernmental scientific and technological agreements.

The major features of globalization of industrial R&D in developing countries could be spelt out as the following: (a) Emergence of an international division of labour in S&T research, (b) emerging flatter technological world regime, (c) growing multinationalization of domestic firms from developing countries, (d) globalization of local R&D from developing countries and (e) growing offshoring industrial R&D services by firms from developed countries.

Emergence of an International Division of Labour in R&D

(a) First Phase of Globalization: The most distinguishing feature of globalization in its initial phase has been the 'scissors phenomenon'. It refers to the tendency among the most developed countries to invest more among themselves, reducing their investment in the developing countries like those from south-east Asia, Africa and Latin America. The tendency also indicates that inter-firm technological strategic alliances/agreements are very few involving firms from less developed and poorer countries as partners. Even long-term research projects such as the fifth-generation computer, the Human Genome and the intelligent manufacturing systems projects, which should involve world cooperation, have not envisaged

involving a research organization or university outside the developed world (Petrella, 1992). Thus, the 'scissors phenomenon' on the one hand entails increasing integration among the most developed countries in matters of technological/R&D alliances and on the other hand it also entails the exclusion of the poor and less developed countries that are of course increasingly segmentalized among themselves. Global 'delinking' may not be an outcome of conscious policy against the less developed countries but a phenomenon inherent in the concentration of capital and technology in the triad countries of western Europe, the United States and Japan. Warrant's (1991) finding has been similar, involving an analysis of geographical distribution of R&D expenditures and personnel that countries of origin remain the preponderant locations of R&D units in Europe, United States and Japan (triad countries). This finding has been corroborated by a survey by Pearce and Singh (1990) concluding that (a) it is very unlikely that an industrial group will delocalize its centre of gravity in R&D abroad and (b) there was a feedback movement to re-concentrate R&D activities at the group centres. Similarly, Patel (1995) also had observed that the increasing globalization (in the past years) of innovation actually reinforces the dominance of the technology districts in the triad countries. Chen (2007) also noted that the geographical limits of the globalization process (or triadization of innovation) has reinforced the concentration of globalized R&D within the existing technology clusters in the triad countries and MNCs would be hesitant to extend their R&D bases to developing countries due to the latter's weak knowledge-based assets, poor infrastructure and limited market. Having treated the process of re-concentration of investment in R&D as 'triadization', Petrella (1992) states that the majority of the developing countries of Africa, Latin America and Asia have no significant roles to play in the global industrial R&D. Their real issue is how to facilitate a successful transfer of knowledge and technology from the triadic world to their own.

(b) Second Phase of Globalization: The latter stage of globalization, unlike the first, witnessed a slow and increasing participation of developing countries in the development of global technologies and R&D, maybe through the process of international subcontracting/outsourcing. Through this participation of developing countries in the global industrial R&D, a new phenomenon has come into existence to be justifiably called the

'new international division of labour', because it is only the low-end, supportive and labour-intensive technical services that are outsourced.

Big MNCs increasingly extend or diversify their field of technological competence through internationally integrated technological networks. These networks in each geographical location try to tap specialized local expertise. Of course, they differentiate their technological capabilities from the geographically separate ones. However, they intend to exploit and add it to their own distinct streams of innovative potentials. Hence, Cantwell and Janne (1999) and Cantwell and Pocatello (2000) observed that European MNCs do not only exploit local knowledge assets from developing countries but also contribute to local technological capabilities. These MNCs by co-developing local technological capabilities together with the local centres of excellence in developing countries set in motion a cumulative causation process that results in the reinforcement of the hierarchy of global technology districts.

There is a growing interest of MNC R&D research in China. It is now well agreed upon that China is becoming the new magnet for global R&D activities with the number of R&D centres increasing from merely 30 in 1999 to 700 in 2005 (Chen, 2007). The case of India would be similar although at a lesser scale. Ever since India signed the GATT in 1993, more than 60 global firms have set up their R&D centres in technology-intensive sectors for using Indian skilled manpower. Prior to 1991, there were only two such centres. MNCs have internationalized their R&D activities and of late have set up their R&D centres in a developing country like India. These centres are either global or regional centres of MNC R&D or are technological alliances possibly with joint venture-R&D projects in India. MNC R&D centres are all in high-tech areas of electronics, biotechnology, pharmaceuticals and chemical technologies. Examples of the first type (FDI as global or regional MNC R&D centres) include the following: the ASTRA-AB Research Centre India (Bangalore), Texas Instruments India (Bangalore), Asia Pacific Design Centre of SGS-Thompson, Motorola VLSI Design Centre (Hyderabad), General Electric (GE) Research Centre India (Bangalore) and Daimler Benz Research Centre India (Bangalore). Examples of the second type (technological alliances/JVC-R&D centres) are the following: AVL Austria with Mahindra and Mahindra; Affymax with Glaxo India; Wockhardt with Sidemark Labs; Novo Nordisk with Dr. Reddy's Laboratories; Cheminor Drugs Limited with Power Resources

International (PRI) Inc.; Airbus with Hindustan Aeronautics Limited; IDEA with TELCO; Mallinckrodt with Natco Pharma; Eli Lilly and Company US and Pfizer with Ranbaxy Laboratories Limited, India; Du Pont US, Abbot Labs US, Park Davis US, SmithKline and Beecham US all with the Indian Institute of Chemical Technology (IICT) Hyderabad; Du Pont US, Akzo Netherlands, GE US all with the National Chemical Laboratory (NCL) Pune; Cadence, International Business Machines (IBM) with the Indian Institute of Technology (IIT) Kanpur and Delhi, respectively; Nokia, Nortel, LG Electronics and Rational software with the Indian Institute of Science (IISc) Bangalore; and Mobil with the Indian Institute of Petroleum. Of course, the basic reason behind this upsurge has been the availability of 'abundant cheap engineers' and the 'large domestic market' (Pattnaik, 2005).

Some of these strategic alliances of R&D between firms are made for purposes such as overseas licensing, global marketing and joint product development. Similarly, the alliances of Indian research institutes have helped them shift their attention to business-led R&D, focusing more on process innovations and product developments in addition to their earlier consultancy works pertaining to troubleshooting, adaptation, testing and problem solving. That apart, the MNC R&D centres are: (a) creating rich employment opportunities for technically qualified manpower in India, (b) revolutionizing the patenting culture in India and posing to integrate themselves horizontally with the Indian industry in the years to come, (c) enabling global firms to place more confidence in India's R&D capabilities and (d) enhancing skills of those managing R&D and innovation (Pattnaik, 2005).

Cantwell and Janne (1999) and Cantwell and Pocatello (2000) have pointed out that even if some developing countries, such as India, Singapore and China, did get integrated into MNC global innovations networks, the innovative activities are usually limited to local market product adaptation and adaptation to lower production cost in host developing countries. Hence, R&D in such developing countries will remain at the bottom of the global innovation locational hierarchy, as it has been exemplified by the product life-cycle theory. This R&D although is internationalized/multinationalized remains limited to low-end jobs (like testing and making components) and are essentially supportive and adaptational in nature (in relation to the parent MNCs) particularly in hi-tech areas of ICT, drugs/pharmaceuticals and biotechnology.

Therefore, as pointed out by Dunning and Narula (1995), the phenomenon of MNC R&D outsourcing from developing countries remained limited primarily to 'asset-exploiting' R&D or 'home-base-exploiting R&D activities'. This is based on the observation from developing countries like China, India and Singapore that the nature of MNC industrial R&D is limited to the home-based exploitation/laboratory type. In addition, 'asset-seeking' or 'home-base-augmenting' R&D activities are assumed to be confined to the triad countries only. This new international division of labour in R&D has been conclusively demonstrated by Kuemmerle (1999).

(c) Third Phase of Globalization: The trends of R&D, however, in countries like China, India and Singapore appear to have again changed in early 2000–01. There was emerging evidence that some industrial R&D in India and China had become 'asset-seeking' or 'strategic/innovative' R&D essentially because of the high uncertainties inherent in the fast-growing large market. The 'triadization of innovation' researchers, like Dunning and Narula (1995), Florida (1997), and Le Bas and Sierra (2002), have found that 'asset seeking' had become a new tendency in the globalization of R&D. Chen (2007) points out that of late there had been a reintegration of 'asset-exploiting' R&D activities (i.e. adaptational, technology transfer, etc., typical to developing countries) and 'asset-seeking' R&D activities (e.g. innovative, technology creating, etc., rare to developing countries) in China. He tries to prove this point with the help of his two empirical observations from China, that is, Motorola and Microsoft. Both Motorola and Microsoft have global innovation networks and are working not only in China but also in other developing countries like Singapore and India. In spite of this, the fact remains that the core R&D of Motorola and Microsoft continues in the United States. Nevertheless, Chen (2007) found ample evidence of 'asset-seeking' behaviour in the R&D activities of Motorola and Microsoft in China (i.e. in technology-driven R&D centres and market-driven R&D centres) and their subsequent reintegration with 'asset-exploiting' R&D centres (i.e. production-driven R&D centres and cost-driven R&D centres). Further, Chen (2006) explains this re-integration of 'asset-seeking' and 'asset-exploiting' R&D activities of the two MNCs in China through their two different upgrading models, that is, (a) bottom-up evolution process (from the vantage of agglomeration of production networks and demanding local customers) and (b) top-down

evolution process (from the vantage of the synergy of Chinese returns and local entrepreneurial culture.). To the author, the current R&D activities of Microsoft and Motorola in India and Singapore would not be much different from that of the counterparts in China.

At the end, the outstanding issue remains, whether China will reap some direct benefits from the plethora of MNCs R&D firms, making China an emerging technological giant or more than just the 'factory to the world'. In his review article, Simon (2007) concluded that China's role in terms of global innovation is slowly growing even if not in a homogeneous manner. There is clearly a series of emerging pockets of excellence appearing in Chinese S&T. The case of India is similar where pockets of excellence with world-class R&D is growing, for example, in software, telecommunication, drugs and pharmaceuticals, etc.

Emergence of Flatter Technological World Regime

Thus, in the latest phase of globalization, the so-called hierarchical pattern of R&D or international division of labour is mostly giving way to a 'flatter' pattern showing that MNCs are now willing to make some substantial bets on China. Because MNCs making fast inroads into China and India to build/extend their global networks for R&D or to upgrade their existing R&D activities are not short-term decisions, knee-jerk responses to new, quick-win opportunities are rather a new set of imperatives driving the intensely competitive market of globalization. This newly emerging world, technologically flat type, is a fast-changing world by virtue of the net addition of five major continental-sized economies like China, India, Russia, Mexico and Brazil to the economic and technological mainstream of global industrial capitalism. Thus, there seems to be slowly emerging a technologically multi-centric world order in place of a technologically uni-centric world order.

Lastly, Simon (2007) offers two perspectives to this unfolding global change in the developing countries: internal and external perspectives. (i) The internal perspective deals with the questions pertaining to historical and cultural impediments to the building of technologically innovative China. Most crucial in this context is the 'sea of acidic attitudes' toward the 'new and the untried', and 'institutional weaknesses' (e.g. slow moving) deeply enrooted in Maoist socialism. So is the case of Russian research scientists and engineers who are desperately trying to overcome the then socialistic Soviet mindset/attitudes, for example, (a) *collectivist messianism* of the then soviet scientific community is now giving way to *individualist messianism* of Western science, (b) the *romantic scientisim* of Soviet science (matched by the cult of service to society of the Soviet scientist) is giving way to *pragmatic scientism* of Western science, (c) the centripetal tendencies of research across the Soviet Union is being replaced by the centrifugal/diffusive tendencies to bring the levelling effect (non-Moscow centric) and tendencies among Russian scientists to migrate even beyond national boundaries, (d) looking at the history of Soviet science, that is, to view prominent Soviet/Russian scientists of the past through the prism of adoration, is being replaced by a critical perspective and (e) perception of Western science and English with suspicion is slowly giving way to perception with open-mindedness, etc. (Yurevich, 2010). (ii) The external perspective reflects the view that the 'opening up policy' practiced by the Chinese government holds the key to the innovation and growth in China. The liberalization of China's economy has substantially expanded the R&D domain and its scale. The strategic role of Chinese diasporas in the United States in guiding MNC FDI and R&D to China is part of the perspective. Further, the strategic recruitment of Western trained Chinese scientists and engineers by MNCs in China as a substitute to domestic trained personnel (plagued by cultural/attitudinal differences) has miraculously converted a traditionally known disadvantage, that is, brain drain, into an advantage (called brain circulation).

In China, a new class of transnational capitalists emerged, which was the vanguard of the ICT revolution in China. These are harbingers of China's Silicon Valley at Zhongguancun district near Beijing, which is in fact overpopulated with thousands of high-tech firms including national and transnational corporations like NEC, Sun, Siemens and Microsoft. The academic mentors of many of these ICT firms are Chinese universities, the Tsinghua University in China leading among them, playing the role of Stanford University in the United States, which was the birthplace of Silicon Valley as some of the first firms were born in Stanford's industrial park in the 1950s. However, the Chinese Silicon Valley Zhongguancun began with an advantage of 'sea turtles' (like the olive green turtles of the high sea returning after years to lay eggs on the beach on which they were born). It is the local nickname for native-born Chinese techies who acquired training and work experience in the United States and returned to start business firms in China.

Chinese techies with years of experience in California in particular have turned entrepreneurs and started business firms in Taiwan, Shenzhen and Zhongguancun that were either manufacturing or trading. These too had strong linkages with Silicon Valley in the United States. Globalization has made it possible. Innovation-driven business in China was being facilitated by informal relations such as those of alumni and professional links of Chinese techies on both sides of the Pacific. Wong (2006) in his study among Silicon Valley Chinese techies noted that many of them maintained multiple passports and thus were 'flexible citizens' with divided loyalty to Silicon Valley as well as their homeland (2006:106). Wong furnished evidence of Chinese techies migrating back to their native country but retaining their Silicon Valley homes (by renting out to other Chinese) and propose to return to Silicon Valley either in a few years or after retirement. In addition, there are many who did not return. Wong reported that in 1993 an estimated 30 per cent Taiwanese migrant engineers who worked in Silicon Valley had returned to Taiwan and in the mid-1990s at least 25 per cent of techies from Hong Kong returned to their land of origin (2006:105) in search of better opportunities. Moreover, the bulk of these professionals turned entrepreneurs made use of the venture capital (VC) funds (e.g. Schoendorf) from Silicon Valley as their Indian counterparts did. This class too maintained its strong transnational linkages with Australia and North Korea and particularly with Silicon Valley, the United States and among themselves (i.e. those in Taiwan, Hong Kong and Mainland China). Wong has also pointed out the existence of strong social capital among these professionals turned capitalists who maintain consciously strong professional networks that they make use of for business purposes. Again, mutual trust and cooperation were key to social capital. Further in China a new class of 'academic entrepreneurs' in the entrepreneurial universities (Chinese UREs) emerged who are more entrepreneurs and less professors as they run profit-making enterprises under universities by virtue of their technical knowledge/innovations. These are of course research universities that emphasize commercialization of their technical knowledge and innovations. Hence, the academic entrepreneurs and the 'sea-turtle'-type entrepreneurs, together, make a sizeable class of new breed capitalists that have revolutionized China's innovation-led business. This class too is an offshoot of China's economic liberalization and subsequently has become part of its globalization.

Likewise in the post-liberalized India, a sizeable section of the Indian technical diasporas had returned home to man the technical positions in the booming ICT industry thriving in the cyber cities of Bangalore and Hyderabad. Besides this, a new class of entrepreneurs has emerged out of the IT revolution in India as it happened in China. As noted by Upadhya (2004) in her study of the Indian software industry, most of the founders of software firms in India were of middle-class origin who had built on their cultural capital of higher education (usually through engineering education) and acquired social capital (knowledge and business networks) through their professional careers. Their social capital enabled them to make use of the trust and cooperation of their professional networks, for making innovative business advancements, because their social capital could reduce the transaction costs, bureaucratic procedures and even corruption. Thus, the combined virtue of cultural and social capital engendered a class of technological capitalists in the Indian IT sector. This class of capitalists is driven not only by high need achievement orientation but also by high level of technological innovations. They could ride innovation-driven business that has flattened the technological world, at least in the IT sector, to a great extent. This class of IT capitalists is also distinguished by its global linkages. Not only that many of them had studied/worked in the United States/abroad but also their business was (a) greatly dependent upon foreign contracts (particularly the United States) and (b) most of them acquired foreign funding either directly or through foreign VCs (which were mostly from the United States, particularly Silicon Valley). This class of entrepreneurs is distinct from the traditional Indian capitalist class (sethjis leading family business) and corporate houses like the Ambanis, Birlas, Godrej, Bajaj, Kirloskar, JKs and Tatas. These are also first-generation entrepreneurs. As the offshoot of liberalization of the Indian policies and globalization of the economy, this bourgeoisie is more adaptable to MNCs. Instead of being wary of the presence of the multinationals in the IT sector, this class could find a synergy with MNCs and have learnt to benefit from their presence. Therefore, this class, first being the offshoot and later being part of the transnational IT business, is the strongest votary of globalization. By virtue of creating enormous employment and profit, this industry has influenced Indian economic and industrial policies greatly during the last two decades. Upadhya (2004) calls them the 'icons of the resurgent India'.

By the mid-1990s, the large Indian IT companies born in the 1980s had 'moved up in the value chain' to offer value-added services even at times turnkey projects. And the 1990s also witnessed the mushrooming growth of IT small- and medium-scale enterprises (SMEs) in Indian metros like Bangalore, Hyderabad, Mumbai and New Delhi as many ambitious techies left large companies to start their own SMEs. This is something akin to what happened then in Silicon Valley as a 'start-up trend'. The easy availability of 'VC' funds after 1995 contributed handsomely to the growth of SMEs in the IT sector. The Indian IT industry then had the following three strong links to the global economy: (a) dependence on foreign contracts (order), (b) dependence on MNC FDI as wholly owned subsidiaries to develop software development centres and (c) dependence on foreign VC funds.

It is noteworthy that in the year 1995 the unfolding liberalization policy allowed the entry of foreign VC funds and institutional capital to India. Further, it is noteworthy that a bulk (80 per cent by 1999) of these funds came from non-resident Indians (NRIs) especially the US-based Indian techies who had done booming business in the United States. Glaring examples of this were the Draper International (headed by Kiran Nadkarni) 1995, the Walden-NIKKO India VC company (in the late 1990s), etc. The US-based VC firms made possible the 'US–India corridor' by which many US-based IT/Software firms had acquired development centres in India. In the dot-com boom period of 1999–2000, there were as many as 100 VC firms (including Indians) that were operating in India but it soon dropped. In 2002, there were 60 foreign VC firms operating in India. In the absence of official figures, unofficial figures show that by the year 2000 the investment from foreign VC was over $1 billion (Upadhya, 2004). Similarly, NRI investment in the IT industry is not exactly available but it is claimed that half of the IT companies in Bangalore since 1999 had some NRI investment component. Saxenian (2002: 43) reported of a relevant survey, which had found that about 50 per cent of Silicon Valley's NRI IT firms had business relations with Indian firms as partial/wholly owned subsidiaries (37 per cent), sub-contractors/supplier of materials/parts (28 per cent) or some sort of joint ventures (16 per cent). Thus, three types of

VC funds poured into the Indian IT industry: (a) Indian (large institutional investors like ICICI, UTI and a few private ones), (b) exclusively foreign VC funding and (c) the ones known as cross-border funding through which the Indian development centres of US-based firms have been funded.

Further, the 'TiE' (The Indus Entrepreneurs), a consortium-type organization of NRI/US-based Indian techies, was an example of funding and fostering entrepreneurship among Indians. TiE was founded in 1992 but soon became successful with 8,000 members in 34 chapters spread across the United States, Canada, Singapore, Dubai, Malaysia, India, etc. TiE became a point of emulation to replicate the Silicon Valley magic in Europe and Asia. In the same decade, pushed by shortage of skilled workers, the United States tripled the number of visas granted to skilled personnel. This new type visa came to be known as H-1B. Nearly half of the H-1B visas granted by the United States was to Indians and most of these technical personnel were in the IT field and a majority of these went to California to participate in the hi-tech gold rush. India's golden diasporas including big achievers like K B Chandrasekhar (founder of Exodus Communication), Vinod Dhan (father of the Practicum chip), Vinod Khosla (co-founder of Sun Microsystems), Kanwal Rakhi (Excelon Corporation), Gururaj Deshpande (co-founder of Sycamore Networks), Rakesh Mathur (Junglee), etc., looked homeward. These Indian diasporas in the United States, like their Chinese counter parts, linked Silicon Valley, the technology hub, with the homeland market (for cheap labour and business opportunities) and created transnational webs. Thus, not only was there an exodus in the 1990s of NRI technical professionals from the United States to India to start IT firms in India but they also had funded the IT industry in India directly as mentors/promoters or through the VC mode.

India had reached the peak of the IT revolution by the late 1990s with the Y2K bug solution and dot-com boom. With the Y2K bug handling Indian techies leaving their footprints the world over, the Indian IT industry began a relationship with the United States. Outsourcing from the United States to India as a new form of collaboration almost exploded. By early 2000, the Y2K work was closed but a new business known as e-commerce emerged on the scene. The fibre-optic lines turned out to be the magic strings. The laying of the fibre-optic lines across the ocean, which connected the mainframes in the United States to Indian IT companies like Wipro, Infosys and Tata Consultancy Services (TCS), was involved in management of the American e-commerce. With the dot-com boom, India had benefited. In addition, the boom laid the cables that connected India with the world. The fibre-optic bubble inflated the business links between the United States/UK and India. More than the boom, the dot-com bust brought enormous benefits to India as it reduced the costs of communication drastically and made transfer of voluminous data from US firms to India possible. This allowed many American firms to outsource Indian knowledge (Friedman, 2006, 133). The dot-com bust further encouraged many Indian engineers to return home, as the techies could find American assignments seated at Bangalore/Bombay/Hyderabad. The big Indian IT firms also came with innovating and value-added proposals to sell their own new products in software and consultancies. They had graduated from maintenance to product IT firms. Their interaction with US firms became increasingly deeper and slowly through business process outsourcing, they started to run the back room of US firms. And thus emerged a mostly technologically flatter world, caused by the ICT revolution and its benefits reaching out to the developing countries.

The post-Y2K recessionary trends in the United States pushed US companies toward outsourcing particularly the offshore type. This outsourcing trend was of course confined to the peripheral tasks of business, what is otherwise known as back room office jobs. This too created opportunities for the Indian IT industry. The Indian IT sector then created within itself a new business sector known as ITES. The ITES industry, a technology-intensive sector (including the medical transcribes, call centres and the BPOs), in India is now a booming sector. Thus, it seems that recession in the United States turned out to be a saving grace for the Indian IT industry.

Growing Multinationalization of R&D in Firms from Developing Countries

The other striking feature of globalization of R&D in developing countries is the internationalization of their own firms, perhaps in limited numbers. And some of the industrial firms from developing countries, like India and China, are even becoming MNCs. So far, we have discussed the MNC FDI flow-based R&D growth in India and China. And now we are to deal with the reverse process of multinationalization of

Chinese/Indian firms. It may be a very recent phenomenon, but it is worth examining as to what drives the Chinese firms to set up their R&D overseas. The hypothesis that becomes relevant in this context entails that Chinese firms internationalize their R&D in order to develop alternate channels of technology sourcing from developed countries (Kuemmerle, 1997). For example, Haier, a Chinese firm, not only has industrial parks in the United States, Jordan and Pakistan but also has 10 listening posts and design centres in various developing countries and many developed countries. In addition, the Chinese automobile manufacturer Dongfeng Motor Corporation has set up four listening posts in the United States, UK, France and Germany for the purpose of being close to major competitors (not markets) and their technological bases. It is observed that where Chinese firms operate in large manufacturing sites in developed countries, as local R&D there, it was seen to have emerged in support of product localization and process innovation (e.g. Haier's R&D centre in south Croatia, the United States and the like). Thus, input-related rationale is the strongest reason for internationalization of R&D by Chinese firms in developed countries. Because of infringements of IPRs at home for using certain Western technologies, Chinese companies are debarred from entering foreign markets. Hence, the local R&D centres of Chinese firms abroad would overcome these difficulties by developing and branding local technologies, which in the process would build new technological competencies for Chinese firms. On the contrary, market- and output-related determinants are said to be the determinants of R&D expansion in developing countries. (For example, Haier's R&D centre at Jordan/Iran developed air conditioners to cope with particularly adverse desert conditions in the Middle East and designed washing machines that could also handle cleaning vegetables in rural Asia.) Apart from Haier, there are other Chinese firms that have internationalized their R&D such as ZTE, 3NOD and Huawei (Zedtwitz, 2005). Similarly, Indian firms that have internationalized their R&D in the recent past, because of liberalization of the restrictions in India are Infosys, Wipro, Satyam, HCL and TCS in the IT sector, Ranbaxy & Dr. Reddy's Laboratories in pharmaceuticals and Tata Steels in metals. Space constraints do not allow me to discuss more about the Indian cases.

Like China and India, Brazil also has been a major destination of R&D-related FDI. Costa (2005) has reported results of a survey conducted by the Economic Intelligence Unit (EIU) in the year 2004 to show that Brazil is the third major destination among developing countries where from MNCs are planning to offshore R&D, whereas the first and the second major destinations remained to be China and India, respectively. I emphasize the technological developments in the later 1990s when some technological upgrading and economic restructuring had taken place in response to competitive market situations. Technological developments in Brazil then were involved in the adaptation of newer technologies, of both product and process and new organizational practices leading to gains in productivity and economic efficiency. The trend even continued further as Costa and Queiroz (2002) reported that, in general, the Brazilian industrial R&D remains at the adaptive level as was the case of industrial R&D in India in the 1960s–1980s. Hence, the very vital question that emerges is if the globalization of R&D by MNCs and flow of FDI in R&D to Brazil would facilitate opening of new opportunities for the Brazilian industry to move beyond adaptive levels of R&D to become innovative?

Querioz et al. (2003), assuming a strong relationship between production capacity, technological capabilities and potentiality to attract MNC R&D, note that foreign affiliates with large and long-established production capacities are in a good position to conduct corporate R&D that results in the accumulation of technological competencies and knowledge. Hence, in developing countries, global product development is mostly observed amongst long-established foreign affiliates that have accumulated technological capabilities in some product or process technologies. In such cases, knowledge embedded in the local R&D team becomes assets for the affiliated MNC. Therefore, Queiroz et al. (2003) noted that MNCs in Brazil are engaged in 'asset-exploiting' R&D activities because these capabilities of local affiliates serve to complement the MNCs themselves. This argument has been substantiated by Queiroz et al. (2003) on the basis of empirical evidence from the Brazilian automobile industry. Car manufacturing MNC affiliates in Brazil like those of General Motors (GM), Fiat, Ford, Volkswagen have accumulated knowledge and skills in styling, prototype testing, laboratories and proving grounds. These MNC affiliates in Brazil have focused their R&D to adapt to local and regional conditions and to develop local derivatives. This process of market-oriented R&D by local affiliates of MNCs is known as *tropicalization*. For instance, in the recent past, GM Brazil proposed to its headquarters the concept of a global derivative (based on the new Corsa) called the *Meriva model*. Similarly, Volkswagen Brazil's TUPI project was a derivative model based on the new Polo platform, the *Fox model*. This enables the local MNC affiliates in Brazil to compete with their sister subsidiaries/affiliates based in other countries for acquiring R&D assignments. The growing competition among the affiliates around the world for central roles in the MNC global network seems to be a significant feature of globalization of R&D.

Globalization of Local Industrial R&D of Firms from Developing Countries

The other significant observation from the globalized R&D in Brazil has been the globalization of the local R&D (through MNC affiliation) that are of course some niche based. The interesting case cited by Galina and Plonski (2002) from the Brazilian telecom sector is worth looking at. Tetax and Batik, both domestic Brazilian firms with strong R&D capabilities on small switches, were acquired by Lucent, an MNC, in the late 1990s during the privatization process. Since the MNC Lucent till then did not have small switches in its product menu, the two Brazilian domestic firms turned into Lucent subsidiaries for small switches manufacturing and became a global centre for technology for small switches. Therefore, it is claimed that high-technological competences in some niches can help local MNC subsidiaries to become part of global R&D network. Perhaps these niches are dependent upon particularities of the technologies and the products. This however appears to be a 'reverse globalization' process. The pharmaceutical industry in Brazil is of course an interesting case but in the opposite direction. Since drugs and pharmaceuticals MNCs have been in Brazil for more than 50 years now, following the earlier argument of Brazilian automobile/telecom industry, one could expect significant R&D activities by local affiliates of the MNCs, maybe of *tropicalization* type (meaning addition of local flavour), or the 'reverse globalization' type. On the contrary, any innovative R&D activities worth mentioning in this sector are almost non-existing. These ageing MNC subsidiaries in the Brazilian pharmaceutical industry have contributed very little to the local technological developments be it product or process type.

Costa (2005) concludes that whatever R&D-related FDI might have flown into Brazil in the second phase of globalization, that is, after the late 1990s, it seems that R&D facilities were established in order to support productive activities mostly. Cases of innovative stand-alone laboratories/firms are almost non-existent. This is partly because of the absence of suitable policy measures in Brazil, although in this later stage of globalization, certain selected sectors of the Brazilian industry as pointed out earlier have made some significant innovative type of R&D that have been the '*tropicalization*'/'asset-exploiting' type of R&D. This is certainly unlike the Chinese case that has taken a big leap forward.

Perhaps under globalization the most significant and profound change in the behaviour of the MNCs, which is of course strategic by nature, has been the breakdown of an 'immutable home country orientation of creative activity (continuously revolutionizing technology for competitive edge) and movement towards globalized programmes in innovation and R&D' (Pearce, 2005). Of course, there has been a stereotype perception that the organizational structures of MNCs are predominantly hierarchical, that is, home country-based R&D headquarters of MNCs strategically control their subsidiary-based R&D activities in different parts of the world. However, this strategic relationship of the controllers with those being controlled has undergone change, pointed out Pearce (2005). The process of globalization has levelled all of them to be parts of the global strategy of R&D. Hence, the earlier intra-MNC relationship has been replaced by 'heterarchy' or as dynamic differentiated networks. With respect to intra-MNC restructuring of relationships, the earlier hierarchical relationship has been replaced by 'heterarchy' on the basis of their current/future potentialities to market. Further, this relationship among an MNC's global network labs/centres is now a dynamic one as their operations in different countries/regions undergo quick change. Of course, this has been possible because of the increasing decentralization of the R&D programmes of MNCs. Their relationship has acquired a dynamic strategic orientation in the MNCs' host countries, particularly those countries that are at the early stage of competitiveness. Strategically, the MNCs now involve their local affiliates with creative resources maybe for the exploitation of local assets or for the localization of global products. Many successful MNCs have benefited from their willingness to respond to local taste differentiations. This has added to the rising significance of their affiliates and subsequently to their heterogeneity.

Offshoring Industrial R&D Services by Firms from Developed Countries

MNCs' FDI and technology transfer are being increasingly interlinked under globalization. Needless to say, MNCs have a very large share in the global R&D activities even today. The largest 700 firms worldwide in the year 2002 had spent $311 billion on R&D (UNCTAD, 2005). Currently under a globalized regime, marked by rapidly changing technologies and shorter product life cycles, MNCs are offshoring increasingly more R&D to different parts of the world through both FDI and technological alliances (non-equity mode). This changing pattern of locating R&D differs radically from that of the past and alters the patterned view that R&D activities by MNCs remain confined mostly to their home countries (almost invariably developed countries). The expansion of R&D beyond the homeland may not be new but significantly the scale of offshoring is rising and its geographical spread in reaching the developing countries is increasing. The spread of R&D-related FDI to developing countries is part of the broader process of offshoring services, which is still rising. With the rising offshoring practices by the MNCs, even the very meaning of the term offshoring is broadened. (Offshoring in this context essentially means locating/transferring R&D activities abroad. It is done often internally by moving assets-/knowledge-based services from the parent MNC to its foreign affiliates involving FDI. Or R&D activities outsourced to the third-party provider abroad, that is, could be a local firm in the host country or could be a foreign affiliate of another MNC in the host country.) Offshoring is different from outsourcing, which always involves a third party, but not necessarily a transfer abroad. Offshoring and outsourcing overlap only when the activities/services are outsourced to third-party service providers.

The offshoring of R&D activities to locations in developing countries has already become a known phenomenon as well-known MNCs like Erickson, GE, IBM, Intel, Microsoft, Motorola, Nokia, Oracle and Texas Instruments have moved into developing countries. Between 1989 and 1999, R&D performance by all foreign affiliates of US MNCs in developing countries increased by nine times as compared to three times increase worldwide. In developing Asia, there was an 18-fold rise in MNC-based R&D expenses. During the same period (1989–1999), Japanese MNCs offshoring R&D to developing country-based foreign affiliates grew by ten times compared to their expenditure worldwide (UNCTAD, 2005). Of course, offshoring of R&D by European MNCs to developing countries is still in its nascent stage.

Indicating the globalization of R&D, foreign affiliates are assuming more important roles in R&D in many host countries. During the period 1993–2002, the R&D expenditure of foreign affiliates worldwide climbed from 10 per cent to 16 per cent of global business R&D (UNCTAD, 2005). Significantly, the expenditures of foreign affiliates in business R&D in the developing world increased from 2 per cent to 18 per cent during 1996–2002 (UNCTAD, 2005). Similar data on greenfield R&D projects initiated during August 2002 to July 2004 indicate that of the 1,000 FDI projects in R&D worldwide, the majority, that is, 739, were located in developing countries. Developing countries in Asia and the Pacific alone accounted for more than half of the total, that is, 563, projects. Prominent among these countries were China and India.

Apart from the FDI route, firms also often use non-FDI forms such as technology alliances, R&D joint ventures, R&D consortiums and university–industry linkages to get access to strategic knowledge abroad. These forms of cooperation can be equity or non-equity based (mostly not considered to be outside FDI definition). As part of their alliances, MNCs are outsourcing some technology development activities from firms and research institutes worldwide, including those located in developing countries. However, R&D by MNCs under globalization reached out to only select developing countries such as China, India, Brazil, Singapore, South Africa and Mexico. However, of late other developing countries like Thailand (Toyota project), Malaysia and even African countries like Kenya (MNCs agricultural technology) have slowly acquired MNC R&D into their folds.

Further, it is noteworthy that the technological change unleashed by globalization impacted the design and organizational patterns of R&D leading to a proliferation and differentiation of corporate R&D units. As products became modular, fragmentation of design and the specialization of knowledge creation in the internal and external networks of MNCs took place. The United Nations Conference on Trade and Development (UNCTAD, 2005) has articulated clearly five types of R&D units that have come into existence (of which three at least are of recent origin): (a) Technology transfer units (typical to developing country firms) linked to manufacturing units, engaged in technology transfer issues and adaptations of products and processes to local conditions, (b) Local technology units (new to developing countries, set up to develop new/improved products for local market when the MNC affiliate identifies distinctive investment opportunities), (c) Regional technology units (set up to develop new-improved products for regional markets share some common features and need for specialized products), (d) Global technology units (single product for global market, when the MNC allocates parts of its product range to specific affiliates and when the MNC opts for a decentralized but integrated R&D programme, for efficiency drive through offshoring and international outsourcing) and (e) Corporate technology units (internationally independent set up to pursue pure/basic research in one area of science and engineering to bring radical product breakthroughs, operating as networks to bring synergetic combination of results, engage in long-term, exploratory research for the parent MNC to protect and enhance future competitive edge).

In addition to this, the emergence of new science-based technologies like micro-electronics, ICT, biotechnology and new materials has had a profound impact on business R&D run by MNCs. These new technologies have created opportunities for developing countries, particularly countries with a tradition of scientific education, without much industrial experience to catch up with the rest. R&D in micro-electronics, biotechnology, pharmaceuticals, software, etc., could be globalized more easily than conventional industries, because these industries are more knowledge intensive and can be presumed even being delinked from production. Further, in these areas of new technologies, R&D activities are so divisible into several modules that certain modules can easily be treated as core and some as 'non-core' activities. Hence, some of the non-core activities can easily be outsourced mostly from developing countries.

The other substantive spillover effects of MNCs' R&D activities in developing countries caused by globalization have been the following: (a) arrival of commercial culture among research scientists and engineers (different from pure academic culture, that is, knowledge for knowledge shake). For example, when FDI-related R&D arrived in India, research scientists and engineers from public-funded laboratories/academic institutions started focusing on patentable research and some of them became entrepreneurs by themselves, (b) emergence of an R&D and innovation culture among local firms, particularly in software, pharmaceuticals and drugs and (c) diverting scarce local R&D resources of host countries, particularly human resources, from local firms and research institutions. The best of R&D manpower may be allured to MNC R&D in developing countries.

Conclusion

These above-discussed trend developments in the globalization of industrial R&D are indicative of the qualitative change in the centre periphery relations. It may be too sweeping to make a derivation about the qualitatively changed relations between the centre and the periphery countries as a general phenomenon emergent of globalization, but it would be otherwise safe to conclude on the basis of the evidence produced above that instead of being a general phenomenon this changed relation between the centre and the peripheral countries is a limited phenomenon confined to the experiences of the handful of developing countries such as China, India, Brazil, Indonesia and the like. It is very much possible that of these a few peripheral countries have changed their relations qualitatively with the countries at the centre keeping the general nature of centre–periphery relations unaltered. The qualitative change in the relations of these few peripheral countries with the centre is of course indicative of their new relations as 'semi-peripheral countries'. It would not be inappropriate to perceive the changing relations of countries like Singapore, China, India and Brazil with the countries at the centre within the 'world-system perspective' of Emanuel Wallerstein, as that of 'centre and semi-periphery' countries. Of course, compared to other countries, Singapore had much earlier entered into the semi-periphery club of countries. However, like Singapore, these other peripheral countries have benefited from their relationship with the countries at the centre, may be within the framework of dependence and have successfully transformed their position from that of periphery to semi-periphery. The case of India as a semi-periphery country may be contested, but it is fast moving into the category, if we take evidence of technological developments in areas of IT, pharmaceuticals and drugs, biotechnology and the subsequent export of industrial products and services.

India also meets the other requirements of becoming a semi-periphery country. Because in the semi-periphery countries, there would be always certain sections of the economy, particularly in the manufacturing sector, that maintain their relevance to the markets in the peripheral countries. These countries produce and export some high-quality finished, semi-finished as well as intermediary products with a relatively low cost (compared to the same in the markets of developed countries). These countries also serve as new sources of industrial capital for peripheral countries. On the contrary, these also import some of the raw materials or semi-processed and intermediary products from the peripheral countries keeping some labour-intensive primary sectors alive there. Thus, the semi-periphery countries do maintain very sound trade relations (devoid of monopoly pricing, restrictive trade clauses, etc.) with the peripheral countries. Of course, these countries are also dependent upon the markets of the developed countries for imports of developed technologies, technological products as well as FDI. Therefore, these maintain bimodal relationships in international trade, that is, dependent peripheral relation with the centre and the reverse type with the peripheries. Thus, the semi-periphery countries placed between the centre and the periphery try to benefit from both the economies of polarities in a world capitalist system. These have been evident to a certain extent in our analysis of the technological developments in India and China. In addition, any further analysis of their exports and imports trends would strengthen this argument.

References

Cantwell, J. and O. Janne. 1999. 'Technological globalization and innovative centers: The role of corporate technological leadership and locational hierarchy', *Research Policy*, 28: 119–44.
Cantwell, J. and L. Pocatello. 2000. 'The location of technology activities of MNCs in European regions: the role of spillovers and local competencies', Working Paper, Reading University Business School, pp. 1–23.
Chen, Y.-C. 2007. 'The upgrading of multinational regional innovation networks in China', *Asia Pacific Business Review*, 13(3): 373–403.

Costa, I. 2005. 'Technological learning, R&D and foreign affiliates in Brazil', in UN (ed) *Globalization of R&D and Developing Countries*. New York and Geneva: UN, pp. 141–54.
Costa, I. and S. Queiroz. 2002. 'Foreign direct investment and technological capabilities in Brazilian industry', *Research policy*, 31: 143–44.
Dunning, J.H. and R. Narula. 1995. 'The R&D activities of foreign firms in USA', *International Studies of Management and Organization*, 25: 39–74.
Florida, R. 1997. 'The globalization of R&D: Results of a survey of foreign affiliated R&D laboratories in USA', *Research Policy*, 26: 85–103.
Friedman, T.L. 2006. *The World Is Flat, The Globalized World in the Twenty-first Century* (Updated and Expanded). Penguin Books, x+, 660pp.
Galina, S.V.R. and G.A. Polnski. 2002. 'Global product development in the telecommunication industry: An analysis of the Brazilian subsidiaries involvement', Ninth International product Development Conference—European Institute for Advanced Studies in Management (EIASM). Sophia-Anipolis, France, May, pp. 12–13. Available at: www.fia.com.br/pgtusp/pesquisas/orq_prorex/subol/SGalina%20%20 9th%20IPDMC. pdf.
Kuemmerle, W. 1997. 'Building effective R&D capabilities abroad', *Harvard Business Review*, March–April: 61–70.
―――, 1999. 'Foreign direct investment in industrial research in the pharmaceutical and electronic industries: Results from a survey of multinational firms', *Research Policy*, 28: 179–93.
Le Bas, C. and C. Sierra. 2002. 'Location versus country advantages in R&D activities: Some further results on multinational's locational strategies', *Research Policy*, 31: 589–690.
Patel, P. 1995. 'Localized production of technology for global markets', *Cambridge Journal of Economics*, 19: 141–53.
Pattnaik, B.K. 2005. 'Impact of globalization on the technological regime in India: Aspects of change', *Perspectives in Global Development and Technology*, 4(1): 63–82.
Pearce, R. 2005. 'The globalization of R&D: Key features and the role of TNCs', UN (ed.) *Globalization of R&D and Developing Countries*. Geneva/New York: UN, pp. 29–42.
Pearce, R. and S. Singh. 1990. 'Internationalization of R&D among the world's leading enterprises; survey analysis motivation, organization and implications', paper presented at the Conference Technology Management and International Business, Stockholm, 17–20 June.
Petrella, R. 1992. 'Internationalization, multinationalization and globalization of R&D, toward a new division of labour in science and technology', *Knowledge and Policy: The International Journal of Knowledge Transfer and Utilization*, Fall, 5(3): 3–25.
Queiroz, S., M. Zanatta and C. Andrade. 2003. 'Internationalization of MNCs' technological activities: What role for Brazilian Subsidiaries?' Paper presented at SPRU conference in the honor of Keith Pavitt, Brighton. pp. 13–15, November. Available at: www.sussex.ac.uk/units/spru/events/kp_conf_03/documents/Queiroz.pdf.
Saxenian, A.L. 2002. 'The Silicon Valley connection: Transnational networks and regional development in Taiwan, China and India', *Science Technology and Society*, 7(1): 117–49.
Simon, D.F. 2007. 'Wither foreign R&D in China: Some concluding thoughts on Chinese innovation', *Asia Pacific Business Review*, 13(3): 471–80.
UNCTAD Secretariat. 2005. 'An overview of the issues', in UN (ed) *Globalization of R&D and Developing Countries*. Geneva and New York: United Nations, pp. 1–25.

Upadhya, C. 2004. 'A new transnational capitalist class? Capital flows, business networks and entrepreneurs in the Indian software industry', *Economic and Political Weekly*, 39(48): 5141–51.

Warrant, F. 1991. 'Deploiment Mondial da la R&D industriells', FAST Research Paper, Commission of the European Communities, Brussels, April.

Yurevich, A.V. 2010. 'Globalization process in contemporary science and scholarship in Russia', in N. Asheulova et al. (ed) vol. *Liberalizing Research in Science and Technology: Studies in Science Policy*. Saint Petersburg: Saint Petersburg Politechnika Press, pp. 94–105.

Zedtwitz, M.V. 2005. 'International R&D strategies of TNCs from developing countries. The case of China', in UN (ed) *Globalization of R&D and Developing Countries*. New York/Geneva: UN, pp. 117–40.

8

Technological Capability, Employment Growth and Industrial Development: A Quantitative Anatomy of Indian Scenario

Lakhwinder Singh and Baldev Singh Shergill*

Introduction

Post-reform spurt in Indian economic growth has been described as 'jobless growth'. The phenomenon of jobless growth during the period of late 20th century and early years of 21st century was not India specific, but was observed across the developed and developing countries alike, except newly industrializing South East Asian countries, including China (Audretsch and Thurik, 1999; Van der Hoeven and Taylor, 2000). The industrial development of the developing economies, including India except Asian newly industrializing countries, has also recorded very slow growth in employment. The industrial sector has been considered as the most dynamic sector of an economy, and therefore, it is expected that this sector should absorb the growing labour force and provide decent livelihood to the growing workforce. Many scholars working in the area of industrial development and its implications for employment have observed dismal scenario so far as employment outcomes of industrial development is concerned (Amsden and Van der Hoeven, 1996; Edquist et al., 2001; Goldar, 2009; Kannan and Raveendran, 2009; Morawetz, 1974; Papola, 2009; Sen, 2008).

---

* This is the revised version of the chapter presented at the UNU-WIDER/UNU-MERIT/UNIDO Workshop Pathways to Industrialization in the 21st Century on 22–23 October 2009 at UNU-MERIT, Maastricht, the Netherlands, and also at Chinese Economic Association, Europe/UK from 12–13 July 2010 at University of Oxford, Oxford. The authors are grateful to participants in the workshop, especially Adam Szirmai, Haider A. Khan and Mulu Gebreeyesus, for their helpful suggestions. The constructive suggestions made by the anonymous referee of the book, which helped in improving the arguments in the chapter, are gratefully acknowledged. However, the usual disclaimer applies.

Economic reforms initiated in the early 1980s and more vigorously since July 1991, both national and international, failed to effect manufacturing employment in the face of excess supply of unskilled labour force in the Indian economy. Indian development strategy has underlined the importance of industrial development with regard to its dynamic characteristics such as capital accumulation and technological capabilities. It is a widely held view that the technological advancement, since the Industrial Revolution, has been concentrated in manufacturing sector and the diffusion of technology takes place from this sector to other economic activities and sectors (Szirmai, 2009). The diffusion of technology across the manufacturing and other sectors not only raised productivity but also generated backward and forward linkages. This process has a capacity to generate special opportunities for a catch-up. Large-scale mass production of manufactured products essentially creates dynamic comparative advantage and triggers change in the industrial structure that generates economic activities based on new knowledge, provides greater employment opportunities and explodes demand for improved variety of products (Audretsch and Thurik, 1999).

Rapid industrial development experience of both developed countries and newly industrialized (East Asian) countries have supported the view that growth can solve the problem of unemployment. However, there has been a growing realization that the gap between the speed at which technological progress is taking place and the capacity to provide new job opportunities has widened dramatically (Commission of the European Communities, 1994; Rifkin, 1995). Policymakers are expected to address the growth–employment dilemma, while understanding the relationship between growth and employment, with a broad spectrum of policy approaches. The mainstream policy response favoured higher dose of market until recent financial meltdown and recession that has engulfed the global economy. Therefore, it is high time to understand and examine the question of industrial output growth implications for employment generation.

In this chapter, an attempt has been made to examine the question of when industrial development provides required dynamism for generating desired employment opportunities for labour force and when it does not. An industrial technological capability-based approach has been adopted to analyze the Indian industrial development experience during the period from 1980 to 2005, which covers the rapid growth period of industrial economy of India and also represents the pre- and post-reform period. Quantitative assessment of industrial employment growth has been made while making use of semi-logarithmic regression analysis and panel data-based models. The discussion is organized into nine sections. Apart from the introductory section, the next section on 'Framework for Analysis' presents the analytical framework of the chapter. Data sources and methodology is discussed in the section 'Database and Methodology'. The changes in the structure of industrial employment are presented in the section 'Changing Structure of Employment in India's Organized Manufacturing'. In the section 'Employment Growth across the Manufacturing Industries in the Pre- and Post-reform Period', the analysis of employment, output and wage trends across the manufacturing industries is presented. The estimates of employment elasticity are presented in the section 'Employment Elasticities across the Indian-organized Manufacturing Industries', and the section 'Determinants of Employment Elasticity across the Manufacturing Industries in the Pre- and Post-reform Period' discusses the determinants of employment elasticity. The employment growth and technological capabilities relationship is examined in the section 'Industrial Technological Capability, Industrial Structure and Employment'. In the final section, concluding remarks and policy implications for other developing countries, which result from the chapter, are presented.

Framework for Analysis

Industrial development, technological capabilities and employment growth are intimately connected. Industrial development generates opportunities for faster capital accumulation and technological capability building that spurs structural transformation in the economy. Historical experience of industrial development of the advanced countries shows that spurts of industrial activities have not only engineered the process of structural transformation but also generated gainful employment opportunities for the workforce. This has led to the shortage of desired workforce in desired skills and initiated the process of either immigration or innovations that has been addressing the problem of labour shortages. Therefore, the evolution of industrial structure in the developed countries has been accompanied by the evolution of technological capabilities addressing the problems encountered by the industrial development process. This co-evolution process remained in operation in the developed countries for a long time, and recent wave of technological revolution further raised the skill requirement of the workforce and raised wage costs along with reduced possibilities of capital–labour substitution.

Towards the last quarter of the 20th century, the emergence of new lower-cost production locations has reduced the competitiveness of European and North American firms. The threat of erosion of traditional comparative advantage has resulted in shifting of production to cheaper locations, laying off workers and reduction of wage cost, which soared the unemployment rates during the 1980s and the 1990s (Audretsch and Thurik, 1999). The continuously rising unemployment rates in the developed countries with moderate growth rates of output have triggered policy debate on inevitable trade-off between employment and wage cost. The recent studies conducted to examine the relationship between innovations, industry evolution and employment suggested that the debate on trade-off between employment and wage cost actually diverted the main issue, that is, alternative to it. The alternative lies in the continuous evolution of the industrial structure and reflects in the shifting of industrial activities from moderate technology industries to newly emerging knowledge-based industries. Thus, the technological capability building approach is path dependent and has a capacity to connect innovation, industry evolution and employment generation.

When India began her march towards modern economic growth after achieving independence, industrial development assumed to play central role for rapid economic growth and structural transformation of the economy. The major emphasis of the Indian government was to lay foundations for faster growth of industrial sector and building technological capabilities that can self-sustain rapid industrialization. It was also envisaged that heavy industrial development strategy will inherently be capital intensive; therefore, special emphasis was made to develop small-scale labour-intensive industries for providing gainful employment to the growing workforce. The catch-up growth model chosen by India strives to develop technological capabilities while doing R&D expenditure in public sector enterprises and institutions. Substantial efforts were made to fulfil the needs of technological requirements through the process of adapting imported technologies and further create technological capabilities to generate new technologies and eventually catching up with the advanced countries (Ray, 2009).

The domestic efforts in terms of R&D expenditure were stepped up from 0.17 per cent in 1958–59 to 0.98 per cent in 1987–88. However, it declined thereafter and hovered around 0.8 per cent. The emphasis has shifted from self-reliant technological capabilities to liberal import of technology. The structure of R&D expenditure has undergone some changes but still remained highly public sector dominated and firm-level capabilities, except some industries usually remained low. The industrial economy of India could not catch up to the frontier of technological knowledge, and industrial productivity still remained quite low, compared not only with the developed countries but also with East Asian countries, especially far behind from China (Papola, 2009). It is significant to know that despite achieving reasonably faster rate of output growth why Indian-organized manufacturing industry could not able to generate desired level of employment growth. The discussion on slow employment growth in the Indian-organized manufacturing industries has been revolved around the inevitable trade-off between output growth and labour cost.

The slow absorption of the labour force in industrial sector even during the liberalization experience put a question mark on industrial development strategy adopted by the Indian policymakers. However, several scholars have investigated and argued that stagnation in employment of labour force in the organized industrial sector of the economy can essentially be attributed to labour security provided by the labour laws. Labour market rigidity was mainly held responsible for decline in the employment elasticity of organized manufacturing industries (Fallon and Lucas, 1993; Goldar, 2009; Hasan et al., 2007). The industrial employment stagnation has resulted from falling employment rates in some set of industries and rising employment rates in other set of industries under the same policy regime, has resulted into overall jobless growth giving credence to the view that supporters of liberalization may not be right to find out factors that has not allowed job creation in the industrial sector. Therefore, some alternative approaches have been put forward to find out a plausible underlined explanation so that right kind of public policy can be formulated to overcome joblessness in the Indian economy (Kannan and Raveendran, 2009; Nagraj, 2000; Papola, 1994; Singh and Gill, 2002).

Alternative to labour market rigidity, the reduction of capital cost and changing pattern of demand of the manufactured products both nationally and internationally put forward as a dominant explanation for decline in the elasticity of employment in the organized Indian manufacturing industry. It is pertinent to argue that the last quarter of the 20th century has witnessed technological revolution, which has introduced automation processes and hence substantially reduced capacity to generate direct employment by the manufacturing sector. Indian industry witnessed structural transformation from primary raw material and metal based to processed intermediates inputs, that is, high-tech processes, which has increased labour productivity and consequently may have reduced the employment potential of output growth (Papola, 1991). The phenomenon of jobless growth of Indian industrial development can be attributed to the pattern of technological change rather than labour market rigidities. The objective of this study is to provide alternative explanation of near stagnation of employment growth and unravel the factors that have led to the jobless growth of the organized manufacturing sector of the Indian economy. To accomplish the above-said objective, we have followed technological capability-impacted evolution of industrial structure approach and accordingly classified manufacturing industries into four groups. These are low-technology, medium-low-technology, medium-high-technology and high-technology manufacturing industries.

Database and Methodology

The purpose of this study is to analyze the long-term pattern of growth and structure of employment at disaggregative level of three-digit manufacturing industries. The study covers the period from 1980–81 to 2004–05, which is a quarter century time period, and data were collected from Annual Survey of Industries (ASI) published by the Central Statistical Organisation. An attempt has been made to develop a consistent data set related to 44 three-digit industries (names of industries and industrial codes are given in the Appendix at the end of this chapter). While developing consistent data set, one faces a problem of frequently changes introduced in the National Industrial Classification (NIC) used by the ASI. The NIC

1970 remained in operation up to the year 1988–89, NIC 1987 up to the year 1997–98, and thereafter, industries were classified on the basis of NIC 1998. Therefore, it is important to construct concordance of the changed NIC 1998 with the earlier two changes introduced in the classifications (NIC70 and NIC87) for developing a consistent data set. After constructing consistent data set for 44 three-digit industries, we have classified industries into four groups based on technological characteristics as low-tech, medium-low-tech, medium-high-tech and high-tech industry groups. The variables other than employment have been corrected with 1993–94 base year wholesale price indices and cost of living indices appropriate for each industry. The whole period from 1980–81 to 2004–05 has been divided into two sub-periods, that is, pre-reform period from 1980–81 to 1991–92 and post-reform period from 1992–93 to 2004–05. We have also made use of the database on research and development expenditure in industry published by the Department of Science and Technology (DST), Government of India. The industry-wise R&D expenditure data has been aggregated into four groups of industries based on technological characteristics (low-tech, medium-low-tech, medium-high-tech and high-tech) to arrive at a matching classification with ASI-based four technological groups. Since the industry-wise database provided by the DST cannot be matched with the three-digit industrial database of ASI, therefore, we have only used DST data for four above-mentioned technology-based industry groups.

To ascertain the long-term trends of the variables, we have estimated trend growth rates based on semi-logarithmic regression equation. The employment elasticity for each industry has been estimated on the basis of percentage change in employment growth for a percentage change in output growth. To estimate the major determinants of employment elasticity, a decomposition analysis, which allows us to compare the trade-off between employment growth and wage growth, has been done. The decomposition analysis is based on the methodology developed by Mazumdar (2003). It enabled to decompose the factors affecting growth rate of real wages for any period and industry into the rate of growth of real value added, the trend share of wages, the rate of growth of employment and the relative price effect. This can be estimated with the help of the following equation:

$$\Delta W = \alpha \Delta V - \alpha L + (\alpha \Delta P_p - P_c),$$

where Δ represents the proportionate rate of change of the variable concerned.

The first term in the right-hand side of the above equation is the output effect, the second the employment effect and the third in parentheses is the price effect (see for a detailed derivation, Mazumdar, 2003). The panel data regression model has also been used to obtain the empirical evidence with regard to ascertaining the impact of technological capabilities on Indian industrial employment growth.

Changing Structure of Employment in India's Organized Manufacturing

Indian industrial development experience, during the import substitution regime, has undergone a substantial structural transformation. During this period, the industrial sector accumulated technological capabilities nurtured and supported by the Indian government while investing in both research and development and tertiary education. While drawing benefits from the capabilities developed during the period of import substitution, the industrial structure has been substantially altered in favour of high-tech industries. According to one estimate, the high-tech Indian industries generated more than 33 per cent of the value added as early as in the year 1980 (Amsden, 2004). This evidence of higher share in value added that originated from high-tech industries has provided India a unique place among the late industrializing countries whose manufacturing industrial sector is dominated by high-tech activities.

The relative shares of employment generation of India's organized manufacturing industries classified on the basis of technological categories are presented in Table 8.1. The analysis of Table 8.1 revealed that a group of 12 low-tech industries in the year 1980–81 had been providing 53.66 per cent of employment of the organized manufacturing industries. Within the low-tech group of industries, there were a very high degree of concentration of employment in two industries (i.e. textiles and food products) and predominantly provided a large proportion (34 per cent) of employment (Figure 8.1). Thereafter, diversification in low-tech industries in terms of employment generation has occurred during the fast pace of

Table 8.1:
Changing structure of employment across Indian industries: 1980–81 to 2004–05 (figures are in percentages)

Industry code	Low technology			Industry code	Medium-high technology		
	1980–81	1992–93	2004–05		1980–81	1992–93	2004–05
151	3.12	1.96	1.79	241	1.17	2.91	2.10
152	0.49	0.88	0.87	252	1.41	1.21	1.97
153	2.27	3.71	3.49	311	2.00	2.25	2.51
154	10.94	9.14	6.86	261	0.92	0.75	0.54
155	3.12	1.96	1.79	290	2.43	3.13	2.10
160	5.12	6.40	4.94	271	7.02	6.35	4.13
171	23.09	19.84	17.87	319–323	2.03	1.61	4.70
201	0.37	0.28	0.11	292	2.60	2.54	5.81
202	1.33	0.55	0.41	293	1.07	0.67	0.27
210	1.67	2.02	1.86	313	0.47	0.52	4.92
221	2.00	1.97	1.18	314	0.26	0.19	4.13
361	0.13	0.09	0.32	315	1.33	0.66	0.83
Subgroup total	**53.66**	**48.79**	**41.49**	341	2.30	2.59	0.83
Medium-low technology				333	0.17	0.28	0.11
182 + 191	0.45	0.80	0.55	359	0.80	1.50	1.37
192	1.38	0.89	1.03		**25.98**	**27.16**	**36.34**
251	1.65	1.36	1.20	**High technology**			
231	0.67	0.49	0.33	223	2.03	1.34	0.02
269	4.00	5.29	4.92	232	2.05	0.39	0.50
272–273	0.65	2.19	1.89	233	0.52	0.00	
281	2.64	3.11	3.33	242	1.17	5.39	5.81
351	3.96	2.48	0.24	300	2.12	0.54	1.20
369	0.39	0.63	1.51	331	0.33	0.40	0.47
371 + 372	0.08	0.09	0.02	332	0.03	0.04	0.07
Subgroup total	**15.86**	**17.33**	**15.03**	243	0.15	0.36	0.27
				Subgroup total	**8.40**	**8.46**	**8.34**

Source: Authors' calculations based on Annual Survey of Industries, various issues.

Figure 8.1:
Low-tech industrial employment shares

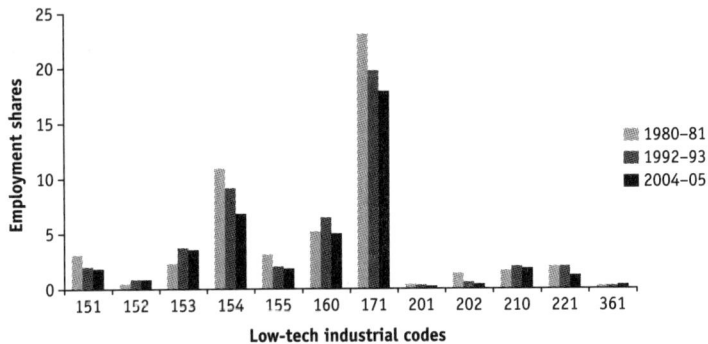

liberalization. The overall shares of employment of the low-tech industrial sector have declined from 48.79 per cent in 1992–93 to 41.49 per cent in the year 2004–05. It is important to note here that during the period of analysis, there was a sharp decline in terms of relative shares of employment provided by low-tech manufacturing industries of the order of more than 12 percentage points.

The medium-tech manufacturing sector employment shares, during the period under consideration, have shown marginal improvement from 1980–81 to 1992–93, but declined in 2004–05 (Figure 8.2). On the whole,

Figure 8.2:
Medium low-tech industrial employment shares

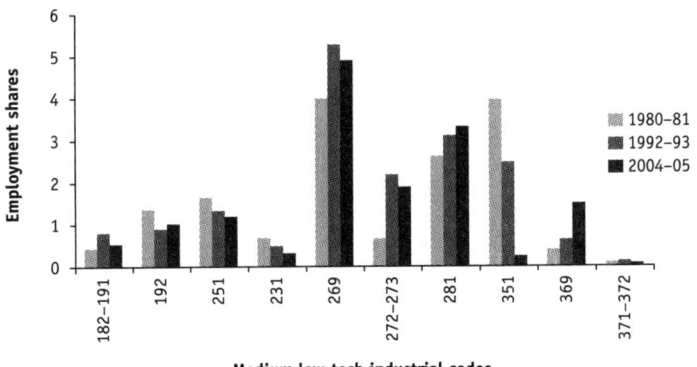

the medium-tech industry continues to maintain its position so far as the shares of employment are concerned. The relative share of labour force employment in high-tech industries has also remained stagnant during the period of analysis. It is quite counter-intuitive result in the sense that in the post-reform period, it is expected that the structure of manufacturing sector should have been driven by domestic and international demand for high-tech products. More so, the growing importance of the operation of multinational corporations is expected to trade both domestically and internationally in high-tech products. The employment outcome of this process tends to show jobless growth of this sector.

The perusal of Table 8.1 reveals that medium-high-tech Indian manufacturing industries have substantially increased its relative shares of employment (Figure 8.3). It is significant to note that in the year 1980–81, the medium-high-tech industries have generated 25.98 per cent of the total industrial sector employment. However, the relative share of employment increased to 27.16 per cent in 1992–93. In the post-reform period, there was a dramatic rise in the relative shares of employment of the medium-high-tech industries. The relative share of employment has improved to 36.34 per cent, which is more than 9 percentage points. If we combine medium-high-tech and high-tech group of Indian manufacturing industries (Figure 8.4), the relative share of employment turns out to be 34.38 per cent in the year 1980–81, which is quite close to the value addition done by these industries (Amsden, 2004). The combined

Figure 8.3:
Medium high-tech industrial employment shares

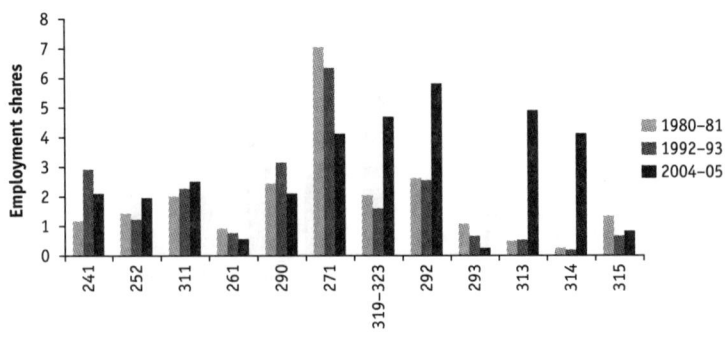

Figure 8.4:
High-technology industrial employment shares

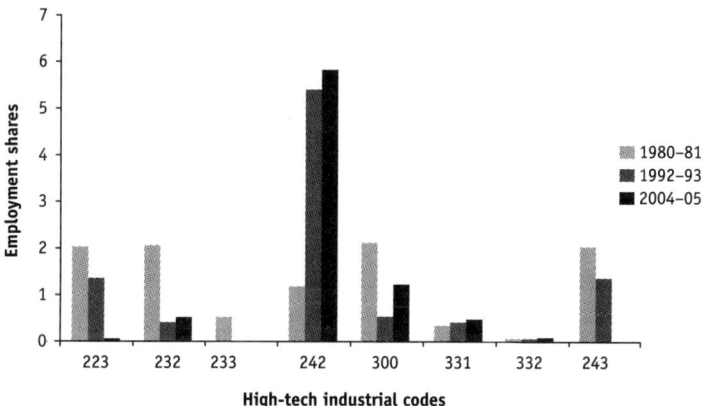

share for the year 2004–05 comes out to be 44.68, which is more than 10 percentage points higher than the initial period. The rising relative importance of high-tech industries in terms of changing proportions of employment sufficiently provides support to the argument that the Indian manufacturing sector has been undergoing a dramatic structural transformation from low-tech manufacturing industries to medium-high-tech industries.

Employment Growth across the Manufacturing Industries in the Pre- and Post-reform Period

The overall employment growth in organized manufacturing sector, both in the pre- and post-reform period, remained quite dismal. The trend growth rate of employment in the pre-reform period (1980–81 to 1991–92) was 0.40 per cent. However, it has marginally improved in the post-reform period (1992–93 to 2004–05) and was 0.63 per cent. The employment elasticity of the overall manufacturing sector also improved marginally. It was 0.06 in the pre-reform period, and in the post-reform period, it was 0.09 (Kannan and Raveendran, 2009). In sharp contrast to this, Goldar (2009) has shown that the estimated labour demand elasticity during the period from 1970–71 to 1990–91 was 0.41, which was declined to 0.27 in the post-reform period (1991–92 to 2003–04). On the basis of labour demand elasticity estimates, the author has concluded that despite dramatic reduction of the tariff rates and dismantling of quantitative restrictions in the post-reform period, the employment demand elasticity results for the organized manufacturing industrial sector are counter-intuitive. Therefore, it is instructive to analyze the pattern of growth of industrial employment across the manufacturing industries for understanding the employment growth enhancing and employment destroying industries. The employment, value added, output and emoluments growth rates across the four groups of industries based on technological categories over the period of quarter century are presented in Table 8.2.

The perusal of Table 8.2 brings out the fact that in the category of low-tech industries, the majority of industries recorded negative trends of growth in both the pre- and post-reform period (Figure 8.5). This clearly shows that the employment growth-creating industries were small in number compared with employment growth-reducing industries. That is why the relative share of low-tech industries declined more sharply in the post-reform period compared with the pre-reform period. The value added and output growth rates remained not only positive but quite high in the low-tech industries, except wood products and publishing. The high growth rates recorded in the majority of low-tech industries provide evidence enough to argue that low-tech industries were partly responsible for jobless growth in the Indian-organized manufacturing industries. The medium-low-tech industries have not only showed stagnation in the relative shares of employment but also generated employment for half the number of industries, resulting in positive growth rates, and equal number of industries recorded negative employment growth rates (Table 8.2 and Figure 8.6). The medium-high-tech industries have recorded positive employment growth rates in as many as eight industries in the pre-reform period (Figure 8.7). However, the employment growth rates have been positive only in five industries in the post-reform period. The high-tech industries have shown better performance so far as employment growth is concerned (Figure 8.8). But the deceleration of employment growth in this group of industries was quite prominent. The job-creating industries in the category of medium-high-tech industries were outnumbered by the job-destroying industries in the post-reform period.

Table 8.2:
Patterns of growth in employment, value added, output and emoluments across Indian industries

Industry code	Employment growth			Value added			Output growth			Emoluments		
	1980–2005	1980–1992	1992–2005	1980–2005	1980–1992	1992–2005	1980–2005	1980–1992	1992–2005	1980–2005	1980–1992	1992–2005
					Low technology							
151	−1.72	−4.58	−0.47	1.60	2.98	3.10	4.24	5.88	8.65	10.22	18.61	2.96
152	3.60	4.60	1.19	10.19	13.98	11.47	7.53	8.75	6.31	16.06	29.70	4.67
153	0.80	−4.14	1.43	6.41	8.14	7.30	5.86	8.46	10.59	15.04	24.92	5.96
154	0.49	−2.30	−0.43	6.74	12.78	1.05	6.75	10.73	3.15	13.67	27.42	2.59
155	3.66	3.55	4.21	6.91	8.69	3.31	7.80	8.24	7.15	16.16	28.31	8.48
160	1.47	1.40	−0.27	6.66	9.12	6.08	3.21	4.61	2.14	12.01	23.06	2.80
171–181	0.08	−1.43	−0.36	5.55	4.94	3.25	7.53	6.55	6.32	9.85	19.55	1.64
201	−6.95	−9.88	−7.19	−9.21	−2.49	−6.83	−4.26	−0.06	4.36	2.18	14.58	−5.24
202	−0.85	−5.02	−1.44	1.95	8.08	−1.15	3.56	7.66	4.93	12.51	22.15	2.82
210	1.63	0.66	0.77	4.94	6.00	2.81	5.40	6.64	4.78	12.63	22.46	3.80
221–222	−1.49	−1.03	−3.88	0.61	−0.47	−2.73	1.80	1.72	−0.70	10.53	21.26	0.32
361	5.50	−2.30	17.05	14.86	1.08	36.96	16.60	1.83	41.55	20.62	19.93	29.65
Subtotal	0.14	−2.33	−0.16	5.31	6.07	3.00	6.55	7.00	5.80	11.30	21.53	2.58

(Table 8.2 Continued)

(Table 8.2 Continued)

Industry code	Employment growth			Value added			Output growth			Emoluments		
	1980–2005	1980–1992	1992–2005	1980–2005	1980–1992	1992–2005	1980–2005	1980–1992	1992–2005	1980–2005	1980–1992	1992–2005
				Medium-low technology								
182–191	1.85	5.92	−2.81	5.45	11.80	−3.56	6.01	8.78	2.12	12.54	27.46	−0.15
192	−0.83	−3.87	2.79	−3.03	−9.67	3.04	−3.38	−10.38	5.33	4.36	9.45	5.21
251	−0.02	−1.76	−0.17	4.59	1.75	5.16	4.81	2.29	5.12	8.56	16.00	2.90
231	−3.48	−3.83	−3.43	0.39	−3.32	1.43	3.30	3.71	3.46	8.65	18.32	1.60
269	1.20	2.11	1.17	7.33	9.74	5.71	7.78	10.63	6.19	12.45	23.21	4.00
272–273	6.41	9.96	0.18	15.74	23.14	9.91	11.97	13.57	11.32	18.14	33.97	4.86
281–289	1.98	1.56	6.38	6.46	4.02	12.27	7.95	5.50	14.85	13.37	23.25	9.95
351	−11.04	−4.44	−22.03	−6.21	1.10	−16.00	−2.22	2.40	−11.40	−1.10	17.65	−19.47
369	3.56	6.95	−0.24	12.10	12.36	10.20	11.86	15.26	10.86	16.31	31.61	5.94
371–372	−6.77	−0.28	−17.60	−7.24	3.98	−20.72	−2.62	4.37	−15.12	0.74	19.56	−17.69
Subtotal	0.58	0.56	0.08	5.88	5.33	5.97	6.47	5.14	8.00	10.58	20.83	2.66
				Medium-high technology								
241	5.37	8.32	−1.12	15.97	23.79	2.49	16.21	24.22	6.40	20.14	36.06	3.13
252	0.03	−2.35	2.06	6.50	−1.86	12.81	7.16	3.56	9.75	8.52	10.43	9.02
311 + 312	−0.47	1.54	−2.95	5.05	7.55	1.57	6.92	7.73	5.52	10.14	25.21	−2.04
261	−0.95	−1.44	−1.33	6.65	8.59	6.99	7.50	9.11	7.53	12.09	23.68	4.06
290	0.31	1.60	−2.44	6.28	5.44	6.79	7.32	7.21	9.40	11.97	24.09	2.85

271	−1.88	−2.13	−2.63	5.95	−0.60	8.44	6.21	4.42	8.56	10.51	18.36	2.80
319–323	9.79	9.14	3.66	16.08	24.26	6.64	20.65	30.06	12.13	22.64	41.49	7.78
292	0.87	−0.91	3.53	7.97	4.60	12.38	8.76	5.92	12.95	13.18	21.24	8.39
293	−4.86	−2.27	−6.65	0.37	7.13	−1.93	2.43	6.58	2.02	5.38	19.81	−2.06
313	3.01	0.75	7.41	5.95	6.11	2.75	7.12	5.99	9.55	12.87	22.43	8.44
314	3.60	−1.52	13.23	9.69	4.10	19.56	8.93	4.61	21.18	14.57	19.55	16.86
315	−6.83	−6.49	−5.83	−1.44	0.23	4.59	−1.07	0.00	4.92	2.73	11.53	−0.51
341	−4.05	0.57	−11.63	4.53	6.45	0.52	8.72	7.79	8.34	9.29	23.81	−3.33
333	0.26	4.67	−5.65	2.98	8.29	−0.70	4.75	12.15	0.04	12.77	28.84	−0.26
359	3.56	6.95	−0.24	12.10	12.36	10.20	11.86	15.26	10.86	16.31	31.61	5.94
Subtotal	0.39	0.35	−0.25	7.73	5.98	7.41	8.68	8.36	9.58	12.19	22.83	3.93
High technology												
223	−11.62	6.89	−36.61	−7.86	22.87	−37.43	−3.77	28.44	−32.84	−1.02	39.00	−33.85
232	−6.53	−14.74	3.51	11.03	9.18	13.12	13.24	12.87	17.03	10.40	12.45	11.12
242	9.47	14.12	1.93	13.09	29.70	5.73	5.44	12.89	5.74	21.34	41.08	4.50
300	−1.73	−2.78	−4.34	6.94	10.33	2.85	10.21	19.46	2.18	10.61	23.10	4.20
331	2.96	1.13	3.11	10.41	7.81	11.45	11.74	8.17	13.65	15.89	22.42	9.43
332	5.73	3.97	3.93	12.42	2.09	13.47	12.68	5.41	6.83	18.81	26.27	8.21
243	3.70	5.07	−2.15				7.70	8.67	−0.36	18.01	37.07	−0.04
Subtotal	2.88	2.68	0.04	11.32	20.05	6.25	8.76	14.45	9.60	16.18	31.07	3.48

Source: Authors' estimates based on data collected from Annual Survey of Industry (various issues).

Figure 8.5:
Low-technology industrial employment growth rates

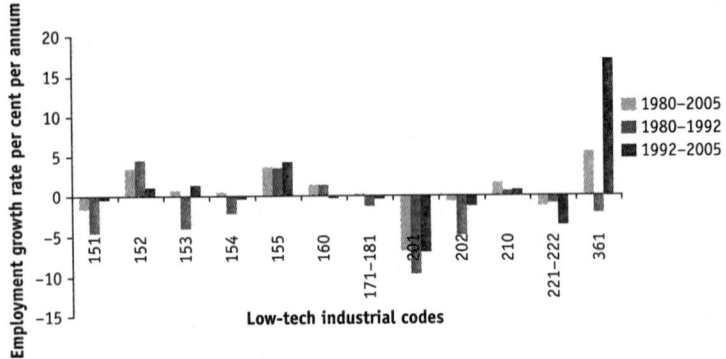

Figure 8.6:
Medium low-technology industrial employment growth rates

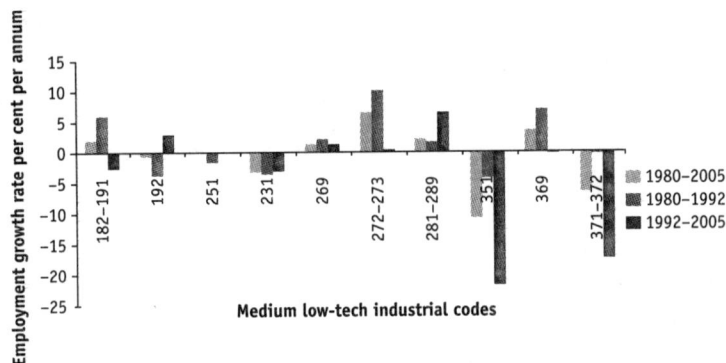

The output (i.e. value added and output) growth rates in the medium-high-tech industries remained quite high compared with the employment growth. The growth rate of emoluments also has shown higher growth rates in the pre-reform period compared with the post-reform period. The deceleration in the growth of emoluments is quite obvious from the pattern of growth of emoluments in the post-reform period (Table 8.2). It is important to note that the high-tech industries have recorded positive employment growth rates in five out of the seven industries in the pre-reform period. Among the five high-tech employment-creating

Figure 8.7:
Medium low-technology industrial employment growth rates

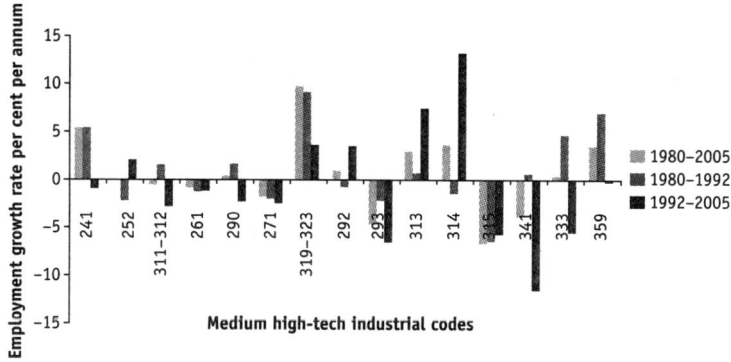

Figure 8.8:
High-technology industrial employment growth rates

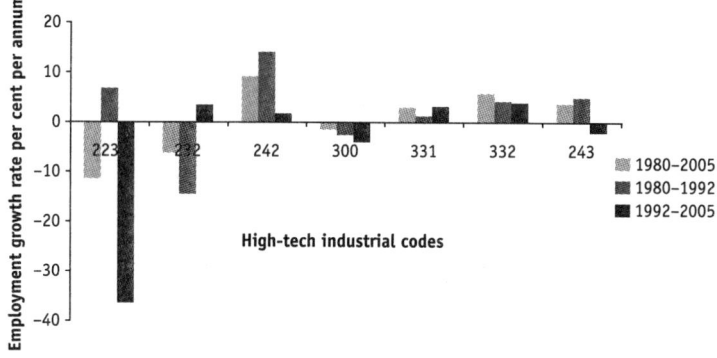

industries, four industries recorded very high growth rates in the pre-reform period. However, in the post-reform period, not only the number of employment-creating industries declined but also the growth rates of job-creating industries observed deceleration (Table 8.2). The high-tech industries have also recorded higher growth rates of value added and output compared to the other industries. The high-tech industries were also high-wage growth industries in the pre-reform period, but deceleration in the growth of emoluments have clearly been occurred in the post-reform period.

Employment Elasticities across the Indian-organized Manufacturing Industries

The employment elasticity helps us in understanding the relationship between employment growth and expansion of output in the manufacturing sector. The low employment elasticity with respect to output signifies that the economic development concentrates in a particular sector and higher growth of manufacturing sector can affect in a limited way the rest of the sectors of the economy (Mazumdar, 2003). On the other hand, higher employment elasticity can generate Hirschman type of linkages with rest of the economy that creates opportunities for economic transformation. The low elasticity of employment results in jobless growth, especially after undertaking economic reforms, and raises serious question with regard to the sharing of benefits of rapid economic development. The enclave-type development reduces opportunities for the labour force in the high-wage sector of the economy and pushes the surplus labour force to find out jobs in the low-wage informal sector of the economy. This process not only generates income gaps but also perpetuates the prevailing disguised unemployment.

The employment elasticities with respect to value added for the Indian manufacturing industries during the period from 1980–81 to 2004–05 are presented in Table 8.3. During the period from 1980–81 to 2004–05, there are four low-tech industries that recorded negative employment elasticity. However, the majority of the low-tech manufacturing industries showed positive but low degree of employment elasticity of output. The values of employment elasticity ranged between 0.01 and 0.53. This shows that capacity to create employment in the low-tech industries during the overall period of analysis remained quite low. When we divide the whole period into two sub-periods, that is, pre-reform and post-reform period, there were eight industries that have observed negative employment elasticities in the pre-reform period. But in the post-reform period, the low-tech manufacturing employment elasticities were recorded negative sign in as many as seven industries. This implies that the employment scenario remained quite grim so far as low-tech manufacturing industries were concerned. It is significant to mention

Table 8.3:
Employment elasticity across organized manufacturing industry groups

Industry code	Low technology			Industry code	Medium-high technology		
	1980–2005	1980–92	1992–2005		1980–2005	1980–92	1992–2005
151	−1.07	−1.54	−0.15	241	0.34	0.35	−0.45
152	0.35	0.33	0.10	252	0.00	1.26*	0.16
153	0.13	−0.51	0.18	311+ 312	−0.09	0.20	−1.88
154	0.07	−0.18	−0.41	261	−0.14	−0.17	−0.19
155	0.53	0.41	1.27	290	0.05	0.29	−0.36
160	0.22	0.15	−0.04	271	−0.32	3.54*	−0.31
171–181	0.01	−0.29	−0.11	319–323	0.61	0.38	0.55
201	0.75*	3.97*	1.05*	292	0.11	−0.20	0.29
202	−0.44	−0.62	1.26*	293	−13.03	−0.32	3.45*
210	0.33	0.11	0.27	313	0.51	0.12	2.70
221–222	−2.43	2.22*	1.42*	314	0.37	−0.37	0.68
361	0.37	−2.12	0.46	315	4.76*	−27.96	−1.27
Subtotal	0.03	−0.38	−0.05	341	−0.89	0.09	−22.39
Medium-low technology				333	0.09	0.56	8.11*
				359	0.29	0.56	−0.02
182–191	0.34	0.50	0.79*	Subtotal	0.05	0.06	−0.03
192	0.27*	0.40*	0.92	High technology			
251	0.00	−1.01	−0.03	223	1.48*	0.30	0.98*
231	−8.90	1.15*	−2.40	232	−0.59	−1.61	0.27
269	0.16	0.22	0.21	233	–	–	–
272–273	0.41	0.43	0.02	242	0.72	0.48	0.34
281–289	0.31	0.39	0.52	300	−0.25	−0.27	−1.52
351	1.78*	−4.04	1.38*	331	0.28	0.14	0.27
369	0.29	0.56	−0.02	332	0.46	1.89	0.29
371–372	0.93*	−0.07	0.85*	243	0.00	0.16	0.03
Subtotal	0.10	0.11	0.01	Total	0.25	0.13	0.01

Source: Authors' estimates based on data collected from Annual Survey of Industry (various issues).

Note: * Represents the negative growth of employment, and value added turns out to be positive elasticity and signifies declining industry.

here that there were at least two low-tech industries that were declining in the pre-reform period, but the number of declining industries increased to three in the post-reform period.

The perusal of the elasticity of employment in medium-low-tech manufacturing industries presented in Table 8.3 shows that there were three declining industries during the overall period under consideration. However, the high negative employment elasticity was noticed in one industry. But the magnitude of positive employment elasticities was quite low during the overall period of analysis. A comparative analysis of pre- and post-reform employment elasticities of the medium-low-tech manufacturing industries clearly brings out the fact that there was a low magnitude of employment elasticities. However, the declining industries increased from two in the pre-reform period to three in the post-reform period. The low employment elasticities of the medium-low-tech manufacturing industries imply that the share of employment has declined contrary to the widely held belief that low-tech industries are less capital intensive and more labour absorbing. It is significant to note that among the medium-high-tech manufacturing industries, the incidence of declining industries was very low. The elasticity of employment in some of the industries was very high in the overall period and also in the two sub-periods.

There were observed wide variations in the employment elasticities in the medium-high-tech manufacturing industries in the pre- and post-reform period. However, there was a rise in employment elasticity in the post-reform period in some industries compared with the pre-reform period. In the pre-reform period, the number of industries with positive employment elasticities was much higher compared with the number of positive employment elasticities industries in the post-reform period. It is important to notice here that this group of industry has increased substantially the share of employment in the overall manufacturing sector of the Indian economy. But this group of industries contained both employment-creating and employment-destroying industries. The variations in estimated employment elasticities in the high-tech manufacturing were quite large (Table 8.3). The employment-creating industries in the high-tech manufacturing sector were more in number compared with the labour-displacing industries. This provided evidence in support of the argument that employment share remained intact in the post-reform period in the high-tech manufacturing industries.

Determinants of Employment Elasticity across the Manufacturing Industries in the Pre- and Post-reform Period

There is a considerable economic literature on the trade-off between wage growth and employment growth (Mazumdar, 2003). In this literature, it has been argued that expansion of employment is constrained by the expansion of output growth. This implies that when employment increases, it has an adverse impact on growth of wages. On the other hand, the rise in wage rate dampens the possibilities of rise in employment. However, this relationship does not work independently from the impact of price rise on wage bill. It needs to be pointed out here that the rise in wage bill does fall behind due to wage-setting rigidities in the face of rise in prices. It is a widely held view that wage setting usually lags far behind the inflationary pressures. The three factors that determine the value of employment elasticity at a given rate of growth of output are the rate of growth of emoluments relative to value of output in current prices, the relative rates of increase in the producer and consumer prices that actually determines the value of emoluments for the labour force and the trade-off between employment expansion and growth in real wages. The decomposition exercise, which segregated the impact of output growth in real wage growth and employment growth along with price effect, has been done to ascertain the actual magnitude of the trade-off between wage growth and employment growth across the Indian manufacturing industries, and the results are presented in Table 8.4.

The analysis of the employment elasticity determinants clearly brings out the fact that price effect is negative in majority of the industries across the board during the period from 1980–81 to 2004–05. However, the intensity of the negative price effect varies widely across the industries and seems to have wiped out moderate growth of output. It is important to note that during the pre-reform period, the price effect was highly positive across the board except few industries and signifies that the output growth was more favourably inclined towards the real wage growth. Thus, given the output growth, the trade-off between real wage growth and employment expansion seems to have been working but varies substantially across the industries. The low-tech industries have lost in terms of real

Table 8.4:
Decomposition of effects on the growth of real wages (1980–81 to 2004–05)

Industry code	1980–81 to 2004–05				1980–81 to 1991–92				1992–93 to 2004–05			
	Real wage growth	Output effect	Employment effect	Price effect	Real wage growth	Output effect	Employment effect	Price effect	Real wage growth	Output effect	Employment effect	Price effect
Low technology												
151	−1.00	−0.01	0.74	−3.63	2.54	−0.43	−1.18	12.60	−0.16	0.06	0.11	1.14
152	−0.96	−0.20	−0.60	−3.13	2.59	0.26	−1.06	10.45	−0.23	−0.21	−1.58	−1.49
153	3.29	0.02	1.42	−2.20	2.92	0.05	6.86	15.22	0.56	−0.30	−0.77	0.23
154	−2.15	−23.02	0.72	27.12	−2.20	−0.22	4.82	12.06	−0.19	−22.53	−1.09	27.01
155	−0.92	0.02	0.28	−3.80	1.51	−0.20	0.78	14.23	−0.01	0.24	0.51	−3.06
160	−0.69	−0.08	−0.04	11.31	0.67	0.07	2.07	29.62	0.19	−0.12	−0.56	15.09
171–181	−0.70	0.01	0.39	−5.26	2.23	0.62	0.70	4.51	0.40	−0.22	−0.41	4.48
201	−2.25	0.09	2.52	−5.66	−10.95	0.10	1.78	−3.21	0.03	−0.89	1.47	1.99
202	−1.99	0.08	2.83	−2.77	−3.44	0.16	1.28	11.66	0.02	0.99	−1.86	−0.44
210	−0.55	−0.10	−0.09	−11.13	3.57	−0.33	0.24	14.14	0.31	0.04	−0.83	−22.84
221–222	−0.56	0.01	−0.52	−11.32	2.64	−0.37	−0.84	15.50	0.04	−0.66	−0.75	−16.14
361	−0.38	0.12	1.86	−3.39	1.47	0.28	1.64	10.10	0.30	−0.18	2.26	7.19
Medium-low technology												
182–191	−0.42	−0.07	−0.92	−3.69	4.55	0.30	1.14	9.29	0.17	−0.28	−0.10	0.67
192	−0.20	−0.10	1.03	−1.07	−3.51	−0.63	−5.55	15.27	0.99	0.39	−0.10	−6.61
251	−0.59	0.11	0.29	−2.58	−1.50	0.02	−0.65	9.11	0.01	−0.16	−1.47	3.33

231	−0.74	−0.20	0.65	1.53	3.52	−0.35	−6.32	15.13	−1.20	−0.22	0.80	5.48
269	−0.57	0.00	−0.37	−3.58	3.34	−0.01	−1.21	12.54	−1.12	−0.04	0.17	−0.61
272–273	5.24	−0.13	4.01	−2.60	21.90	0.22	16.78	14.12	0.67	0.06	−0.01	4.78
281–289	−0.57	−0.23	0.05	−0.74	1.69	1.48	0.90	17.07	−0.71	1.28	26.24	2.30
351	−1.01	0.05	−5.67	−4.47	0.12	6.59	2.01	13.13	0.12	0.32	−0.18	−3.60
369	−0.21	0.11	0.83	−4.26	2.47	0.25	2.71	9.81	0.39	−0.14	−0.08	2.89
371–372	−1.29	−0.60	−17.07	−1.09	0.35	−0.38	−0.91	14.55	0.35	−0.04	−13.48	6.32
Medium-high technology												
241	−0.31	−0.02	−0.96	−4.99	6.27	0.36	3.36	15.95	0.42	−0.04	−2.50	−1.32
252	−28.46	0.12	−20.69	−2.84	−4.76	−0.70	−4.24	12.08	−52.98	−0.13	−33.05	2.82
311 + 312	−3.24	0.04	−0.86	−6.49	1.60	0.31	2.35	1.61	−7.06	−0.08	1.52	2.23
261	−1.23	−0.22	0.40	2.31	0.23	−0.70	3.62	36.09	−0.23	0.03	1.66	1.27
290	−0.27	0.02	−1.64	−2.32	1.71	−0.22	−1.46	12.03	2.51	0.11	−2.13	1.57
271	−0.51	−0.09	−0.11	−2.27	−1.57	−0.68	−3.12	15.31	−2.59	−0.30	−0.64	1.41
319–323	−3.62	0.17	17.17	−5.13	−9.99	0.87	11.03	8.92	−3.05	0.17	4.23	0.95
292	−1.06	0.19	1.46	−2.19	2.36	−0.04	−0.20	12.17	−1.71	0.47	4.87	2.19
293	0.20	−0.03	−0.90	−2.88	2.40	−0.26	−5.36	12.28	4.37	0.32	−1.61	−0.45
313	−2.37	0.52	1.64	−4.79	1.65	−0.03	0.32	12.31	−4.97	1.89	5.27	−6.65
314	−0.34	0.62	2.91	−2.68	−0.15	−0.40	1.01	17.59	1.71	1.95	7.04	1.74

(Table 8.4 Continued)

(Table 8.4 Continued)

Industry code	1980–81 to 2004–05				1980–81 to 1991–92				1992–93 to 2004–05			
	Real wage growth	Output effect	Employment effect	Price effect	Real wage growth	Output effect	Employment effect	Price effect	Real wage growth	Output effect	Employment effect	Price effect
315	0.59	0.42	0.99	−3.08	0.90	−0.83	−11.92	12.77	3.57	1.89	4.14	1.69
341	−0.35	−0.36	−2.54	−2.69	1.76	−0.42	−0.66	11.45	0.84	0.00	−0.65	1.06
333	−1.37	−0.07	−1.88	−3.37	1.13	0.12	1.12	10.94	−2.92	−0.70	−2.90	1.55
359	−0.30	0.12	−1.35	−2.61	2.69	−1.16	−0.68	20.55	1.41	0.05	−1.84	0.58
					High technology							
223	−2.95	−0.46	−63.15	−4.15	−9.71	0.77	108.77	8.94	0.19	−0.41	−10.16	1.44
232	−0.89	0.05	4.27	4.29	5.87	4.10	−29.31	−43.92	0.47	−0.11	−1.54	0.87
233												
242	−0.74	−0.19	−0.72	−2.25	1.51	0.03	3.68	7.12	0.72	0.03	−0.78	1.85
300	−3.71	0.20	10.72	−3.41	−9.83	0.01	65.57	10.47	−2.86	0.43	8.74	3.21
331	−0.05	0.05	−0.01	−2.52	3.60	−0.06	−1.70	12.85	−0.34	−0.37	−2.20	2.88
332	0.02	−0.22	−0.29	2.20	9.29	0.16	3.17	30.15	1.26	−0.06	−0.26	1.69
243	−0.84	−0.25	−1.42	−2.79	7.18	0.53	−0.66	13.98	0.88	−0.47	−2.09	3.28

Source: Authors' estimates based on data collected from Annual Survey of Industry (various issues).

wage growth. The medium-high-tech industries have gained in terms of expansion of employment during the pre-liberalization period. The analysis of the decomposition exercise in the post-liberalization period shows that there were wide variations of price effect across the industries. The negative price effect was substantial in the case of low-tech industries. The medium-high-tech and high-tech industries recorded positive price effect, except three industries in the medium-high-tech industries where price effect turns out to be negative (Table 8.4).

The distribution of output growth among the medium-high-tech industries more favourably inclines towards real wage growth, and employment effects largely turn out to be negative. However, the fall in the real wage growth in the medium-high-tech industries seems to have positive employment effects. This shows that there has occurred a trade-off between real wage growth and employment expansion. In the case of high-tech industries, the magnitude of the trade-off differs across the industries. But the real wage growth has positive gains in the majority of industries, except two industries where real wage growth has actually declined. It needs to be mentioned here that the moderation of the trade-off between real wage growth and employment has been done by the other factors such as price effects. Thus, the decomposition procedure adopted has allowed us to quantify the relative importance of the factors determining the share of wages and price effects and enables us in understanding the labour market outcomes. There are other alternative factors in operation that has played a significant role in deviating the interests of employment growth and real wage growth. Therefore, it is important to understand employment outcomes of the economic growth process beyond the inevitable trade-off between employment growth and wage growth explanation.

Industrial Technological Capability, Industrial Structure and Employment

It is a widely recognized fact that industrial development and technological capabilities are highly correlated. The evolution of industrial structure in the developed countries shows that innovative capabilities have played an important role in stimulating change in the industrial structure. Employment outcomes of industrial development have remained highly dependent on the technological capabilities. It has been argued in the literature on economics of innovation that product innovations are employment creating, but the process innovations are employment destroying (Edquist et al., 2001). The net increase or decrease in employment outcomes of industrial development will largely be determined by the relative strength of the effects of product versus process innovations. It is important to note here that the technological capabilities of most of the developing countries are either very weak or related to adaptation of the innovations generated in the developed countries. Therefore, the technological capabilities of the developing countries are generally related to the process innovations and improvements in the technologies imported from the developed countries. Consequently, the employment implications of industrial development of the developing countries are quite dismal.

According to the United Nations Conference on Trade and Development (UNCTAD) innovation capability index, which consists of technological activity index and human capability index, India has been ranked among the low technological capability developing countries. In the year 1995, the ranking of India with regard to innovation capability index was 81 with the index score point 0.287. However, India's global position declined to 83 with score point 0.285 in the year 2001 (UNCTAD, 2005). It is pertinent to point out here that among the Indian industries, only pharmaceutical and information and communications technology (ICT) industries have been considered to be possessing substantial technological capabilities, but the majority of industries are having weak technological capabilities and are highly dependent on technology imports (Ray, 2009). Therefore, the relationship between technological capabilities and employment growth of the industrial development is expected to be quite weak. The estimates of this relationship are presented in Table 8.5.

The analysis of Table 8.5 reveals that the coefficient of research and development expenditure turns out to be negative but non-significant during the period from 1980–81 to 2004–05. However, the elasticity of technological capability with respect to employment was negative but statistically significant for both the pre- and post-reform period. This implies that technological capability, measured in terms of research and development expenditure incurred, has shown dramatic decline during the period of liberalization. The output and employment elasticity have been positive and significant, but the magnitude was quite low. This may

Table 8.5:
Estimated fixed effects models (dependent variable is log of employment)

Name of the variables	1980–81 to 2004–05	1980–81 to 1991–92	1992–93 to 2004–05
Log of R&D expenditure	−0.019	−0.240*	−0.180*
	(0.651)	(−3.60)	(−3.64)
Log of value added	0.014*	0.062***	0.0005
	(3.25)	(1.35)	(−0.004)
Log of wages	0.068**	0.263*	0.619*
	(1.49)	(4.212)	(2.804)
AIC	−2.675	−3.165	−3.330
Adjusted R^2	0.961	0.982	0.976
Autocorrelation	0.000	0.000	0.000

Source: Authors' estimates based on data collected from Annual Survey of Industry (various issues) and R&D expenditure, Department of Science and Technology, Government of India, New Delhi (various issues).

Note: * Represents statistically significant at 1 per cent level; ** represents statistically significant at 15 per cent level; *** represents statistically significant at 20 per cent level.

be because of two reasons. One, the workers in the organized sector, and more specifically working in the large and public sector enterprises, have a substantial bargaining power. Two, during the liberalization period, the private sector instead of hiring new workers remained dependent on using existing workforce more intensively. This has led to a rise in the real wage rate in the organized industrial sector of the Indian economy (Nagraj, 2006). However, the value of the magnitude has declined substantially in the post-reform period and also turned insignificant. The relationship between employment and output turned out to be positive and significant during the period of analysis (Table 8.5). This implies that the aggregate demand plays an important role to increase employment in the manufacturing sector of the Indian economy and thus justify the Keynesian explanation holds true rather than neoclassical one.

To ascertain the relationship between the technological capability index[1] and employment growth, we have estimated the correlation across the industries classified on the basis of technological characteristics and are presented in Table 8.6. The magnitude of the correlation coefficients across the technology-based industrial groups between technological

Table 8.6:
Correlation coefficients between technology capability index and growth rate of employment (1992–93 to 2004–05)

Name of industrial groups	Correlation coefficient
Low-technology industries	0.21
Medium-technology industries	–0.35
Medium-high technology industries	–0.40
High-technology industries	–0.29
All industries	–0.21

capability index and employment growth is quite small. The coefficient of correlation turns out to be positive in the case of low-technology industries. This implies that there is a positive connection between technological capabilities and employment growth. But the non-significant coefficient could not allow us to ascertain definite conclusions. However, the direction of the correlation indicates a positive relationship.

The correlation across the three industrial categories, that is, medium-technology, medium-high-technology and high-technology industries, between employment growth rates and technological capability index, turns out to be negative, and the magnitude of the correlations was very low. It needs to be noted here that Indian industrial technological capabilities remained quite weak; therefore, implications of this relationship for employment need to be interpreted with some caution. The South Korean industrial technological capability development experience is quite instructive. As the technological capabilities have increased at a faster rate during the period of 1980s and 1990s in the South Korean industrial sector (Lee, 2009), the employment growth slowed down in the 1980s and turned negative in the 1990s (Singh, 2004).

Concluding Remarks

In this chapter, an attempt has been made to analyze the long-term trends of employment growth across the Indian manufacturing industries classified on the basis of technological categories over the period of quarter century. The phenomenon of jobless growth of the organized manufacturing has put to test at disaggregative level, and for this purpose, a consistent data set for 44 three-digit industries based on ASI have been constructed. The analysis of the changing structure of industrial employment brings out the fact that low-tech industries have shown signs of fatigue, and the majority of low-tech industries have lost their relative importance in the industrial economy of India. The changing pattern of employment structure has allowed us to identify the medium-high-tech industries, which have shown dynamism in terms of increasing their relative share in total employment of the organized manufacturing industries. However, the relative shares of employment in the medium-low-tech and high-tech industries have remained stable over the quarter century.

The pattern of employment growth has shown that some industries have generated employment and others have destroyed jobs. The employment elasticities across the technological groups of industries have shown wide variations across the industries. The positive employment elasticities have shown the ability of the industries to create new jobs, but the majority of the industries have shown negative elasticities, implying jobless growth. In the post-reform period, estimates of employment elasticities across the industries have shown rise in the number of industries recorded positive employment elasticities compared with pre-reform period. These results imply that the employment-creating industries in the high-tech manufacturing sector were more in number compared with the labour-displacing industries. The changing structure of employment elasticities underlined the importance of emerging dynamic high-tech industries in the Indian industrial sector. The labour market regulation view put forward by various scholars supporting the liberalization policies could not stand the scrutiny of clear demarcation among job-creating and job-destroying industries under the same circumstances over the quarter century period examined here.

The decomposition of the determinants of the employment elasticity procedure clearly brings out the fact that the labour market outcomes have shown the importance of the factors such as wage shares and price effects in leaning out the interests of employment growth and wage growth. The relationship between industrial technological capabilities and employment growth turns out to be ambiguous. This implies that weak technological capabilities adversely affect the employment growth, and heavy dependence on imported technological know-how from the developed countries is labour displacing. Therefore, it is suggested that developing countries should invest in both institutions and industrial firms to develop technological development that suits to resource endowment, specificities of local conditions and the stage of industrial development. There is a dire need to explore alternative paths of industrial and technological capability development to sustain economic transformation process for achieving prosperity and reducing the time for catch-up development.

Appendix

Names of industries classified on the basis of technology characteristics and industrial codes

Low technology	Industry code	Medium-high technology	Industry code
Production, processing and preservation of meat, fish, fruit vegetables, oil and fats	151	Manufacture of basic chemicals, fertilizers and nitrogen compounds	241
Manufacture of dairy product	152	Manufacture of plastic products	252
Manufacture of grain mill products, etc., and animal feeds	153	Manufacture of electric motors, generators, transformers and control apparatus	311 + 312
Manufacture of other food products	154	Manufacture of glass and glass products	261
Manufacture of beverages	155	Machinery	290
Manufacture of tobacco products	160	Metal	271
Spinning, weaving and finishing of textile + other textiles + knitted and crocheted fabrics and articles, wearing apparel, except fur apparel and tailoring	171–181	TV, radio and video	319–323
Saw milling and planning of wood	201	Special-purpose machinery	292
Manufacture of products of wood, cork, straw and plaiting materials	202	Domestic appliances	293

(Table Continued)

(Table Continued)

Low technology	Industry code	Medium-high technology	Industry code
Manufacture of paper and paper products	210	Insulated wire and cables	313
Publishing and printing and service activities related to printing	221–222	Manufacture of accumulator, primary cells and battery	314
Manufacture of furniture	361	Manufacture of electrical lamps and lighting equipment	315
Medium-low technology	**Industry code**	Manufacture of motor vehicles	341
Dressing and dyeing of fur, manufacture of articles of fur, tanning and dressing of leather, manufacture of luggage handbags, saddlery and harness	182–191	Manufacture of watches and clocks	333
Manufacture of footwear	192	Manufacture of transport equipment n.e.c. (not elsewhere classified)	359
Manufacture rubber products	251	**High technology**	**Industry code**
Manufacture of coke oven products	231	Reproduction of recorded media	223
Manufacture of non-metallic mineral products n.e.c.	269	Manufacture of refined petroleum products	232
Manufacture of basic precious and non-ferrous metals and casting of metals	272–273	Manufacture of pesticides, paints, varnishes and similar coating and pharmaceuticals and other chemicals	242
Manufacture of structural metal products, tanks, reservoirs and steam generators	281–289	Manufacture of office, accounting and computing machinery	300
Building and repairing of ships and rails	351	Manufacture of medical instruments	331
Other manufacturing	369	Manufacture of optical and photography instruments	332
Recycling of metal waste and scrap and recycling of non-metal waste and scrap	371–372	Manufacture of man-made fibres	243

Note

1. Technological capability index for the three-digit manufacturing industries is taken from Pradhan and Puttaswamaiah (2008). The authors have constructed composite technology index using principal component approach while using four indicators, that is, R&D intensity, foreign disembodied technology intensity, foreign capital goods intensity and domestic capital goods intensity. The data for indicators of technology used to develop composite technology index was based on the Prowess database of the Centre for Monitoring Indian Economy (see for a detail, Pradhan and Puttaswamaiah, 2008).

References

Amsden, A.H. 2004. 'Import substitution in high-tech industries: Prebisch lives in Asia!', *CEPAL Review*, 82(April): 75–89.

Amsden, A.H. and R. Van den Hoever. 1996. 'Manufacturing output, employment and real wages in the 1980s: Labour's loss until the century's end', *Journal of Development Studies*, 32(4).

Audretsch, D.B. and A.R. Thurik. 1999. *Innovation, Industrial Evolution and Employment*. Cambridge: Cambridge University Press.

Commission of the European Communities. 1994. *Growth, Competitiveness, Employment: The Challenges and Way Forward into the 21st Century—White Paper*. Luxembourg: Office for Official Publications of the European Communities.

Edquist, C., L. Hommen and M. McKelvey. 2001. *Innovation and Employment: Process versus Product Innovation*. Cheltenham: Edward Elgar.

Fallon, P.R. and R.E.B. Lucas. 1993. 'Job security regulations and the dynamic demand for industrial labour in India and Zimbabwe', *Journal of Development Economics*, 40: 241–275.

Goldar, B.N. 2009. 'Trade liberalisation and labour demand elasticity in Indian manufacturing', *Economic and Political Weekly*, XLIV(34): 51–58.

Hasan, R., D. Mitra and K. Ramaswamy. 2007. 'Trade reforms, labour regulations and labour-demand elasticities: Empirical evidence from India', *Review of Economics and Statistics*, 89(3): 466–81.

Kannan, K.P. and G. Raveendran. 2009. 'Growth sans employment: A quarter century of jobless growth in India's organised manufacturing', *Economic and Political Weekly*, XLIV(10): 80–91.

Lee, K. 2009. 'How can Korea be a role model for catch-up development? A capability-based view', Research Paper No. 2009/34, World Institute for Development Economic Research, United Nations University, Helsinki.

Mazumdar, D. 2003. 'Trends in employment and the employment elasticity in manufacturing, 1971–92: An international comparison', *Cambridge Journal of Economics*, 27: 563–82.

Morawetz, D. 1974. 'Employment implications of industrialisation in developing countries: A survey', *Economic Journal*, 84(335): 491–542.

Nagraj, R. 2000. 'Organised manufacturing employment', *Economic and Political Weekly*, 35(38).

Nagraj, R. 2006. *Aspect of India's Economic Growth and Reforms*. New Delhi: Academic Foundation.
Papola, T.S. 1991. 'Industry and Employment: Recent Indian Experience', ISID Foundation Day Lecture, Institute for Studies in Industrial Development, New Delhi.
────── 1994. 'Structural adjustment, labour market flexibility and employment', *Indian Journal of Labour Economics*, 37(1).
────── 2009. 'India: Growing fast. But also needs to industrialise!', *Indian Journal of Labour Economics*, 52(1): 57–70.
Pradhan, J.P. and S. Puttaswamaiah. 2008. 'Trends and patterns of technology acquisition in Indian organised manufacturing: An inter-industry exploration', *Indian Journal of Economics*, XXXIX(353): 269–313.
Ray, A.S. 2009. 'Emerging through technology capabilities? Analysis of India's technological trajectory', in Manmohan Agarwal (ed.) *India's Economic Future: Education, Technology, Energy and Environment*. New Delhi: Social Science Press.
Rifkin, J. 1995. *The End of Work*. New York: G.P. Putnam's Sons.
Sen, K. 2008. 'International trade and manufacturing employment outcomes in India: A comparative study', Research Paper No. 2008/87, UNU-WIDER, Helsinki, Finland.
Singh, L. 2004. 'Technological progress, structural change and productivity growth in the manufacturing sector of South Korea', *World Review of Science, Technology and Sustainable Development*, 1(1): 37–49.
Singh, L. and A. Gill. 2002. 'Post-reform employment growth in the organised manufacturing sector of India: A disaggregative analysis', *Manpower Journal*, 30, (1): 87–98.
Szirmai, A. 2009. 'Industrialisation as an Engine of Growth in Developing Countries 1950-2005', UNU-MERIT Working Paper No. 2009-010, Maastricht Economic and Social Research and Training Centre on Innovation and Technology, United Nations University, Maastricht, The Netherlands.
UNCTAD. 2005. *World Investment Report: Transnational Corporations and the Internationalization of R&D*. New York: United Nations.
Van der Hoeven, R. and L. Taylor. 2000. 'Introduction: Structural adjustment, labour markets and employment—Some considerations for sensible people', *Journal of Development Studies*, 36(4).

9

Intellectual Property Protection, Innovation and Medicines: Lessons from the Indian Pharmaceutical Industry

Dinesh Abrol

Introduction

This chapter examines the Indian experience of innovation making during the post-Trade-Related Aspects of Intellectual Property Rights (TRIPS) Agreement period under the influence of strong intellectual property rights (IPRs) system in the pharmaceutical sector. It is largely an investigation undertaken with a view to connect the experience of implementation of strong IPR system in the pharmaceutical sector in the presence of the ingredient of open trade in the policy on technology strategy with the relevant IPR writings that Professor Robert E. Evenson did in the period of last two decades (Evenson et al., 1989; Evenson and Singh, 1997; Evenson, 2001, 2003). The aim is to draw ultimately policy-related lessons for the development of an optimal intellectual property (IP) protection system in the case of emerging economies to take care of the impact of strong IPRs on public health.

Analysis of the discourse on post-TRIPS period Indian experience of pharmaceutical innovation would have greatly interested Professor Robert E. Evenson. At the time of General Agreement on Tariffs and Trade (GATT) negotiations when the debate on international harmonization of trade-related aspects of IP protection was still hot and kicking, Evenson was actively writing on the issue of how the design of IPR policy for the developing countries should be tackled. The economic impact of strengthening of IPRs in emerging economies still remains a controversial issue. Recent papers have analyzed the impact of stronger IPRs and of the TRIPS Agreement and find mixed evidence. There is no clear-cut evidence that strong IPRs have a positive impact on domestic innovation, trade activities and foreign direct investments (FDIs) in emerging countries (Basheer, 2005; Delgado et al., 2008; McCalman, 2001; Montobbio et al., 2010; Qian, 2008; UNCTAD-ICSTD, 2005).

Today the post-TRIPS system of patent protection, as adopted by India, is again the subject of intense debate, particularly as it has impact on public health and access to essential medicine in the developing world. In 2008, the member countries came together in the World Health Organization (WHO) forum to adopt a resolution entitled 'Global Strategy and Plan of Action on Public Health, Innovation and Intellectual Property' that recognized that to achieve the goals of public health, the world community cannot rely on the system of strong IPRs and will have to promote the model of open-source drug discovery (OSDD). In the case of developing countries, Evenson was concerned about the absence and presence of R&D and innovation (Evenson and Westphal, 1995). The issue of directions of R&D and innovation was not so important for him. Evenson did not ever directly engage in his scholarly writings with the theme of the connection between strong IP system and pharmaceutical innovation, particularly in respect of the impact on public health, but the writings he did on the subject of IPR policy during the period of GATT negotiations certainly contributed to the building of public understanding on the issue of IPR protection in general and pharmaceutical patents as they exist today in India.

Speaking about the work of Evenson, in terms of his world view, he had a distinct viewpoint on the role of IPRs. His view can be characterized as a position favouring the introduction of strong IPRs, which need only to be tempered by the stage of development of the country's technological capabilities. Recently in a lecture Evenson delivered in 2003 while speaking on Africa, he articulated himself again in favour of the same policy in Africa.

Kanwar and Evenson (2001, 2003) empirically examined again the same question of connection between strong IP system and innovation in their scholarly contribution entitled 'Does Intellectual Property Protection Spur Technological Change?'. They wrote, 'Given fairly recent changes in the international policy ethos where a regime of stronger IP protection has become a fait accompli for most developing countries it is of some significance to ask whether more stringent protection of IP does indeed encourage innovation'. After the examination of cross-country panel data gathered by them on R&D investment, patent protection and other country-specific characteristics spanning the period 1981–90, once again their conclusion was that 'Evidence unambiguously indicates the significance of intellectual property rights as incentives for spurring innovation'. The econometric modelling approach used by Evenson and his colleagues in the case of working out the impact of strong IPRs as such has also been evaluated and the method and conclusions have been commented upon and questioned[1] (Beattie, 2006; Hall, 2007).

However, for Evenson, it would also be appropriate to note that the issue of optimal design of patent system remained of key interest. Evenson was opposed to the developed countries imposing an unequal agreement on IPRs in the case of developing world. In the view of Evenson, 'the economic case for international agreements to recognize the rights of foreigners by treating them on a par with nationals is only clear for countries at similar levels of technological development'. When the negotiations were taking place between member countries on the issue of international harmonization of standards and norms of IP protection, Professor Robert E. Evenson was in favour of the adoption of the principle of special and differential treatment. Evenson believed that this kind of understanding would be sufficient to enable the developing countries to design the system of IP protection in a balanced manner as per their own requirements in consonance with their specific stage of development of technological capabilities.

Technology Strategy and Developing Countries

The ideas that Evenson deployed to discuss the challenge of formulation of technology strategy for the developing world are described here below in brief to allow us to define the context of our own analysis of the experience of innovation making and strong IP system in the case of Indian pharmaceutical industry. Evenson and Westphal (1995) discuss the issue of achievement of balance in the system of IP for the developing world in their seminal contribution entitled 'Technological Change and Technology Strategy', which was published as a chapter in the *Handbook of Development Economics*, Volume IIIa, ed. J. Behrman and T.N. Srinivasan, pp. 2209–99, North Holland, Amsterdam. In this article, Evenson explicitly states his own understanding on the connection between innovation making and IP protection in the following way: 'Stronger intellectual property rights would be a powerful instrument for encouraging many forms of investment at all levels of technological development if they are sufficiently focused on promoting those forms of investment that are respectively important at each level'. His own position, in the case of a developing country like India which he, like many others, saw advancing towards NIC-hood, was that India has a weak system of IPR protection and would need to adopt a stronger system of IP protection. Evenson and Westphal (1995) suggested that India, defined as a level 2b country, had a weak IPR system and this system reflected the previous dominance of international concerns to the detriment of domestic interests.

Consistent with conventional wisdom, in this chapter Evenson also took a view that as the developing countries needed to encourage FDI and technology transfer, it would be better for them to strengthen the system of IP. Of course, Evenson also noted in the case of emerging economies that the imposition of sufficient penalties can now seemingly be considered a fact of life. Further, Evenson also held that 'the recognition of foreign IPRs is only half of what is needed' (Evenson and Westphal, 1995). Evenson and Westphal (1995) were clear that the developing countries also need to recognize the importance of IPRs in stimulating domestic inventive effort and refashion the IPR systems. More imagination than has previously been given to their design is clearly in order (Evenson and Singh, 1997).[2] Of course, in order to exemplify the gains to creativity in this area, in this seminal contribution entitled 'Technological Change and Technology Strategy', they suggested to the developing countries to adopt the less strong systems of breeder rights and utility models. Their rationale was that such kind of IPR systems would be able to stimulate the kinds of minor, adaptive inventions that are important in the early to middle phases of technological development. Their understanding was that positive technological distance between advanced and developing countries is often optimally overcome by adaptive technological effort. In order to capture the diversity of adaptive efforts based on formalized R&D capabilities, they distinguish minor adaptations that involved changes in the technique from inventive adaptations, making use not of the technique but of the knowledge that underlies it. Evenson and Westphal (1995) even wrote, 'As a general rule, minor adaptations cannot overcome great technological distances, only inventive adaptation has the potential of doing so'.

But at the same time, as we see that even the limited requirement of inventive adaptation also did not require him to discuss the challenge of formulation of technology strategy for the development of new drugs. As we show later, the problem of catching up in the pharmaceutical sector, in which formal technology transfer to developing countries for manufacturing medicines, and particularly active ingredients, is scant or non-existent, is complex. It needs the emerging economies to get the directions of R&D and innovation right to encourage strategic collaboration with advanced countries for technology transfer and greenfield FDI. The challenge of how the system of strong IP protection would be able to tackle the problem of coordination of multiple property rights in the pharmaceutical sector remained unaddressed in the writings of Evenson.

In both, agriculture and pharmaceuticals, the model of innovation is rapidly changing. Important sectoral differences that had to be recognized were not appropriately examined in the writings. Notwithstanding the order of difficulty involved in the case of drug discovery work being high, the regulatory challenge of clinical development too makes the pharmaceutical sector quite unique. In the pharmaceutical sector, the processes of search even demand a higher level of coordination among the multiple agents. Today in the pharmaceutical sector, the processes of search include the use of tools of rational design and genomics-based drug discovery involving multiple technologies and disciplines and different types of economic agents as well. New set of actors in private sector (e.g. CROs, clinical research organization) have come to populate the innovation-making space. Public–private partnerships (PPPs) for early stages of drug discovery work are common. Further, Evenson also chose to ignore the issue of market power that global pharmaceutical firms could exercise when the developing world would try entering the area of pharmaceutical innovation.

Again consistent with conventional wisdom, Evenson took the view that the much needed activities of adaptive search for innovation in the case of industrial inventions was mostly the forte of private sector firms, and therefore the system of strong IPRs would alone be appropriate for industrializing countries. Evenson took into account mainly the requirements of adaptive research where even by practising discrete model of innovation making, individual private sector firms could have easily succeeded as inventors. Further, by failing to confront the important issue of how, in spite of being technologically weak, the private sector in an emerging economy can possess greater influence on the directions of public sector R&D and, consequently, impact in a big way on the achievement of appropriate systemic character in respect of innovation making, for example, domestic firms being incentivized to develop inventive activity for home market when there also exist opportunities for higher profit-making in the regulated markets. Evenson underestimated how important can ultimately be the role of directions of inventive activities under perusal in the domestic and foreign firms. All these questions remained unattended in the work of Evenson. Being largely engaged with agricultural innovation, there was an insufficient attention from his side to the nature of challenge posed in respect of the formulation of technology strategy in the case of pharmaceutical sector.

Evenson's pro-strong IP stance remained unshaken even when they came across contrary evidence arising out of the survey research that had given only weak support to the proposition that patent protection stimulates R&D. They wrote, 'Much of it is attitudinal evidence comparing patents with incentives in settings where patent protection has typically been available for long periods. Furthermore, the evidence does not generally address the question of what would happen if the patent system were eliminated. Clearer, stronger evidence of its importance is found in the fact that all developed countries have been strengthening their own patent systems over time'.

Today, scholars worldwide are far more clued into the problems of building of innovation systems in the emerging scenario where innovation making in almost every sector is far more embedded in the world of technological convergence involving the interplay of different types of information technologies and biotechnologies and needs the IP system to address the problem of coordination of multiple economic agents. It is not therefore completely surprising that several scholars are now beginning to focus on the theme of how the system of patents as it exists in developed countries also urgently needs a major reform. Recently, Richard Nelson, George Blumenthal Professor of International and Public Affairs, Business, and Law, Columbia University, suggested in his keynote address that in developing their own patent law, developing countries need to recognize that patent law and practice, in some cases, overshot, and need to be reformed. Developed countries have made serious mistakes committed in constituting over-protectionist IP regimes, and these mistakes should not be committed in the case of developing countries. The above-said view is the burden of the recent National Academy of Sciences/National Research Council (NAS/NRC) report on 'A Patent System for the 21st Century'. Therefore, recently, again many scholars have shifted their attention to the theme of optimal design of system of protection of IPRs.[3] Although the emerging literature is still not ready to conclude on the specifics of the reform needed in respect of IPR system, but they are advancing towards the position that TRIPS as it stands is against the interests of developing countries, and needs reform (Correa, 2008; Reichman, 2009).

Reichman (2009) stated this concern explicitly, 'Developing countries, particularly the BRIC countries of Brazil, Russia, India, and China, should accommodate their national systems of innovation to the worldwide intellectual property (IP) regime emerging after the adoption of the Agreement on Trade-Related Aspects of Intellectual Property Rights in a way that maximizes global welfare in the foreseeable future'. Further, continuing with the theme, Reichman clearly states that the challenge is for emerging economies to capture the benefits of IP without importing the serious problems that developed countries have themselves failed to solve. Emerging economies can attain this balance by pursuing a policy of counter-harmonization in which they take advantage of existing exemptions in international agreements governing IP to establish regional, local and international practices that promote more innovative, flexible uses of IP. Such practices include a research exemption for experimental uses of IP: government-imposed non-exclusive licensing, anti-blocking provisions, an essential facility doctrine and compulsory licenses. Additional tools include an ex ante regime of compensatory liability rules for small-scale innovation and sensible exceptions, particularly for science as well as general fair use provisions.

It is time to review the Indian experience of post-TRIPS changes in Indian pharmaceutical industry. Almost one and a half decade has now passed; the first change in Indian Patent Act, 1970 was made in 1995 through an ordinance issued by the government on the subject of exclusive marketing rights (EMRs). The second amendment was made from 2000 to 2002. This amendment was introduced to make the patent legislation compatible with the TRIPS Agreement signed by the government as a part of the larger World Trade Organization (WTO) package in 1994.

The third and last amendment to the Indian patent legislation made the product patent in pharmaceuticals enforceable from 2005. Although by making these changes the country has already fulfilled her international obligations adequately, there also exists scope for making those changes that its own specific circumstances and ambition required and are yet to be undertaken.

However, before we even go into the details of the issue of how what kind of changes are feasible and needed to encourage the domestic and foreign companies to undertake innovation making in directions favouring public health, we would like to critically examine the experience of the post-TRIPS technology behaviour of the domestic and foreign pharmaceutical companies on the basis of evidence gathered on the quality of FDI, technology transfer and overseas R&D and on the directions of R&D and innovation along with their implications for the future of domestic pharmaceutical companies in the Indian home market.

Revisiting the Making of IPR Policy on the Basis of Post-TRIPS Evidence

Contrary to the predictions of the above-discussed academic viewpoint of Evenson and many others like him who believed in the beneficial impact of strong IPRs, evidence of the post-TRIPS behaviour of Indian pharmaceutical industry confirms the apprehensions of all those groups who suggested that there also exist other options before the firms of developed countries, namely, increased imports rather than technology transfer, expansion of market power by them for increased control of the Indian market by using FDI and overseas R&D strategically, and were sceptical of the positive impacts accruing from the introduction of TRIPS for the development process and the realization of social welfare of the people in India. As far as the question of increase in exports of off-patent drugs and pharmaceuticals to regulated markets is concerned, this benefit to the domestic firms is not a result of the acceptance of product patent and strong IPR system. It is a result of the space available for generic business in the US and European markets after the introduction of Hatch–Waxman Act of 1984, and the capabilities made possible by the pre-TRIPS industrial policy regime.

Analysis of the experience of Indian pharmaceutical industry during the period of last 15 years indicates that the system of strong IPRs has not been able to favour India with the benefits claimed in respect of increased access to good-quality FDI, technology transfer, overseas product R&D and stimulation of domestic investment in R&D for product innovation for local needs. Analysis also shows that how, as of today, both domestic and foreign pharmaceutical firms do not have any big plans to invest in R&D on the development of medicines related to local needs of India, which is an issue of directions of R&D and innovation stimulated in the case of domestic pharmaceutical firms. But their incorporation into the emerging international division of labour within the global pharmaceutical industry is already on the horizon. And this is also leading the existing clusters of pharmaceutical production and the linked innovation systems to move even further away from the goal of development of medicines needed by developing countries for the improvement of the health conditions of their own people.

Foreign Direct Investment

As far as the benefit of rise in the quantity and quality of FDI is concerned, contrary to the expectations of pro-TRIPS policymakers, the performance of pharmaceutical sector has been the worst among the sectors expected to be positively impacted in terms of FDI inflows on account of the acceptance of strong IPRs. The total amount of FDI received in the pharmaceutical sector is US$1,458.78 million. It happens to be a meagre 1.69 per cent of the total FDI attracted by India during the period 2000–09. This amount puts the sector at 14th rank as compared to the sectors that have been receiving FDI. Compared to the services sector, which received US$19,828.95 million, that is, 22.96 per cent of the total, the amount received by the pharmaceutical sector is certainly paltry. A bigger time frame of 15 years from 1991 to 2005 indicates that compared to the other sectors, the FDI in pharmaceuticals has been not only lagging but also falling in rank. Over the period of these 15 years, the amount of FDI received by this sector amounts to US$936 million compared to the amount received by the electrical equipment(s) sector, which is US$4,266 million for the same time period, and the amount is ranked eighth. Its

Table 9.1:
Statement on sector-wise FDI inflows from April 2000 to April 2009

S. no.	Sector	Amount of FDI inflows (in millions)	% Age of total FDI inflows
1	Services sector	19,828.95	22.96
2	Computer software and hardware	9,019.59	10.46
3	Telecommunications	6,956.34	8.21
4	Housing and real estate (including cineplex, multiplex, integrated townships and commercial complexes, etc.)	5,872.98	6.72
5	Construction activities	5,327.13	6.00
6	Automobile industry	3,434.86	4.02
7	Power	3,227.28	3.73
8	Metallurgical industries	2,732.49	3.03
9	Petroleum and natural gas	2,565.29	2.90
10	Chemicals (other than fertilizers)	2,153.76	2.54
11	Cement and gypsum products	1,674.89	1.93
12	Electrical equipment	1,519.45	1.76
13	Trading	1,561.45	1.75
14	Drugs and pharmaceuticals	1,458.78	1.69
15	Ports	1,558.75	1.67
	Total	92,157.54	100.00

Source: Fact Sheet on FDI, August 1991 to April 2009, Annex B, Pages 8–10, Department of Industrial Promotion and Policy, Government of India.

rank has declined from 8th position to 13th position in 2009. See Table 9.1 for the evidence of how the sector of drugs and pharmaceuticals has a much lower position in respect of attracting FDI in the period following the adoption of TRIPS.

Further, when we come to the patterns of investment by global pharmaceutical firms in the Indian pharmaceutical industry, it is to be noted with concern that a large part of the newer investments of foreign firms in manufacturing activity have largely been for the expansion of formulation activity. Newer investments in the bulk drug were few and far between. The post-1999 situation is certainly now far more permissive environment for imports. Global pharmaceutical firms have been able to increase their operating freedom. They are able to shift to import-based production for a number of product segments (Abrol, 2005). Second, their preference for the establishment of new operations through the incorporation of wholly owned subsidiaries is also now a well-confirmed tendency.

During the post-TRIPS period, quite a large part of the new FDI in pharmaceuticals have been devoted for the benefit of mergers, acquisitions and takeovers to facilitate the parent firms to increase their control over the operations located in India (Abrol, 2003). Global mergers have affected the foreign pharmaceutical industry on familiar lines. Even today, the motive of stronger control over the ownership of investments continues to be the main driver of merger and acquisition activity for the pharmaceutical multinational corporations (MNCs) in India. Bhaumik et al. (2003) confirm the same motive being the driver of investment for pharmaceutical industry in their survey of FDI in India when they suggest that MNCs investing in the pharmaceutical sector prefer greenfield investment to joint venture. The government has been made to relax its laws with regard to the control of FDI. For example, earlier the Indian government used to grant permission for the establishment of 100 per cent wholly owned subsidiaries only on the condition that the industry would be willing to take up the production of pharmaceuticals right from the basic stage of manufacture of bulk drugs involved. This is no longer a requirement. Today the motive for acquisitions has moved, and the dominant global pharmaceutical companies want to have a strong presence in the generic business in addition to the sales of patented drugs.

Analysis of Activity-wise FDI

While we can see in Table 9.2 that R&D activity accounted during the period 2003–2009 for the highest number of projects from the United States with a total of 36, representing 43 per cent of investment projects, we will be misleading ourselves if the issue of directions of R&D is ignored and left unexamined. Analysis undertaken for the purposes of FDI for R&D shows that a large number of foreign R&D investment projects are focused on the development of facilities for phase III clinical trials and other such modules that only integrate the Indian talent and facilities into the global objectives of foreign pharmaceutical firms. As such these R&D

Table 9.2:
Industry analysis: No. of US FDI projects by activity

Business activities	2003	2004	2005	2006	2007	2008	2009	Total	Average annual growth
Research and development	2	4	10	5	8	5	2	36	44.5%
Manufacturing	3	8	6	3	3	5		28	n/a
Sales, marketing and support		2	2	3	1	1	1	10	n/a
Design, development and testing		1	1		2	1		5	n/a
Business services			1					1	n/a
Headquarters				1				1	n/a
Logistics, distribution and transportation						1		1	n/a
Retail		1						1	n/a
Overall total	5	16	20	12	15	12	3	83	42.0%

Source: FDI Markets Intelligence.

projects have little to do with the development of much needed capabilities for drug discovery and development. Investments in R&D are for clinical trials of drugs discovered and developed in advanced countries. Further, these investments have also little relation with the research needs of local population. Thus by any stretch of imagination, the quality of FDI being attracted into pharmaceutical R&D cannot be characterized as really capability enhancing. Although, prior to the signing of TRIPS Agreement and the adoption of stronger IPRs, there existed R&D units of MNCs that explicitly catered to some of the needs of local population, even the nature of R&D activities under implementation cannot be characterized as of high quality. More evidence on the quality of FDI in R&D is provided in the latter parts of this chapter.

Evidence on Technology Transfer from MNCs to India

During the pre-TRIPS era, foreign pharmaceutical firms exhibited in India often an almost near complete aversion to technology transfer in respect of bulk drug production. Evidence collated on the recent patterns of technology transfer from foreign firms to domestic companies shows that in respect of technology transfer, the results are again not very encouraging for pharmaceuticals. Evidence obtained on the intensity of R&D and royalty payments made by the domestic pharmaceutical and the foreign pharmaceutical firms to their own parents and local sources is also quite clear and shows that royalty figures have been extremely small for the domestic firms until now. Table 9.3 gives the sample characteristics for which knowledge accumulation expenditure was undertaken by the authors. For more details on R&D and royalty paid, look Table 9.4 for a comparison of R&D and royalty payment expenditure of domestic and foreign firms. There is also a confirmation that foreign firms are still spending much less on R&D as compared to domestic firms.

Table 9.3:
Sample characteristics of selected pharmaceutical companies

Total number of listed companies in Prowess database of CMIE	Number of selected companies in the sample	Total sales of 134 pharmaceutical companies	Total sales of selected companies	Percentage
134	(Domestic + foreign) 50	55,214.06	50,397.46 (50 companies)	91.27
134	(Domestic) 41	55,214.06	39,745.47 (41 companies)	71.98
134	(Foreign) 9	55,214.06	10,651.99 (9 companies)	19.29

Source: CMIE Prowess Database, 2009.

Table 9.4:
Intensity of R&D, royalties and marketing and advertising (2006–08)

CMIE rank	Companies	R&D intensity	Marketing and advertising	Royalties paid
	Foreign companies			
	Integrated companies			
3	Ranbaxy Laboratories Ltd	12.65	11.08	0.01
35	Merck Ltd	0.40	5.51	3.07
	Formulation			
16	Aventis Pharma Ltd	0.43	4.21	0.00

(Table 9.4 Continued)

(Table 9.4 Continued)

CMIE rank	Companies	R&D intensity	Marketing and advertising	Royalties paid
8	GlaxoSmithKline Pharmaceuticals Ltd	0.47	4.61	0.01
20	Pfizer Ltd	3.23	6.26	1.56
24	Abbott India Ltd	0.56	3.12	0.00
28	Novartis India Ltd	0.19	7.91	1.97
	Bulk drug			
10	Matrix Laboratories Ltd	10.70	1.49	0.04
	Domestic companies			
	Integrated companies			
2	Dr. Reddy's Laboratories Ltd	9.87	7.01	0.00
4	Lupin Ltd	6.90	8.30	0.00
5	Aurobindo Pharma Ltd	4.75	0.81	0.00
6	Sun Pharmaceutical Industries Ltd	8.98	4.23	0.00
12	Ipca Laboratories Ltd	3.88	6.32	0.01
44	Natco Pharma Ltd	0.00	1.64	0.00
45	Fresenius Kabi Oncology Ltd	10.66	7.50	0.00
49	Marksans Pharma Ltd	1.36	3.07	0.00
50	Wanbury Ltd	3.03	3.88	0.00
	Formulation			
1	Cipla Ltd	5.11	9.33	0.00
7	Piramal Healthcare Ltd	4.48	4.68	0.03
9	Cadila Healthcare Ltd	9.42	9.33	0.04
11	Wockhardt Ltd	9.87	3.99	0.21
15	Alembic Ltd	4.45	6.51	0.03
17	Ankur Drugs and Pharma Ltd	0.00	0.00	0.65
19	Glenmark Pharmaceuticals Ltd	5.24	5.64	0.00
25	J.B. Chemicals and Pharmaceuticals Ltd	2.19	12.11	0.27
26	Unichem Laboratories Ltd	4.15	10.84	0.00
27	Elder Pharmaceuticals Ltd	0.72	5.24	0.00
29	Strides Arcolab Ltd	8.17	2.49	0.00
30	FDC Ltd	1.78	5.56	0.00
32	Ind-Swift Ltd	1.56	3.37	0.00

(Table 9.4 Continued)

(Table 9.4 Continued)

CMIE rank	Companies	R&D intensity	Marketing and advertising	Royalties paid
34	Plethico Pharmaceuticals Ltd	4.28	5.03	0.00
40	Twilight Litaka Pharma Ltd	0.08	0.92	0.00
41	Indoco Remedies Ltd	5.17	6.09	0.00
42	Ajanta Pharma Ltd	6.50	6.14	0.00
47	Granules India Ltd	0.81	1.88	0.00
	Vaccine			
22	Panacea Biotec Ltd	11.21	3.93	0.09
	Fine chemical/biotech			
18	Biocon Ltd	6.06	2.03	0.10
	Bulk drug			
13	Divi's Laboratories Ltd	1.46	0.29	0.00
14	Orchid Chemicals & Pharmaceuticals Ltd	6.36	2.07	0.07
21	Nectar Lifesciences Ltd	0.00	0.84	0.00
23	Surya Pharmaceutical Ltd	3.40	0.41	0.00
31	Ind-Swift Laboratories Ltd	13.16	1.10	0.00
33	Shasun Chemicals & Drugs Ltd	5.50	2.40	0.12
36	Dishman Pharmaceuticals and Chemicals Ltd	4.21	0.00	0.00
37	Sharon Bio-Medicine Ltd	0.00	0.00	0.00
38	Aarti Drugs Ltd	1.24	0.89	0.00
43	Neuland Laboratories Ltd	9.09	1.41	0.00
46	SMS Pharmaceuticals Ltd	4.47	0.80	0.00
48	Themis Medicare Ltd	0.82	2.13	0.00

Source: Compiled from the CMIE in respect of sample firms analyzed in 2009.

Directions of Overseas R&D in India

Evidence is also clear that the TRIPS Agreement has not been able to succeed in inducing the foreign firms to take up overseas R&D for the discovery and development of drugs where the Indian markets could be large. In those cases where some MNCs had located part of their global

R&D outfit in India, activities have been on the decline. Barring Hoechst and Astra, who do limited drug discovery work here, others have closed down the units that had the mandate to develop products for the benefit of local markets. Earlier Ciba-Geigy had larger presence in R&D; its R&D centre is now closed in India. Hoechst has also been reducing its involvement in R&D in India. Their current strategy is to reduce the locally oriented in-house R&D investment in India. They are now building on the work done at these centres on natural products in European laboratories.

Upadhyay et al. (2002) indicated the same in their R&D survey that higher-stage R&D activities of MNC subsidiaries were found to be minimal. And whatever little R&D they had undertaken, it reflected more thrust on formulation R&D (or product development) compared to bulk drug R&D (or process development). Some adaptive R&D was undertaken for troubleshooting. Their focus remained on conventional dosage forms. Although few of them manufactured novel drug delivery system, no research on new drug delivery system (NDDS) was being undertaken at the subsidiaries. See Tables 9.5 and 9.6 for the details of contribution made to the pattern of innovative activities undertaken for the benefit of domestic market by the foreign pharmaceutical firms from the Indian soil.

India does not seem to figure much in the increased strategic R&D alliance activity of the global biopharmaceutical and biotechnology firms. An explanation for this trend is simple because in biopharmaceutical research, the distribution of capabilities is the major determinant of the partner and the mode of alliance. Visalakshi et al. (1995) confirmed this on the basis of a survey of biopharmaceutical R&D undertaken by the Department of Biotechnology, the same that there are hardly any foreign firms engaged in biotechnological research in India. Visalakshi and Sandhya (2000) and Visalakshi (2009) confirm once again the same conclusion regarding biotechnological research by foreign firms working in India after the implementation of TRIPS. Presently under the route of wholly owned subsidiary, AstraZeneca is the only example of drug discovery operations for TB, a type II disease. Of course, here too one needs to keep this in mind that these operations were started when Astra was an independent company. In fact, the Indian government induced Astra to start its operations as a joint venture with the government to work on TB-related drug discovery and diagnostic work. After its merger with Zeneca, the Indian operations are now taking place under the direct control of AstraZeneca. This is still

Table 9.5:
Directions of innovative activities of foreign firms (1999–2009)

	Foreign companies	CMIE rank	Compounds commercialized	Process patents	NDDS	NCE	MOT	New forms of substances	Other products
1	Ranbaxy Laboratories	3	71	88	20		4	109	239
2	GSK Pharmaceutical	8	4						
3	AstraZeneca Ltd	59	4						
4	Pfizer Ltd	28	3						
5	Shantha Biotechnics	137	5						
6	Novo Nordisk			1		2			1
7	Alfred								1
8	Hindustan Lever/Unilever			30			18	18	363
9	Johnson & Johnson								3
	Grand total		87	119	20	2	22	127	607

Source: Data of commercialization and launched compound collected from news archive search of individual pharma companies from 1999 to 2009, and emerging patterns of pharmaceutical innovations (process, product, NDDS, NCE, dosage/formulation/composition, salt/polymorphs/derivative) data collected from USPTO of 1992–2007.

Abbreviation: NDDS, new drug delivery system; NCE, new chemical entity.

Note: Other products included skin products, cosmetics, oral dental care, toiletries products, antifungal, antibacterial, antimicrobial products.

Table 9.6:
Disease type-wise product R&D activities of foreign firms active in India (1999–2009)

Foreign companies	1999–2001			2002–2004			2005–2007			2008–2009			Total
	I	II	III	I	II	III	I	II	III	I	II	III	
Ranbaxy Laboratories Ltd	2						2	1		3			8
Abbott India Ltd	3			2			5			3			13
AstraZeneca India Ltd										6			6
Pfizer Ltd										14	1		15
GSK Pharmaceutical Ltd							1			14		2	17
Aventis Pharmaceutical										13			13
Merck Ltd										17	1		18
Novartis India Ltd										28			28
Grand total	5			2			8	1		98	1	3	118

Source: Source data collected from individual website and latest annual report of individual pharma companies and Clinical Trials Registry India (CTRI).
Disease type: type I, type II, type III.
Type I: diabetes, cancer, metabolic diseases, hepatitis, influenza, cardiovascular, infectious diseases, inflammatory diseases, allergy and respiratory diseases.

an isolated case; foreign firms are unlikely to establish integrated drug discovery facilities for the diseases that affect disproportionately India. The latest is that even the AstraZeneca has taken a decision to close its Bengaluru R&D centre undertaking research for tuberculosis.

Therefore, the policy design-related question is that whether the MNCs should be allowed to use India merely as a cheap source of science and technology (S&T) manpower and patients and as a 'listening post'. Foreign pharmaceutical firms are unlikely to develop India as a location for the development of system integration capacity. Since in the new drug discovery paradigm, the system integration capacity is going to finally count[4] and if the development of this capacity cannot be expected to take place automatically, it is obvious that the foreign subsidiary mode in which the MNCs are now restructuring their investments should not be encouraged at all by the government in India.

But the current expectations of global pharmaceutical firms are clear. Global pharmaceutical firms will prefer to selectively invest in the selected R&D operations, namely, bioinformatics and clinical research, where by relocation, it is possible for them to cut down the R&D costs without increasing information spillovers. Available evidence from India suggests that in many cases, the MNCs appear to have preferred the route of outsourcing of R&D from fully dedicated companies to reduce costs in respect of clinical trials and bioinformatics-related R&D work. Presently, only for the health care management and pharmaceutical services, the choice of MNCs has been to establish fully owned R&D subsidiaries. Establishment of operations for the implementation of clinical trials, data management and biostatistics by Quintiles, a leading pharmaceutical service provider, is an example.

Technology acquisition is done by domestic firms through the routes of contract research and manufacturing services (CRAMS) and export of generic medicines. Again the claimed benefit of increased technology transfer to the domestic firms is also yet to accrue in the case of India. Foreign technical collaborations have not been important for export, only many small- and medium-scale firms have entered into collaborations with foreign firms mostly to cater to the domestic market. Kumar and Pradhan (2000) too had shown using empirical evidence on the patterns of relationship of IPRs with FDI, technology transfer through licensing and overseas research and development (R&D) that technology licensing continues to be on the decline; imports and low-quality FDI are the preferred routes for foreign companies for whom the control over ownership and markets has become far easier with liberalization.

Expectations from the route of contract manufacturing are also clear in the case of India. Exploitation of contract manufacturing would not be able to improve the prospects for technology transfer by itself because there are no new technologies being transferred. Production capabilities can certainly get better on account of the enforcement of good manufacturing practices (GMP) in the case of some firms. Analysis indicates that though the players like Matrix Laboratories, Divi or Shasun Chemicals or Cadila have made much use of this opportunity to grow, their technological capabilities have not been upgraded through the provision of contract manufacturing services. Recently, Matrix Laboratories got warning about their manufacturing practice from the US Food and Drug Administration (USFDA). Apart from Ranbaxy and Cipla, which were earlier warned by the FDA, Matrix was the third drug company working from India for the US market to get warning from the regulatory authorities of United States.[5]

There is also evidence that as far as the terms and conditions of contract manufacturing of bulk drugs are concerned, in the post-TRIPS scenario, deals being entered in to by the Indian firms are far from being equal. Ranbaxy Laboratories and Lupin Laboratories were among the first Indian companies to bag manufacturing contracts from MNCs—Ranbaxy from Eli Lilly and Lupin from Cyanamid. In pre-TRIPS scene, contracts for manufacturing came through when Ranbaxy developed an alternative process for manufacturing Eli Lilly's patented drug cefaclor because the American company sensed that it would lose its markets to Ranbaxy's low-cost substitute in countries that did not recognize product patents. Eli Lilly offered a manufacturing contract to Ranbaxy for producing 7-Amino-3-Chloro Cephalosporanic (7-ACCA), intermediate for cefaclor, to make the best of a bad situation. Of course, Ranbaxy is no more an Indian company. It has been sold to Daiichi Sankyo, a Japanese MNC. Today the situation is changed due to the implementation of TRIPS. Take the example of Nicholas Piramal; it entered a joint venture (49:51) with Allergan Incorporated, USA, to earn business for the manufacturing of bulk drugs. The same is true for the negotiations that it is carrying out with UK-based Baker Norton to earn business in the form of contract manufacturing. And it seems that the growth in contract manufacturing will come from the efforts of companies such as Divi, Shasun and Nicholas Piramal India (now taken over by Abbott Laboratories, USA), which have been willing to accept even 'subordinate relationships' in their collaborations for contract manufacture. See Table 9.7 for a glimpse into the pattern of CRAM activities being undertaken by the large domestic pharmaceutical firms since the time of the adoption of TRIPS Agreement in India.

In this context, it needs to be kept in view that the pharmaceutical fine chemicals (PFC) industry is also at the moment experiencing a period of development driven by changes in the demands of its major pharmaceutical clients and the desire to depart from the low-profit areas of basic fine chemical manufacture. PFC companies have responded to recent pressures through a series of high-value acquisitions, increasing their size, core capabilities, geographical reach and the ability to buffer the pitfalls associated with late pharmaceutical withdrawals. Expansion of companies' core competencies and their geographical reach, in addition to the acquisition of new technologies and facilities, appear to have been the objectives of a number of recent acquisitions. PFC companies

Table 9.7:
Pharmaceutical companies in CRAM activities in India

Companies in contract research (excluding clinical trials)	Clinical trials
Nicholas Piramal	Clinigene (Biocon)
Aurigene (Dr. Reddy's)	Jubilant Clinsys (Jubilant Organosys)
Syngene (Biocon)	Wellquest (Nicholas Piramal)
GVK Biosciences	Synchron
Jubilant Organosys	Vimta Labs
Divi's Laboratories	Lambada
Suven Life Sciences	SIRO Clinpharm
Dr. Reddy's Laboratories	Reliance Life Sciences
Vimta Labs	Asian Clinical Trials (Suven Life Sciences)

Source: Annual report and IDMA (International Disease Management Alliance) news, 2007.

are also thus involved in a longer-term process of restructuring that makes them to vacate the low-profit areas of basic fine chemical manufacture for the benefit of Asian companies and to undertake a shift to the higher-margin territories involving innovative technologies. Consequently, the big PFC-producing firms have preferred to shift these low-margin areas to Asia, which however involves very little gain in respect of competence building for the Indian firms.

For the big Indian companies, namely, Ranbaxy (now sold to Daiichi Sankyo), Cipla, Reddy's, Lupin and Wockhardt, the option of export of generics to the regulated markets of United States and Europe remains till this day the main route for the realization of growth of markets. Technology upgrading has been undertaken to support the same route. Domestic companies continue to invest much of their money into generic market with the understanding that they should be skimming maximum out of the generic export opportunity when the market competition is low and the margins are high. As this scenario is possible only immediately after the drugs come off-patent, they are putting their full might and filing four to five abbreviated new drug applications (ANDAs) every year to be first in the market and exploit the period of exclusivity available under the US drug regulation laws. Stimulation of the effort in respect of generic exports is of course also driven by the taxation policy benefit enacted and continued at home by the successive governments. In this policy move, the role of acceptance of the TRIPS Agreement is incidental.

However, the Indian experience of last 15 years also indicates that the road ahead for the export of generics to regulated US market remains like always tedious and full of hurdles. First of all, from being mere producers of broad-range generics, Indian companies have had to learn to challenge the US patent law and go through costly litigation involving risk of losing and making the investment made in process development and clinical data generation redundant. Both Ranbaxy and Dr. Reddy's Laboratories have already burned a lot of money in this business. Hurdles are only growing due to the steps taken in the form of adoption of the policy of allowing parent firm-authorized generics and creating fresh IP barriers for biogenerics/biosimilars. Both the United States and European Union (EU) are every day taking steps that can help the MNCs to maintain monopoly in the area of pharmaceuticals even beyond 20 years. In the United States, this kind of pharmaceutical-specific re-engineering of patent length and breadth has been made possible with the involvement of federal drug regulation authority (USFDA), the body responsible for drug regulation. The constraints of a unitary patent system have led the United States to extend the period of market exclusivity for drugs by adjusting the system with the help of drug regulators. And the generic route is its target; it is being seriously affected by such an adjustment of the drug regulation.

If we have to be more specific, in the United States, under the Hatch–Waxman Act, already the government has a system of patent term 'restorations' that can extend the monopoly of original patentee for a maximum of five years, in addition to the initial patent term. In the EU too, there exists a scheme for supplemental protection certificate. In the United States, no ANDA can be submitted until five years after the referenced brand name product gets its first FDA approval if the originator product was the first drug product to contain that active ingredient to obtain approval. Similarly, an ANDA cannot be submitted for three years if an originator's new drug application or supplemental application is supported by new clinical investigations conducted by the applicant and essential to approval (normally for a new indication). As of 1997, the United States now allows for an additional six months of exclusivity as a reward for studying drugs in children. In the United States, the first version of an orphan drug is entitled to seven years of exclusivity, preventing approval of an ANDA.

The United States also allows, as a reward, 180 days exclusivity to the first generic manufacturer to file a successful paragraph IV certification alleging that a listed patent is invalid or not infringed. Thus, as far as the question of export of generics is concerned, it faces important IPR-related hurdles today in the markets of EU and United States.[6]

For the purpose of illustration below, we give some examples to bring on surface the implications of patent law changes brought in by the developed countries for the export of generics. Take the example of Augmentin—a powerful and expensive antibiotic produced by SmithKline. It was initially expected to come off-patent in 2002. However, by securing patents covering other properties of the drug, Augmentin will now remain covered until 2017, fully 15 years more than expected. The new patent was not granted for innovative research on a new drug, but for work conducted in the early 1970s. Take the case of anti-anxiety drug Buspar; its makers secured a new patent covering the absorption of Buspar just one day before a generic competition was set to begin distribution of its pill that would have given consumers a 25 per cent discount. A listing automatically delays FDA approval of a generic product by 30 days under US law. Bristol-Myers Squibb has been charged with also falsely listing a patent claim for Buspar in the USFDA's orange book. In the case of Hoechst Marion Roussel (Aventis) and Andrx Corporation in 2001, the former sued the latter for patent infringement. Such a lawsuit has the effect of preventing generic manufacturers from seeking FDA approval. Aventis had paid Andrx several million dollars to delay the introduction of a generic version of the drug Cardizem CD (Horton, 2002, p. 181).

Being aware that their drug cannot come off-patent when there is an ongoing patent litigation, the original property right holder files suits against generic manufacturing claiming infringement on one or more layers of patents subsequently filed on various, and often insignificant, elements of the drug. To date, such cases are going on in the United States for Claritin, Buspar, Cardizem and Prozac. Under the Uruguay Round Agreement Act (URAA), the United States provided an extension of three years as gratis. This way, Zantac of Glaxo gained an additional 19 months of protection. Claritin got an extra 22 months of exclusivity. Claritin was able to extend its patent life further by six months for an estimated US$3 million paediatric trial, but for the company (Schering-Plough), the gain made through this alone was US$1 billion. Estimates exist that through these and other legal tricks, total gains for Claritin have been of the order of extra four and a half years of patent life, which is going to give an additional US$13 billion to Schering-Plough. It should therefore be clear that the Indian pharmaceutical firms cannot assume the traditional pharmaceutical generics opportunity to continue to easily fall in their lap due to the United States and Europe being in position to strengthen and link their own patent law with drug regulation and control the entry of Indian companies into their own markets against which Indian can do little.

As available evidence seems to tell that in the area of biogenerics, a tough fight is in waiting for the Indian pharmaceutical industry in the Indian market. The recombinant products market has been led so far by the imports of established global brands and marketing of the products either by local subsidiaries (SmithKline Beecham, Novo), or through marketing arrangements as it happened in the case of Nicholas Piramal and Roche. Though certainly changes have come in due to the recent introduction of local firms such as Shantha, Bharat, Panacea and Wockhardt into the Indian market for products like hepatitis B vaccine, interferon-alpha, insulin and erythropoietin (EPO), nevertheless the situation has changed quite radically after 1 January 2005. As already discussed in the earlier section, the Indian policymakers should expect the litigations to grow in the case of biogenerics in the US market. The Indian industry is getting a taste of this at an early stage; almost all the export-oriented Indian firms have faced recently this challenge in the United States.

Domestic Firms' R&D Directions

Studies differ in their degree of optimism in respect of the positive effects of stronger patents on product development by local firms based on disclosed foreign patents and on additional R&D efforts. Looking at the domestic sector today, only a handful of firms have been able to increase their R&D investments. Some of these have earlier demonstrated that they can devise with the help of public sector research their expertise in the creation of new processes for patented products. Dr. Reddy's Group was the first domestic company to file the first two-product patent applications for anti-cancer and anti-diabetes substances in the United States. But it is also clear that Dr. Reddy's Group does not want to engage autonomously in drug development. It is interested in selling its rights to the partners abroad for the reason that it does not have the capacity to invest further and stopping after the stage of drug discovery work. Examples of Wockhardt joining hands with Rhein Biotech GmbH, Germany, Ranbaxy shaking hands with Eli Lilly for development work, Cipla undertaking custom synthesis, collaborations with Japanese and Swiss firms indicate the limitations of and opportunities available to Indian firms.

Even a cursory glance thrown at the evidence available in Tables 9.8–9.10 appears to clearly show that all the important developments that we see in respect of the creation of R&D capabilities for drug discovery and development within the Indian firms have a global market favouring R&D orientation. Today as the situation stands, their pharmaceutical research is largely directed to the needs of the Western markets and much

Table 9.8:

Disease type-wise product-specific R&D activities of domestic firms active in India (1999–2009)

Domestic companies	Disease type												
	1999–2001			2002–04			2005–07			2008–09			
	I	II	III	I	II	III	I	II	III	I	II	III	Total
Orchid Pharmaceuticals Ltd				2			6			2			10
Sun Pharmaceutical Ltd							2			7			9
Biocon Ltd				2			4			6			12
Glenmark Pharmaceuticals Ltd				1			5	1		7			14
Bharat Biotech Ltd							1	1		3	2		7
Alembic Ltd													–
Dr. Reddy's Laboratories Ltd				7			2	1		15			25
Lupin Ltd	1			1			4	4		4	1		15
Cadila Healthcare Ltd							3	1		9			13
Piramal Healthcare Ltd							7			5			12
Wockhardt Ltd							1			2			3
Ipca Laboratories Ltd										2	2		4
Aurobindo Pharmaceutical Ltd													–

(Table 9.8 Continued)

(Table 9.8 Continued)

Domestic companies	Disease type												Total
	1999–2001			2002–04			2005–07			2008–09			
	I	II	III	I	II	III	I	II	III	I	II	III	
Torrent Pharmaceuticals										1			1
Ajanta Pharma										7			7
Natco Pharma										2			2
Granules India Ltd										1			1
SMS Pharmaceutical										10			10
Shantha Biotechnics							3	2		10	1		16
Panacea Biotec												2	2
Matrix Laboratories										3			3
Grand total	1	–	–	12	1	–	37	7	4	96	3	5	166

Source: Source data collected from individual website and latest annual report of individual pharmaceutical companies and Clinical Trials Registry India (CTRI).
Disease type (type I, type II, type III).
Type I: diabetes, cancer, metabolic diseases, hepatitis, influenza, cardiovascular, infectious diseases, inflammatory diseases, allergy and respiratory diseases.
Type II: HIV/Aids, tuberculosis and malaria.
Type III: leishmaniasis, trypanosomiasis, lymphatic filariasis, leprosy and diarrhoea (neglected diseases of the developing world).

Table 9.9:
Clinical Phases of Compound for Various Diseases by Foreign and Domestic Pharmaceutical Industry during 2007–2009

Company	Disease type			Status of trial/phases			
Domestic firms (16 companies)	Type I	Type II	Type III	Phase I	Phase II	Phase III	Phase IV
	65	3	2	5	20	35	9
Foreign firms (9 companies)	Type I	Type II	Type III	Phase I	Phase II	Phase III	Phase IV
	110	3	3	12	23	12	9

Source: Data collected from individual website and latest annual report of individual pharmaceutical companies and Clinical Trial Registry India (CTRI).
Disease type: type I, type II, type III.
Type I: diabetes, cancer, metabolic diseases, hepatitis, influenza, cardiovascular, infectious diseases, inflammatory diseases, allergy and respiratory diseases.
Type II: HIV/Aids, tuberculosis and malaria.
Type III: leishmaniasis, trypanosomiasis, lymphatic filariasis, leprosy and diarrhoea.
Status of involvement of domestic and foreign firms in the trials (phase I, phase II, phase III, phase IV).

Table 9.10:
Therapeutic area-wise estimation of pharmaceutical projects and patents and the pattern of matches with the national burden of disease (1992–2007)

S. no.	Major therapeutic areas/ disease/health conditions	Share in the total burden of disease (%)	Domestic pharmaceutical project (%)	Foreign pharmaceutical project (%)	Domestic companies pharmaceutical patents percentage (%) of total domestic patents	Domestic companies pharmaceutical patents percentage (%) of total patents	Foreign companies pharmaceutical patents percentage (%) of total foreign patents	Foreign companies pharmaceutical patents percentage (%) of total patents
1	Diabetes	0.7	17.15	16.36	5.94	5.91	20	0.084
2	Cancer	3.4	10.05	8.81	5.6	5.57		
3	Tuberculosis	2.8	1.18		0.50	0.50		
4	Malaria	1.6	2.36		0.93	0.92		
5	Metabolic disease	–	7.36	0.9	6.79	6.76	20	0.084
6	HIV/Aids	2.1	0.59	0.23	0.84	0.84		
7	Inflammatory diseases		3.55	0.67	5.6	5.57		
8	Infectious diseases/injuries	16.1	8.28	4.54	38.96	38.79		
9	Respiratory diseases	1.5	4.73	5.61	1.1	1.09		
10	Arthritis	–						
11	Bone disease	–		6.63	1.27	1.26		
12	Brain disorders	8.5	4.73	0.56	10.18	10.14	40	0.16
13	Ulcer	–			0.5	0.50		
14	Psoriasis	–			0.33	0.33		
15	Cardiovascular	10.0	0.59		2.63	2.78	20	0.084

16	Maternal and prenatal problems	11.6	1.34		0.25	0.25
17	Diarrhoea	8.2	1.77		0.08	0.084
18	Heart disease	–			0.93	0.92
19	Depression	–			3.56	3.55
20	Hypertension	–		10.12	4.49	4.48
21	Allergy	–			1.78	1.77
22	Hepatitis	–		1.81	0.16	0.16
23	Leprosy	0.1				
24	Childhood disease	5.4				
25	Otitis media	0.1				
26	Blindness	1.4				
27	Oral diseases	0.5				
28	Prosthetic hyperplasia	–			1.01	1.014
	Others	25.4	30.17	18.18	6.45	6.42

Source: USPTO from 1992 to 2007, company websites and data available on the Burden of Disease from GoI.

less to undertaking R&D for neglected diseases of the poor in developing countries. Under the emerging conditions of competition in the 'global' pharmaceutical industry, locally bred firms of developing countries are likely to be lured by the MNCs to work for the Western markets. Furthermore, outsourcing markets in clinical trials, R&D and production are also becoming accessible to the locally bred firms of countries like India. Because of very many short-term benefits, it is obviously quite tempting to direct the industry totally or mainly for these markets in countries like India.

The dynamic of biotechnology in India is also apparently dependent on the overall movement of internationalization of R&D. The examples of Dr. Reddy's Lab and Biocon are especially useful for discussion on the conditions for gains to accrue from the contract work being undertaken by these two companies. Both these companies have created several entities, each of them corresponding to a different strategy. Dr. Reddy's Lab is involved in the development of recombinant DNA-based products and has an internal programme of biotechnology-based drug targets discovery. It has also set up a company named Molecular Connections, and a contract research company named Aurigene, involved in chemical and biological research for drug discovery. Similarly, Biocon too, whose core activity is the manufacturing of industrial enzymes, has set up a contract research subsidiary named Syngene, and a CRO named Clinigene.

However, as far as the contribution of these domestic firms to meeting the product development challenge for neglected diseases is concerned, analysis is clear that the current level of opportunities which limit Aurigene, GVK Bio and Syngene to cloning the genes and getting the genes to express would not allow these companies to build an industry capable of doing cutting-edge biotechnology research. At the moment, their mother companies do not have any intention of interfering with their subsidiaries because of the agreements of confidentiality signed by them with the partners who have outsourced the part of drug discovery or clinical research to them. This means that no technological information can circulate between the company in charge of contract research work and the parent company involved in its own research. From the standpoint of priorities of public health protection, the moot question is one of how it would benefit the country in terms of promotion of indigenous drug discovery and development efforts.

Emerging Relations of Public Sector R&D with Domestic Firms

Coming to the issue of emerging relations of public sector R&D with industry in the case of drugs and pharmaceutical sector in India, the main challenge is that the public sector R&D institutions maintain a long-term vision and strategy directed by public health priorities of the Indian nation whose citizens have a first claim on their outcomes. See Table 9.11 for the current status of matches and mismatches of R&D priorities under

Table 9.11:
Comparison with disease burden of public sector projects from 1992 to 2007

S. no.	Major therapeutic areas/disease/ health conditions	Share in the total burden of disease (%)	IMR projects (%)	EMR projects (%)	Public sector patents as percentage (%) of total patents
1	Diabetes	0.7	2.08	8.29	5.96
2	Cancer	3.4	12.71	19.21	13.1
3	Tuberculosis	2.8	8.30	12.66	6.37
4	Malaria	1.6	10.38	5.24	9.87
5	Metabolic disease	–			4.73
6	HIV/Aids	2.1	8.43	10.26	9.85
7	Inflammatory diseases				2.05
8	Infectious diseases/injuries	16.1			24.27
9	Respiratory diseases	1.5		1.74	2.26
10	Bone disease	–		2.35	1.4
11	Brain disorders	8.5		4.71	2.26
12	Ulcer	–			
13	Psoriasis	–			
14	Cardiovascular	10.0	1.43	2.18	4.11
15	Maternal and prenatal problems	11.6	5.96	3.02	5.25
16	Diarrhoeal diseases	8.2	0.26	1.39	0.20

(Table 9.11 Continued)

(Table 9.11 Continued)

S. no.	Major therapeutic areas/disease/ health conditions	Share in the total burden of disease (%)	IMR projects (%)	EMR projects (%)	Public sector patents as percentage (%) of total patents
17	Heart disease	–			
18	Depression	–			0.41
19	Hypertension	–			2.26
20	Allergy	–			
21	Hepatitis	–	3.37	5.02	2.44
22	Leprosy	0.1	4.15	3.93	2.24
23	Childhood disease	5.4	2.52	1.21	0.41
24	Otitis media	0.1			
25	Blindness	1.4			0.2
26	Oral diseases	0.5			0.3
27	Prosthetic hyperplasia	–			
28	JE		3.11		0.61
29	Dengue		3.11	0.43	0.41
30	Leishmaniasis		9.86	4.80	3.29
31	Others	25.4	23.48		12.1

Source: Project-specific database built by the authors from the public databases on R&D projects and patenting activities being undertaken by the public sector R&D organizations in India, 2009.

perusal with the priorities of burden of disease in the public sector. It appears that while there are too many mismatches to be taken care of and reflect a clear systemic failure that is seemingly connected with, on the one hand, the systemic determination of disciplinary priorities of the Indian scientific community in the West and the decisions of the government to subject the public sector to the short-term demands of private sector in the post-TRIPS period.

All over the world, the PPPs is the latest new buzzword in the system of health research and technology development. In India, New Millennium Indian Technology Leadership Initiative (NMITLI) of Council of Scientific and Industrial Research (CSIR), Drugs and Pharmaceuticals Research Programme (DPRP) and Technology Development Board (TDB) of Department of Science and Technology (DST) and Small Business

Innovation Research Initiative (SBIRI) of Department of Biotechnology (DBT) constitute the main examples of public private partnerships. A large experience has been gathered through these schemes in respect of the determinants of success in implementation of PPPs. A large number of NMITLI PPPs have preferred to catalyze health innovations as a vehicle for the domestic industry to attain mainly global leadership position in selected niche area by synergizing the best competencies of publicly funded R&D institutions, academia and private industry. In the last six years, NMITLI has supported 42 R&D in various fields, including new targets, drug delivery systems, bioenhancer and therapeutics for psoriasis, *Mycobaterium tuberculosis*, pain management in osteoarthritis, insulin sensitization in diabetes mellitus type II and process of Tamiflu and so forth with about 287 partners, 222 in public sector and 65 in private sector, with an estimated outlay of over Rs. 300 crore. Analysis of SBIRI efforts (37 cases till May 2008) shows that there is not much focus for diseases of Indian interest, though a couple of cases pertain to malaria and typhoid. Similarly, in the case of Drug Promotion Research Programme (DPRP), it is also known that the government had to add a special grant-in-aid programme for the promotion of research on neglected diseases because in the earlier years, the programme was unable to attract domestic companies to work on these areas.

Conceived in 2003, the Golden Triangle partnership is also now receiving special budgetary support for an integrated technology mission focused on the development of Ayurveda and traditional medical knowledge that synthesize modern medicine, traditional medicine and modern science. In this way, efforts on traditional medicine have also picked up momentum. The CSIR and Indian Council of Medical Research (ICMR) are working with the Department of Ayurveda, Siddha, and Homeopathy to bring out safe, efficacious and standardized classical products for identified disease conditions. New Ayurvedic and herbal products for diseases of national/global importance are also being pursued. Innovative technologies are being used to develop single- and poly-herbal mineral products, which have the potential for IP protection and commercial exploitation by national/multinational pharma companies.

Since the priorities of private sector have much support of the political and bureaucratic apparatus, even the scientific community has been willing to subject the priorities of public sector R&D organizations to the short-term priorities of the domestic industry. Leadership of the scientific community is also far more willing to give a higher priority to the R&D work to be undertaken on the problems of ageing disorders, psoriasis, rejuvenation and so on, rather than putting in money into products for neglected diseases (type III). But since with the intervention of public groups, there is somewhat better response from the Indian public sector agencies, and the situation can change and head for better. It is therefore essential to plan, monitor and evaluate the public sector R&D institutions on the basis of the above-described public health priorities. Here also, it seems that the results would come forth for the benefit of public health only if the leadership of the science agencies is willing and determined to pursue the road map for the development of products that are required locally and have the support of public health system. Even the perusal of OSDD by CSIR clearly illustrates the importance of public policy, providing directions to R&D and, in the process, shaping the dynamic of IP regime. India might be able to lead the fight in respect of Global Strategy and Plan of Action (GSPOA) in the forum of WHO in the Consultative Expert Working Group on IP and financing of R&D for the priorities that matter to the developing world. Hopefully with the dynamic generated by OSDD, India would be able to pave way for the emergence of better conditions for IP management and move the world to a more balanced international regime in patents.

There is also evidence that the Government of India has not been interested to use the TRIPS flexibilities in full. Under pressure from the US government, most of the provisions of compulsory licensing remain unutilized till this day. Lack of interventions on the front of stimulation of local demand-oriented innovation activity and local competition has much to do with the belief in the promise of promotion of innovation through stronger IPRs. It is clear that since the market-friendly institution of stronger IPRs is unable to stimulate technological change, the assumptions of Evenson about IPRs would need to be strongly tempered. Evidence of the inability of institution of stronger IPRs to correct the directions of R&D and innovation by itself is growing. Clearly, the need of the hour is to have a wider focus and bring in to play the policy instruments other than IPRs. Policymakers should introduce a measure of healthy competition and achieve for the domestic firms a measure of competitiveness in the case of local pharmaceutical industries. Pharmaceutical innovation-related capability building processes need a strong dose of public sector intervention in translational R&D. By enforcing contract law, India is capable of assuring confidentiality of the results of contract research being undertaken for clinical trials. Investments in clinical trials do not need strong IPRs in the case of pharmaceuticals in India.

The Road Ahead

The Indian Patent Act, 1970 was an enabling instrument for both domestic market creation and capability building in the case of Indian pharmaceutical industry because it promoted a balanced IP regime. India utilized it in conjunction with a whole range of industrial policy changes. In the area of pharmaceuticals, India should define her national interest in innovation. Policymakers need to decide it by taking into account the interests of all the sections. Of course, the public interest should ultimately matter more to the policymakers. The challenge of deciding on what all flexibilities that India must use to protect has to be tackled by appropriately defining the national interest. Demand articulation and well-targeted investment subsidies can do a better job in respect of the promotion of R&D in private sector. Investment of public sector in translational R&D and public procurement should be used to support the development of domestic market in line with the immediate and future health needs.

As earlier mentioned, available TRIPS flexibilities have not been fully utilized on the basis of the interests of all the sections. Amendments made to the Indian Patent Act, 1970 were only geared to satisfy partly the immediate interests of large domestic firms. Still there exists much scope to utilize the TRIPS flexibilities to benefit the forces of competition and good technological citizenship. The government should be designing the national patent regime to create an enabling environment for the development of new chemical entities (NCEs). Policymakers should support the efforts in process innovation being undertaken by the domestic enterprises through the institutionalization of a policy package aimed at strengthening the sectoral system of innovation as a whole. The opening available through the need for a discussion in the parliament on the recommendations of the Technical Expert Group on the issue of patentability of pharmaceutical substance and microorganisms should be used not only to seek all the relevant changes at one go in the amended patent act by bringing a Fourth Patent (Amendment) Bill in one of the forthcoming parliamentary session of 2007 but also to introduce a suitable supporting technological, industrial and health policy package. A truly selective IP protection policy that is capable of restricting the protection to only those products that are truly innovative is the most urgent need of the developing nations. Nevertheless, the selective IP protection policy would need to be accompanied by a path of calibrated promotion of an industrial policy framework to achieve ultimately the goals of competitiveness of domestic enterprises and the promotion of health needs in the case of developing nations.

Notes

1. Hall states 'The Study by Kanwar and Evenson (2003) looks at the variation across country in R&D spending as a function of the Ginarte–Park index over the 1981–2005 period and finds results to be suggesting strong IP protection being related to higher R&D intensity. Although well done in many respects, this study makes no attempt to explore the potential endogeneity of the relationship nor does it control for the level of development of countries, which arguably drives both R&D and the development of IP institutions'.
2. Evenson and Singh (1997) provide the rationale for strong IPRs in terms of the institution, enabling appropriation of benefits and acting as incentive. They write, 'In the recent past, it has been growlingly felt that the companies that engage in the development of new technologies (be it for manufacturing of new goods, improved performance of existing products, or cheaper production processes) face difficulties in appropriating the fruits of their labor. If innovation is a principal engine of economic growth and agents innovate to capture or hold a share of market they would not retain otherwise, then the protection of intellectual property rights (IPRs) may boost long-run economic growth'.
3. In the opening remarks of the proceedings prepared for the international seminar, held in Maastricht on 23–24 September 2005 on the theme of 'Contributions to the Development Agenda on Intellectual Property Rights', Castro and Bohrer (2008) highlight that there is now near consensus among informed observers that 'TRIPS as it stands is against the interests of developing countries, and needs reform. In developing their own patent laws, developing countries need to recognise that patent law and practice, in some cases, overshot, and need to be reformed. That is the burden of the recent NAS/NRC report on 'A Patent System for the 21st Century'.
4. Nightingale (2000) emphasizes this by suggesting that the learning of system integration skills is a precondition of further competition in the development of innovative drugs in the global pharmaceutical industry today.
5. When it comes to manufacturing, India ranks only second to the United States in the number of global drug master filings (DMF) every year. DMF is essentially a permission to enter the US bulk actives market with the objective of supplying to either a large US generics player or a captive consumption. DMFs by Indian companies rose

to 19 per cent of the world filings in 2003 compared to 2.4 per cent in 1991. For the April–June quarter 2003, India accounted for 34 per cent of the world's filings.
6. Glasgow identifies five main ways that pharmaceutical firms employ to lengthen the patent life of their drugs in order to benefit from monopoly rents longer. One, these firms are using legislative provisions and loopholes to apply for a patent extension. Two, these firms are suing generic manufacturers for patent infringement. Three, these firms are merging with direct competitors as patent rights expire in an effort to continue the monopoly. Four, these firms are recombining drugs in slightly different ways to secure new patents and layering several patents on different aspects of the drug to secure perennial monopoly rights. Five, these firms are using advertising and brand name development to increase the barrier to entry for generic drug manufacturers (Glasgow, 2001, pp. 227–58).

References

Abrol, D. 2003. 'Report on international conference on healthcare and food: Challenges of Intellectual Property Rights, Biosafety and Bioethics, NISTADS, New Delhi, India', *International Journal of Biotechnology*, 5(2): 186–212.
——— 2005. 'Conditions for the achievement of pharmaceutical innovation for sustainable development: Lessons from India', in the *World Review of Science, Technology and Sustainable Development*.
Castro, A.C. and M.B.A. Bohrer. 2008. 'International Seminar—Contributions to the Development Agenda on Intellectual Property Rights', in *Economica*, Rio de Janeiro, 10(2): 11–14, December.
Basheer, S. 2005. 'India's tryst with TRIPs: The patents amendment act 2005', *The Indian Journal of Law and Technology*, 1: 15–43.
Beattie, P.M. 2006. 'The Intellectual Property Law and Economics of Innocent Fraud: The IP and Development Debate', be press Legal Series, law.bepress.com.
Bhaumik, S.K., P.L. Beena, L. Bhandari and S. Gokarn. 2003. Survey of FDI in India, Centre for New and Emerging Markets, London Business School, April. *Business World*, 30 January 2010, Available at: http://www.businessworld.in/bw/.
Correa, C.M. 2008. 'Designing patent policies suited to developing countries needs', *Economica*, Rio de Janeiro, 10(2): 82–105, December.
Delgado, M., M. Kyle and A.M. McGahan. 2008. 'The influence of TRIPS on Global Trade in Pharmaceuticals, 1994–2005', University of Toronto Working Paper.
Evenson, R.E. 2001. *Comment: Intellectual Property Rights and Economic Development*. Maskus, K.E. (Ed), Cleveland, Ohio: Case Western.
——— 2003. 'Making Science and Technology work for the Poor', UN Economic Commission for Africa, Institute for Natural Resources in Africa, United Nations University, October 7–10, 2003, UNU/INRA-UNECA Reserve, *Journal of International Law*, 33(2001): 187.
Evenson, R.E. and L. Singh. 1997. 'Economic Growth, International Technological Spillovers and Public Policy: Theory and Empirical Evidence from Asia', Economic Growth Center, Yale University Center, Discussion Paper No. 777, September 1997.
Evenson, R.E. and L.E. Westphal. 1995. 'Technological Change and Technological Strategy', in Behrman, J. and T.N. Srinivasan (eds), *Handbook of Development Economics, Volume III*. Amsterdam: Elsevier Science B.V.

Evenson, R.E., J. Putnam and S. Kortum. 1989. 'Invention by Industry', Unpublished Manuscript. New Haven, Connecticut: Yale University.
Glasgow, L.J. 2001. 'Stretching the limits of Intellectual Property Rights: Has the pharmaceutical industry gone too far?', *IDEA The Journal of Law and Technology*, 41(2): 227–58.
Hall, B.H. 2007. 'Patents and Patent Policy', OxREP article, December 2007.
Horton, R. 2002. 'Countering delays in introduction of generic drugs', Editorial, *The Lancet*, 359(9302): 181.
Kanwar, S. and R.E. Evenson. 2001. 'Do Intellectual Property Rights Spur Technological Change?', Yale Economic Growth Center, Discussion Paper No. 831.
———— 2003. 'Does intellectual property protection spur technological change?', *Oxford Economic Papers*, 55: 235–64.
Kumar, N. and P. Jayaprakash. (2000). *Reforms, TRIPs and Indian Pharmaceutical Industry*. New Delhi: Research and Information System for Developing Countries.
McCalman, P. 2001. 'Reaping what you sow: An empirical analysis of international patent harmonization', *Journal of International Economics*, 55: 161–86.
Mercurio, B.C. 2010. 'Reconceptualising the debate on intellectual property rights and economic development', *Law and Development Review*, 3(1): Article 3.
Montobbio, F., A. Primi and V. Sterzi. 2010. 'Meet Me after the TRIPS: Do IPRs Reinforcement Facilitate International Technological Cooperation', OECD Directorate for Science, Technology and Industry, 2010.
National Working Group on Patent Laws. 1989. *Conquest by Patent: On Patent Law and Patent Policy*. Delhi.
Nightingale, P. 2000. 'Knowledge, technology and organization: Understanding the dynamics of pharmaceutical innovation process', SPRU, Available on University of Sussex website, http://www.druid.dk/summer2000/conf-papers/Nightingale.pdf.
Qian, Y. 2008. 'Are National Patent Laws the Blossoming Rains?—Evidences from Domestic Innovation, Technology Transfer and International Post Patent Implementations in the Period 1978–2002', in *The Development Agenda: Global Intellectual Property and Developing Countries*. Oxford, UK: Oxford University Press.
Reichman, J.H. 2009. 'Intellectual property in the twenty-first century: Will the developing countries lead or follow?', *Houston Law Review*, 46(4): 1115–1185, 31 January, 2009.
Srinivas, S. 2008. 'Intellectual property rights, innovation and healthcare: Unanswered questions in theory and policy', *Economica, Rio de Janeiro*, 10(2): 106–146.
UNCTAD-ICSTD. 2005. *Resource Book on TRIPS and Development*. Cambridge, UK: Cambridge University Press.
Upadhyay, V., A. Ray and P. Basu. 2002. 'A Socio-economic Study of the Role of In-house R&D in Indian Industry: A Case Study of Pharmaceutical and Tyre Sectors', Department of Science and Technology, March 2002.
Visalakshi, S. 2009. 'Transferring biotechnology in India: Experiences and lessons', *International Journal of Technology Transfer and Commercialisation*, 8(2/3).
Visalakshi, S. and G.D. Sandhya. 2000. 'R&D capabilities in pharma industry in India in the context of biotechnology commercialisation. *Economic and Political Weekly*, XXXV(4): 4223–33.
Visalakshi, S., S. Gautam and D. Abrol. 1995. 'Assessment of R&D and production capabilities in pharmaceutical industry in India in the context of commercialization of BT', Report no REP-156/95, NISTADS (CSIR), Delhi.

Appendix

Selected Publications of Robert E. Evenson

A. Books and Monographs (Authored and Co-authored)

Science for Agriculture: A Long-term Perspective (Second Edition) (with W. Huffman). Ames, Iowa: Blackwell Publishing Professional, 2006.
Measuring the Impact of Climate Change on Indian Agriculture (with A. Dinar, R. Mendelsohn, J. Parikh, A. Sanghi, K. Kumar, J. McKinsey and S. Lonergan). World Bank Technical Paper No. 402. Washington, DC: The World Bank, 1998.
Rice Research in Asia: Progress and Priorities (with R. Herdt, M. Hossain). Wallingford, UK: CAB International, 1996.
The Development of U.S. Agricultural Research and Education: An Economic Perspective (with W. Huffman). Ames, Iowa: Iowa State University, 1994.
Adjustment and Technology: The Case of Rice (with C. David). OECD, 1993.
Science for Agriculture: A Long-term Perspective (with W. Huffman). Ames, Iowa: Iowa State University Press, 1993.
Research and Productivity in Asian Agriculture (with Carl Pray et al.). Ithaca, New York: Cornell University Press, 1991.
Research, Productivity and Incomes in Brazilian Agriculture: A Study of the EMBRAPA Program (with E.R. da Cruz, J. Strauss, M.T.L. Barbosa and D. Thomas). Brasilia: EMBRAPA, 1991.
Agricultural Research Productivity in Pakistan (with Q.T. Azam and Erik Bloom). Islamabad: Pakistan Agricultural Research Council, 1991.
National and International Agricultural Research and Extension Programs (with J.K. Boyce). New York: Agricultural Development Council, 1975.
Agricultural Research and Productivity (with Y. Kislev). New Haven, Connecticut: Yale University Press, 1975.

B. Books and Monographs (Edited and Co-edited)

Bantilan, M.C.S., U.K. Deb, C.L.L. Gowda, B.V.S. Reddy, A.B. Obilana and R.E. Evenson (eds). *Sorghum Genetic Enhancement: Research Process Dissemination and Impacts*. (Forthcoming).

Evenson, R.E. and P. Pingali (eds), 2006. *Handbook of Agricultural Economics, Volume 3*. Amsterdam, The Netherlands: Elsevier-North Holland Publishers.

Evenson, R.E. and P. Pingali (eds), 2007. *Handbook of Agricultural Economics, Volume 4*. Amsterdam, The Netherlands: Elsevier-North Holland Publishers. (Forthcoming).

Evenson, R.E. and V. Santaniello (eds), 2004. *Consumer Acceptance of Genetically Modified Foods*. Wallingford, UK: CAB International.

Evenson, R.E. and V. Santaniello (eds), 2004. *The Regulation of Agricultural Biotechnology*. Wallingford, UK: CAB International.

Evenson, R.E. and D. Gollin (eds), 2003. *Crop Variety Improvement and Its Effect on Productivity: The Impact of International Agricultural Research*. Wallingford, UK: CAB International.

Evenson, R.E., V. Santaniello and D. Zilberman (eds), 2002. *Economic and Social Issues in Agricultural Biotechnology*. Wallingford, UK: CAB International.

Evenson, R.E. and D. Gollin (eds), 2001. *Crop Variety Improvement and Its Effect on Productivity: The Impact of International Agricultural Research*. Wallingford, UK: CAB International.

Evenson, R.E., V. Santaniello, D. Zilberman and G.A. Carlson (eds), 2000. *Agriculture and Intellectual Property Rights*. Wallingford, UK: CABI Publishing.

Agricultural Values of Plant Genetic Resources (edited with D. Gollin and V. Santaniello), 1998. Wallingford, UK: CABI Publishing.

Invention-Innovation Input-Output Measures: Special Issue Economic Systems Research (edited with D. Johnson), April 1997.

Science and Technology: Lessons for Development (edited with Gustav Ranis), 1990. Boulder, Colorado: Westview Press.

Rural Household Studies in Asia (edited with C. Florencio, B. White and H.P. Binswanger), 1979. Singapore: University of Singapore Press.

Philippine Household Economics (ed), 1978. *Philippine Economic Journal*, Symposium Volume.

C. Publications: Economics of Agriculture

1. Journal Papers

Huffman, W.E. and R.E. Evenson. 2006. 'Do formula or competitive grant funds have greater impacts on state agricultural productivity?', *American Journal of Agricultural Economics*, 88(4): 783–798.

Evenson, R.E. 2004. 'Food and population: D. Gale Johnson and the Green Revolution', *Economic Development and Cultural Change*, 52(3) (April): 543–570.

'Assessing the impact of the Green Revolution, 1960–2000' (with D. Gollin), *Science*, 300: 758, 2003.

'Structural and productivity change in U.S. agriculture, 1950–1982' (with W. Huffman), *Agricultural Economics*, 24, 2001.

'The effect of agricultural extension on farm yields in Kenya' (with G. Mwabu), *African Development Review*, 2001.

Appendix 311

'How far away is Africa? Technological spillovers to agriculture and productivity', (with D.K.N. Johnson), *American Journal of Agricultural Economics*, 82 (August): 743–749, 2000.

'R&D spillovers to agriculture: Measurement and application' (with D.K.N. Johnson), *Contemporary Economic Policy*, 17(4): 432–456, 1999.

'Technological spillovers in Southern Cone agriculture' (with E. Rodrigues da Cruz), *Economia Aplicada*, 1(4): 709–730, 1997.

'The impact of T&V extension in Africa: The experience of Kenya and Burkina Faso' (with V. Bindlish), *World Bank Observer*, 1997.

'Genetic resources, international organizations, and rice varietal improvement' (with D. Gollin), *Economic Development and Cultural Change*, 45(3): 471–500, 1997.

'Global warming impacts on Brazilian agriculture: Estimates of the Ricardian model' (with A. Sanghi, D. Alves and R. Mendelsohn), *Economia Aplicada*, 1(1): 7–33, 1997.

'Science for agriculture: International perspectives. *Asian Journal of Agricultural Economics*, 2(1): 11–38, 1996.

'Productivity change and technology transfer in the Brazilian grain sector', (with A. Flavio Avila), *Revista de Economia Rural*, 34(2), November 1995.

'Measurement of the total factor productivity of Spanish agriculture: 1962–1989', (with M. Carmen Fernandez and A. Casmiro Herruza), *Oxford Agrarian Studies*, 23(1), 1995.

'Efficiency in agricultural production: The case of peasant farmers in eastern Paraguay', (with B. Bravo-Ureta), *Agricultural Economics*, 10(1), January 1994.

'Agricultural technology: International dimensions' (with Emphasis on Rice), *Technology Forecasting and Social Change*, 43: 337–351, 1993.

'Agricultural productivity growth in Pakistan and India: A comparative analysis' (with M. Rosegrant), *The Pakistan Development Review*, 32(4) (winter), 1993.

'Agricultural productivity and sources of growth in South Asia' (with M. Rosegrant), *American Journal of Agricultural Economics*, 74(3): 757–761, August 1992.

'The contribution of public and private science and technology to U.S. agricultural productivity' (with W. Huffman), *American Journal of Agricultural Economics*, 74(3): 751–756, August 1992.

'Research and extension in agricultural development', *Forum Valuazione*, 2, November 1991.

'The economic impact of agricultural extension: A review' (with D. Birkhaeuser and G. Feder), *Economic Development and Cultural Change*, 39(3), April 1991.

'Spillover benefits of agricultural research', *American Journal of Agricultural Economics*, 71(2), 1989.

'Supply and demand functions for multi-product U.S. cash grain farms' (with W. Huffman), *American Journal of Agricultural Economics*, 71(3), August 1989.

'Forestry Research: A Provisional Global Inventory' (with F. Mergen, M. Ann Judd and J. Putnam), *Economic Development and Cultural Change*, 37(1), 1988.

'Balancing basic and applied science: The case of agricultural research' (with Glenn Fox and V.W. Ruttan), *Bio Science*, 37(7), July/August 1987.

'Institutional change in intellectual property rights' (with J. Putnam), *American Journal of Agricultural Economics*, 69, 1987.

'Methods for agricultural policy analysis: An overview' (with C.F. Habito, A.R. Quisumbing and C.S. Bantilan), *Journal of Philippine Development*, XIII, 1986.

'Regional total factor productivity change in Philippine agriculture' (with M.L. Sardido), *Journal of Philippine Development*, XIII, 1986.

'Infrastructure, output supply and input demand in Philippine agriculture: Provisional estimates', *Journal of Philippine Development*, XIII, 1986.
'Impact multiplier policy models for the agricultural sector: An application to India', *Journal of Philippine Development*, XIII, 1986.
'Equity implications of public policy in an unstable agricultural economy: Discussion', *American Journal of Agricultural Economics*, 67(2), May 1985.
'Output supply and input demand effects of high yielding rice and wheat varieties in north Indian agriculture', *Indian Journal of Quantitative Economics*, 1(1), 1985.
'The influence of international research on the size of national research systems' (with C. Pray and G. Scobie), *American Journal of Agricultural Economics*, 67(5), 1985.
'Investing in agricultural supply' (with M.A. Judd and J.K. Boyce), *Economic Development and Cultural Change*, 35(1), 1985.
'The political economy of agricultural research and extension: Grants, votes, and reapportionment' (with S. Rose-Ackerman), *American Journal of Agricultural Economics*, 67(1), February 1985.
'Intellectual property rights and agribusiness research and development: Implications for the public agricultural research system', *American Journal of Agricultural Economics*, 65(5), December 1983.
'Observations on Brazilian agricultural research and productivity', *Revista de Economia Rural*, 19(4), October–December 1981.
'Benefits and obstacles to appropriate agricultural technology', In *Technology Transfer: New Issues*.
'New analysis' (A.W. Heston and H. Pack, eds), *The Annals of the American Academy of Political and Social Science*, 458, November 1981. Also Chapter 24 in *Agricultural Development in Third World* (Carl K. Eicher and John M. Staatz, eds). Baltimore: Johns Hopkins University Press, 1984.
'Economic benefits from research: An example from agriculture' (with V.W. Ruttan and P. Waggoner), *Science*, 205, 14 September 1979.
'Technology access and factor markets in agriculture', *Philippine Economic Journal*, 1–2, 1979.
'Social returns to rice research in the Philippines: Domestic benefits and foreign spillover' (with P. Flores-Moya and Y. Hayami). *Economic Development and Cultural Change*, 26(3), April 1978.
'Productivity in Philippine agricultural regions' (with M. Sardido and M.C. Antonio), *Human Resource Development Journal*, March 1977.
'Research and factor productivity in agriculture: An inter-country study' (with Y. Kislev), *International Journal of Agrarian Affairs* (supplement), 1, 1976.
'International transmission of technology in the production of sugarcane', *Journal of Development Studies*, 12(2), January 1976.
'The intensive agricultural districts program in India: A new evaluation' (with R. Mohan), *Journal of Development Studies*, 1976.
'Investment in agricultural research and extension: An international survey' (with Y. Kislev), *Economic, Development and Cultural Change*, 23(3), April 1975.
'Agricultural technology transfer', *Productivity* (New Delhi), XVI(2), July–September 1975.
'The Green Revolution in recent development experience', *American Journal of Agricultural Economics*, 56(2), May 1974.
'International diffusion of agrarian technology', *Journal of Economic History*, XXXIV(1), March 1974.

'Research and productivity in wheat and maize' (with Y. Kislev), *Journal of Political Economy*, 81(6): 1309–1329, November–December 1973.

'The contribution of the agricultural research system to agricultural production in India', (with D. Jha), *Indian Journal of Agricultural Economics*, XXVIII(4), October–December 1973.

'Production quota systems with production uncertainty', *International Journal of Agrarian Affairs*, 1972.

'Economic factors in research and extension investment policy', *FAO FCA Monthly Bulletin*, 1972.

'The contribution of agricultural research to production', *Journal of Farm Economics*, 49(5), December 1967.

2. Other Publications

'Objectives and Methodology for Country Studies', Chapter 21, in *Crop Variety Improvement and Its Effect on Productivity: The Impact of International Agricultural Research*, R. Evenson and D. Gollin (eds). Wallingford, UK: CAB International, 2003.

'Crop Genetic Improvements on Indian Agriculture' (with J. McKinsey), Chapter 19, in *Crop Variety Improvement and Its Effect on Productivity: The Impact of International Agricultural Research*, R.E. Evenson and D. Gollin (eds). Wallingford, UK: CAB International, 2003.

'Brazil' (with F. Avila, S. DeSilva and F.A. de Almeida), Chapter 20, in *Crop Variety Improvement and Its Effect on Productivity: The Impact of International Agricultural Research*, R.E. Evenson and D. Gollin (eds). Wallingford, UK: CAB International, 2003.

'Production Impacts of Crop Genetic Improvement', Chapter 22, in *Crop Variety Improvement and Its Effect on Productivity: The Impact of International Agricultural Research*, R.E. Evenson and D. Gollin (eds). Wallingford, UK: CAB International, 2003.

'The Economic Consequences of Crop Genetic Improvement Programmes' (with Mark Rosegrant), Chapter 23, in *Crop Variety Improvement and Its Effect on Productivity: The Impact of International Agricultural Research*, R.E. Evenson and D. Gollin (eds). Wallingford, UK: CAB International, 2003.

'From the Green Revolution to the Gene Revolution', In Evenson, R.E., V. Santaniello and D. Zilberman (eds), *Economic and Social Issues in Agricultural Biotechnology*. Wallingford, UK: CABI Publishing, 2002.

'The Value of Plant Biodiversity for Agriculture' (with B. Wright), in *Agricultural Science Policy, Changing Global Agendas*, J. Alston, P.G. Pardey and M.J. Taylor (eds). IFPRI, Baltimore: Johns Hopkins University Press, 2001.

'IARC "Germplasm" Effects on NARS Breeding Programs', Chapter 21, in *Crop Variety Improvement and Its Effect on Productivity: The Impact of International Agricultural Research*, R.E. Evenson and D. Gollin (eds). Wallingford, UK: CAB International, 2001.

Productivity and Agricultural Growth in the Second Half of the 20th Century: Survey of Food and Agriculture. Rome: FAO, 2000.

'Economics of Intellectual Property Rights for Agricultural Technology', Chapter 6, in *Agriculture and Intellectual Property Rights*, V. Santaniello, R.E. Evenson, D. Zilberman and G.A. Carlson (eds). Wallingford, UK: CABI Publishing, 2000.

'Biotechnology Inventions: Patent Data Evidence' (with A. Zohrabyan), Chapter 12, in *Agriculture and Intellectual Property Rights*, V. Santaniello, R.E. Evenson, D. Zilberman and G.A. Carlson (eds). Wallingford, UK: CABI Publishing, 2000.

'Agricultural Technology Spillovers', Chapter 11, in *Public–Private Collaboration in Agricultural Research: New Institutional Arrangements and Economic Implications*, K.O. Fuglie and D.E. Schimmelpfennig (eds). Ames: Iowa State University Press, 2000.

'Global and local implications of biotechnology and climate change for future food supplies', *Proceedings of the National Academy of Science*, 96: 5921–5928, May 1999.

'Biotechnology and Genetic Resources', Chapter 19, in *Agricultural Values of Plant Genetic Resources*, R.E. Evenson, D. Gollin and V. Santaniello (eds). Wallingford, UK: CABI Publishing, 1998.

'The Economic Value of Genetic Improvement in Rice', Chapter 18, in *Sustainability of Rice in the Global Food System*, N.G. Dowling, S.M. Greenfield, and K.S. Fischer (eds). Davis, California: Pacific Basin Study Center, and Manila, Philippines: International Rice Research Institute, 1998.

'Rice Varietal Improvement and International Exchange of Rice Germplasm', in *The Impact of Rice Research*, P.L. Pingali and M. Hossain (eds). Bangkok, Thailand: Development Research Institute, and Manila, Philippines: International Rice Research Institute, 1998.

'Optimal Collection and Search for Crop Genetic Resources' (with S. Lemarié), in *Farmers, Gene Banks and Crop Breeding: Economic Analyses of Diversity in Wheat, Maize, and Rice*, M. Smale (ed). Boston, MA: Kluwer Academic Publishers, 1998.

'Plant Breeding: A Case of Induced Innovation', in *Agricultural Values of Plant Genetic Resources*. Wallingford, UK: CABI Publishing, 1998.

'An Application of Hedonic Pricing Methods to Value Rice' (with D. Gollin), in *Agricultural Values of Plant Genetic Resources*. Wallingford, UK: CABI Publishing, 1998.

'Varietal Trait Values for Rice in India' (with K.P.C. Rao), in *Agricultural Values of Plant Genetic Resources*. Wallingford, UK: CABI Publishing, 1998.

'Modern Varieties, Traits, Commodity Supply and Factor Demand in Indian Agriculture', in *Agricultural Values of Plant Genetic Resources*. Wallingford, UK: CABI Publishing, 1998.

'Crop-loss Data and Trait Value Estimates for Rice in Indonesia', in *Agricultural Values of Plant Genetic Resources*. Wallingford, UK: CABI Publishing, 1998.

'Breeding Values of Rice Genetic Resources' (with D. Gollin), in *Agricultural Values of Plant Genetic Resources*. Wallingford, UK: CABI Publishing, 1998.

'Biotechnology and Genetic Resources', in *Agricultural Values of Plant Genetic Resources*. Wallingford, UK: CABI Publishing, 1998.

'The Economic Principles of Research Resource Allocation', in *Rice Research in Asia: Progress and Priorities*. Wallingford, UK: CAB International, 1996.

'Priority Setting Methods', in *Rice Research in Asia: Progress and Priorities*. Wallingford, UK: CAB International, 1996.

'An Application of Priority-Setting Methods to the Rice Biotechnology Program', in *Rice Research in Asia: Progress and Priorities*. Wallingford, UK: CAB International, 1996.

'Rice Research Priorities: An Application' (with M.M. Dey and M. Hossain), in *Rice Research in Asia: Progress and Priorities*. Wallingford, UK: CAB International, 1996.

'Economic Valuation of Biodiversity for Agriculture', in *Biodiversity, Biotechnology, and Sustainable Development in Health and Agriculture: Emerging Connections*. Scientific Publication No. 560, Washington, DC: Pan American Health Organization, 1996.

'Sources of Agricultural Productivity Growth in India' (with M. Rosegrant and C. Pray), to be published as *Research Report—International Food Policy Research Institute*, 1996.

'Extension, Technology, and Efficiency in Agriculture in sub-Saharan Africa', in *Achieving Greater Impact from Research Investments in Africa*. SAA/Global 2000/CASIN, 1996.

'Total Factor Productivity and Sources of Long-Term Growth in Indian Agriculture' (with Mark W. Rosegrant). IFPRI: EPTD Discussion Paper No. 7, April 1995.

'The Economic Contribution of Agricultural Extension to Agriculture and Rural Development', in *Improving Agricultural Extension: A Reference Manual*. FAO, 1995.

'Research and Extension Impacts on Food Crop Production in Indonesia', in *Upland Agriculture in Asia*, J.W.T. Bottema and D.R. Stolz (eds). CGRPT Center, 1994.

'Analyzing the Transfer of Agricultural Technology', in *Agricultural Technology: Policy Issues for the International Community*, J.R. Anderson (ed). Wallingford, UK: CAB International, 1994.

'Research, Extension and Productivity in Asian Agriculture: Implications for Viet Nam', Chapter 4, in *Impact of Agricultural Policies: Experiences from Asian Countries and Implications for Viet Nam*, Randolph Barker (ed). FAO, 1994.

'Commercial Forestry and Rural Development: Looking at the Next Fifty Years' (with W.F. Hyde), in *Forest Resources and Wood-Based Biomass Energy as Rural Development*, W.R. Bently and M.R. Geron (eds). Oxford, 1994.

'The Effects of R&D on Farm Size, Specialization and Productivity' (with W. Huffman), Chapter 3, in *Industrial Policy for Agriculture in the Global Economy*, S.R. Johnson and S.A. Martin (eds). Ames, Iowa: Iowa State University Press, 1993.

'India: Population Pressure, Technology, Infrastructure, Capital Formation, and Rural Income', in *Population and Land Use in Developing Countries*. National Research Council, National Academy of Science, 1993.

'Evaluation of T&V Extension in Burkina Faso' (with V. Bindlish and M. Gbetibouo), World Bank Technical Paper Number 226, African Technical Department Series, 1993.

'Education, Extension and Farmer Productivity' (with D.T. Jamison, M. Lockheed and L.J. Lau), in *Encyclopedia of Educational Research*. New York: Macmillan for Educational Research Association, pp. 395–494, 1992.

'Genetic Resources: Assessing Economic Value', in *Valuing Environmental Benefits in Developing Economies*. Special Report 29, Michigan State University, Agricultural Experiment Station, September 1991.

'Genetic Research: Assessing Economic Value', in *Managing Global Genetic Resources: Agricultural Imperatives*. Washington: NAS, 1993.

'Evaluation of the Performance of T&V Extension in Kenya' (with V. Bindlish), *World Bank Agricultural and Rural Development Series #7*. World Bank, 1993.

'Education, Extension and Farmer Productivity' (with D.T. Jamison, L.J. Lau and M.E. Lockheed), in *Encyclopedia of Educational Research*, Sixth Edition. New York: Macmillan, 1991.

'Evaluating Agricultural Research and Extension in Peru' (with V. Ganoza, G. Norton, C. Pomereda and E. Walters), in *Methods for Diagnosing Research System Constraints and Assessing the Impacts of Agricultural Research Vol. II*. The Hague: ISNAR, 1990.

'The Economics of Extension', in *Investing in Rural Extension: Strategies and Goals*, Gwyn E. Jones (ed). Elsevier Applied Science Publishers, August 1989.

'U.S. Agricultural Competitiveness: Evidence from Invention Data', in *Biotechnology and the New Agricultural Revolution*, J. Molnar and H. Kinnuae (eds), AASA Selected Symposium 108, 1989.

'Agricultural Technology, Supply and Factor Demand in Asian Economies', in *Conference on Direction and Strategies of Agricultural Development in the Asia-Pacific Region, January 5–7, 1988*. Taipei: The Institute of Economics, Academia Sinica, 1988.

'Research, Extension, and U.S. Agricultural Productivity: A Statistical Decomposition Analysis', in *Agricultural Productivity: Measurement and Explanation*, S.M. Capalbo and J.M. Antle (eds). Washington, DC: Resources for the Future, 1988.

'Technological Opportunities and International Technology Transfer in Agriculture', in *The Agro-Technological System Towards 2000*, G. Antonelli and A. Quadrio-Curzio (eds). North-Holland: Elsevier Science Publishers B.V., 1988.

'The IARCs and Their Impact on National Research and Extension Programs', in *Agricultural Extension Worldwide*, Wm. H. Rivera and Susan G. Schram (eds). New York: Croon Helm, 1987.

'The Importance of Agricultural Research during a Period of Farm Surpluses', in *Critical Concerns of U.S. Agriculture*. The Philadelphia Society for Promoting Agriculture, 1987.

'The International Agricultural Research Centers: Their Impact on Spending for National Agricultural Research and Extension', Study Paper No. 22, Consultative Group on International Agricultural Research, Washington, DC: The World Bank, 1987.

'Investment in Agricultural Research and Extension' (with M.A. Judd and J.K. Boyce), in *Policy for Agricultural Research*, V.W. Ruttan and C.E. Pray (eds). Boulder: Westview Press, 1987.

'Private Sector Agricultural Invention in Developing Countries' (with D.D. Evenson and J.D. Putnam), in *Policy for Agricultural Research*, V.W. Ruttan and C.E. Pray (eds). Boulder: Westview Press, 1987.

'Resources for the Production of Agricultural Growth in the Pacific Basin Region' (with M.A. Judd), in *Food, Agricultural, and Development in the Pacific Basin: Prospects for International Collaboration in a Dynamic Economy*, G.E. Schuh and J.L. McCoy (eds). Boulder: Westview Press, 1986.

'The Pretechnology Agricultural Sciences', in *The Agricultural Scientific Enterprise: A System in Transition*, L. Busch and W.B. Lacey (eds). Boulder: Westview Press, 1986.

'Discussion, Price Policies in Developing Countries', in *The Role of Markets in the World Food Economy*, D. Gale Johnson and G. Schuh (eds). Boulder: Westview Press, 1984.

'Economics of Agricultural Growth: The Case of Northern India', in *Issues in Third World Development*, K.C. Nobe and R.K. Sampath (eds). Boulder: Westview Press, 1984.

'Labor Demand in North Indian Agriculture' (with H. Binswanger), in *Contractual Arrangements, Employment and Wages in Rural Labor Markets in Asia*, H. Binswanger and M. Rosenzweig (eds). New Haven: Yale University Press, 1984.

'Economic Benefits from Research', in *Twentieth Century Agricultural Science*. Beltsville, Maryland: National Agricultural Library, 1983.

'Agriculture', in *Government and Technical Progress*, R.R. Nelson (ed). Elmsford, New York: Pergamon Press, 1982.

'Economic Issues in Tropical Forest Policy', in *Tropical Forests: Utilization and Consumption*, F. Mergen (ed). New Haven: Yale School of Forestry and Environmental Studies, 1981.

'Research Evaluation: Policy Interests and the State of the Art', in *Methodology for Evaluation of Agricultural Research*, B. Sundquist and G. Norton (eds). University of Minnesota: Agricultural Experiment Station Bulletin, 1981.

'Productivity Growth in U.S. Agriculture: An Historical Perspective on Causes, Consequences and Prospects', in *Increasing Understanding of Public Problems and Policies*, J. Hildreth (ed). Oak Brook, Illinois: Farm Foundation, 1980.

'An Evaluation of Methods for Examining the Quality of Agricultural Research' (with B. Wright), *An Assessment of the United States Food and Agricultural Research System, Vol. II*. Office of Technology Assessment, Washington, DC: U.S. Government Printing Office, 1980.

'Risk and Uncertainty as Factors in Crop Improvement Research' (with R. Herdt, J. O'Toole, W.R. Coffman and H. Kaufman), in *Risk, Uncertainty, and Agricultural Development*, J. Roumasset, et al. (eds). Philippines: College Laguna, SEARCA, 1979.

'The Organization of Crop and Animal Improvement Research in Low-Income Countries', in *Distortions of Agricultural Incentives*, T.W. Schultz (ed). Bloomington: Indiana University Press, 1978.

'Agricultural Research and Education in Asia', *Second Asian Agriculture Survey*. Survey Supplementary Papers Vol. II, Asian Development Bank, 1978.

'Agricultural Research and U.S. Comparative Advantage', in *The Competitive Threat from Abroad: Fact or Fiction*. Lafayette, Indiana: Purdue University Press, 1978.

'Social Returns to Rice Research' (with P. Flores-Moya), in *Economic Consequences of New Rice Technology in Asia*, Y. Hayami (ed). Los Banos: International Rice Research Institute, 1978.

'Technology Transfer and Research Resource Allocation' (with H.P. Binswanger), in *Induced Innovation: Technology, Institutions, and Development*, H.P. Binswanger and V.W. Ruttan (eds). Baltimore: Johns Hopkins University Press, 1978.

'Comparative Evidence on Returns to Investment in National and International Research Institutions', in *Resource Allocation and Productivity in National and International Agricultural Research*, T. Arndt, D.G. Dalrymple and V.W. Ruttan (eds). Minneapolis: University of Minnesota Press, 1977.

'Cycles in Research Productivity and International Diffusion Patterns in Sugarcane, Wheat, and Rice', in *Resource Allocation and Productivity in National and International Agricultural Research*, T. Arndt, D.G. Dalrymple and V.W. Ruttan (eds). Minneapolis: University of Minnesota Press, 1977.

'Technology Generation in Agriculture', in *Agriculture in Development Theory*, L.G. Reynolds (ed). New Haven: Yale University Press, 1975.

'Science and the world food problem', *Connecticut Agriculture Experiment Station Bulletin*, 758, 1975.

'Consequences of the Green Revolution', *Proceedings of the VI Pacific Trade Congress*, Mexico City, July 1974.

'Research, Extension and Schooling in Agricultural Development', in *World Yearbook of Education 1974*. London: Evans Bros. Limited, 1974.

'Agricultural Trade and Shifting Comparative Advantage', in *Trade and Agricultural Development*, G.W. Tolley (ed). New York: Ballinger, 1973.

'Economic Aspects of the Organization of Agricultural Research', in *Resource Allocation in Agricultural Research*, W. Fishel (ed). Minneapolis: University of Minnesota Press, 1971.

'Technical Change and Agricultural Trade: Three Examples—Sugarcane, Bananas and Rice' (with V.W. Ruttan and J.P. Houck), in *The Technology Factor in International Trade*, R. Vernon (ed). New York: Columbia University Press, 1970.

D. Publications: Economics of Invention

1. Journal Papers

Evenson, R.E. 2004. 'GMOs: Prospects for productivity increases in developing countries', *Journal of Agricultural & Food Industrial Organization*, 2(2): Article 2.

'Economic growth, international technological spillovers and public policy: Theory and empirical evidence from Asia' (with Lakhwinder Singh), *Indian Management Studies Journal*, 4, 2000.

'Foreign technology licensing in Indian industry' (with K.J. Joseph), *Economic and Political Weekly*, July 3, 1999.

'Industrial productivity growth linkages between OECD countries 1970–1990', *Economic Systems Research*, April 1997.

'Innovation and invention in Canada' (with D. Johnson), *Economic Systems Research*, April 1997.

'Intersectoral technology flows: Estimates from a patent concordance with an application to Italy' (with J. Putnam), *Innovazione e Materie Prime*, 1993.

'Patents, R&D, and invention potential: International evidence', *American Economic Review*, 83(2): 463–468, 1992.

'Imitation and the new industrialists', *Foreign Service Journal*, 68(11), November 1991.

'Technological opportunities and international technology transfer in agriculture', *International Journal of Development Banking*, 9(2), July 1991.

'Invention intended for use in agriculture and related industries: International comparisons', *American Journal of Agricultural Economics*, 73(3), 1991.

'Technology production and technology purchase in Indian industry: An econometric analysis' (with A. Deolalikar), *Review of Economics and Statistics*, 1989.

'R&D, innovation and the total factor productivity slowdown' (with A.S. Englander and M. Hanazaki), *OECD Economic Studies*, 11(Autumn), 1988.

'Patents, innovation and competition in Australia', *Prometheus*, 3(2), December 1985.

'Intellectual property rights and the Third World', *European Intellectual Property Review*, 5(12), December 1983.

'Invention in Philippine industry' (with K. Mikkelson and E. Medalla), *Journal of Philippine Development*, IX(1–2)(16), 1982.

'Scale economies, elasticities of substitution and productivity change in the agro-based industries in India' (with M.A. Oomen), *The Asian Economic Review*, XIX(1), April 1977.

'A stochastic model of applied research' (with Y. Kislev), *Journal of Political Economy*, 84(2): 265–282, April 1976.

2. Other Publications

Evenson, R.E. *Agricultural Development and Intellectual Property Rights*. Cambridge: Cambridge University Press, 2004.

Evenson, R.E. 'Induced Adaptive Invention/Innovation and Productivity Convergence in Developing Countries', In *Technological Change and the Environment*, A. Grübler, N. Nakicenovic, W.D. Nordhaus (eds). Washington, DC: RFF Press, 2002.
Evenson, R.E. 'Economic Impacts of Agricultural Research and Extension', In B. Gardner and G. Rausser (eds), *Handbook of Agricultural Economics, Vol. 1*. Amsterdam: Elsevier Science, BV, 2001.
'Technological Specialization in International Patenting' (with J. Eaton, S. Kortum, P. Marino and J. Putnam), In *Essays in Honor of Gustav Ranis*. Ann Arbor, Michigan: University of Michigan Press, 1997.
'Technological Change and Technology Strategy' (with L. Westphal), *Handbook of Development Economics, Vol. 3*, T.N. Srinivasan and J. Behrman (eds). Amsterdam: North Holland Publishing Company, 1994.
'Human Resources and Technological Development', In *International Investment in Human Capital*, Crauford D. Goodwin (eds). IIE Report No. 24, 1993.
'Global Intellectual Property Rights in Perspective', In *Global Dimensions of Intellectual Property Rights*. National Research Council, NAS, 1993.
The Role of Patents and Intellectual Property Rights for Development of Agribusiness Markets. Alabama: International Fertilizer Development Center, 1993.
'Patent Data by Industry: Evidence for Potential Exhaustion of Invention', In *Technology and Productivity: The Challenge for Economic Policy*. Paris: OECD, 1991.
'Human Capital and Agricultural Productivity Change', In *Agriculture and Governments in an Interdependent World*. Proceedings of the 20th International Conference of Agricultural Economists, Buenos Aires, Argentina, January, 1990.
'Strengthening Protection of Intellectual Property in Developing Countries: A Survey of the Literature' (with W. Siebeck, M. Lesser and C. Primo-Braga), World Bank Discussion Paper 12, 1990.
'Intellectual Property Management' (with J. Putnam), In *Agricultural Biotechnology: Opportunities for International Development*, G.J. Persley (ed). Wallingford, UK: CAB International, 1990.
'Introduction', Chapter 1 (with G. Ranis), In *Science and Technology: Lessons for Development Policy*, R.E. Evenson and Gustav Ranis (eds). Boulder: Westview Press, 1990.
'Private Inventive Activity in Indian Manufacturing: Its Extent and Determinants' (with A. Deolalikar), In *Science and Technology: Lessons for Development Policy*, R.E. Evenson and Gustav Ranis (eds). Boulder: Westview Press, 1990.
'Intellectual Property Rights, R&D, Inventions, Technology Purchase, and Piracy in Economic Development: An International Comparative Study', In *Science and Technology: Lessons for Development Policy*, R.E. Evenson and Gustav Ranis (eds). Boulder: Westview Press, 1990.
'Intellectual Property Rights and the Land Grand University', *Protecting Intellectual Property Rights in an Academic Environment*. Minnesota: University of Minnesota, 1990.
'Invention and inventors in India', *Journal of Development Economics*, 1989.
'Technology, Productivity Growth and Economic Development', In *The State of Development Economics*, Gustav Ranis and T. Paul Schultz (eds). New York: Basil Blackwell, 1988.
'The Potential for Transfer of U.S. Agricultural Technology' (with C. Pray and J. Putnam), Office of Technology Assessment, 1986.
'International Invention: Implications for Technology Market Analysis', In *R&D, Patents, and Productivity*, Z. Griliches (ed). Chicago: University of Chicago Press, 1984. (Also EGC Discussion Paper No. 419.)

'Legal Systems and Private Sector Incentives for the Invention of Agricultural Technology in Latin America' (with D. Evenson), In *Technological Progress in Latin American Agriculture*, E. Trigo and M. Pineiro (eds). San Jose, Costa Rica: Inter-American Institute for Cooperation on Agriculture, 1982.

'Productivity Measurement in the Developing Economies: The Indian Case', In: *On the Measurement of Factor Productivity*, Franz-Lothar Altman, Oldrech Kyn and Hans-Jurgen Wagener (eds). Goettingen: Vandenhoech & Ruprecht, 1971.

E. Publications: Economics of Rural Households

1. Journal Papers

'Gender and agricultural extension in Burkina Faso' (with Michele Siegel), *Africa Today*, 46(1): Winter, 1999.

'Household composition and expenditures on human capital formation in Kenya' (with Germano Mwabu), In *Research in Population Economics*, 8, 205–232, 1996, JAI Press.

'Marriage in rural Philippine households', *Journal of Philippine Development*, 1995.

'Measuring food production (with reference to South Asia)' (with Carl E. Pray), In *Journal of Health Development Economics*, 44, Elsevier Science, 1994.

'Population pressure, technology, infrastructure, capital function and rural incomes in Asia', *Southeast Asian Journal of Agricultural Economics*, 1994.

'Markets, institutions and family size in rural Philippine households' (with J. Roumasset), *Journal of Philippine Development*, XIII, 1986.

'The allocation of women's time: An international comparison', *Behavior Science Research*, 17(3–4): spring–summer, 1983.

'Food policy and the new home economics', *Food Policy Journal*, 6(3), August 1981.

'Introduction', *The Philippine Economic Journal*, XVII(1–2)(36), 1978.

'Fertility, schooling and home technology in rural Philippine households' (with K. Banskota), *The Philippine Economic Journal*, XVII(1–2)(36), 1978.

'Production and consumption of nutrients in Laguna households: An exploratory analysis' (with S. Ybanez-Gonzalo), *The Philippine Economic Journal*, XVII(1–2)(36), 1978.

'Time allocation in rural Philippine households', *American Journal of Agricultural Economics*, 60(2), May 1978.

'Fertility, schooling and the economic contribution of children in rural India' (with M. Rosenzweig), *Econometrica*, 45(5), July 1977.

'On the new household economics', *Journal of Agricultural Economics and Development*, VI, January 1976.

2. Other Publications

'Health, Climate and Development in Brazil: A Cross-Section Analysis' (with E. Rosenberg, D. Alves and C. Timmins), Research Network Working Paper No. R-386, Inter-American Development Bank, August 2000.

'Ability, Schooling and Occupational Choice in Rural Philippine Households', In *Choice, Growth and Development (Essays in Honor of Jose Encarnacion)*, E.S. de Dios and R.V. Fabella (eds). Philippines: University of the Philippines, 1996.

'India: Population Pressure, Technology, Infrastructure, Capital Formation, and Rural Incomes', In *Population and Land Use in Developing Countries*. National Research Council, National Academy Press, 1994.

'Population, Technology and Rural Poverty in the Philippines: Rural Income Implications from a Simple CGE Impact Multiplier Model' (with A. Quisumbing and C. Bantilan), *Perspectives on Poverty in the Philippines*. Philippines: University of the Philippines, 1994.

'Institutions and Rural Poverty in Asia', In *Rural Poverty in Asia: Priority Issues and Policy Options*, M.G. Quibria (ed). Oxford, UK: Oxford University Press, 1993.

'Fertility and Poverty', Chapter 12, In *Poverty Theory & Policy: A Study of Panama*, G.S. Sahota (ed). Baltimore: Johns Hopkins University Press, 1990.

Women in Development: Philippines. The World Bank, 1990.

'Population Growth, Infrastructure and Real Incomes in North India', In *Population, Food, and Rural Development*, R.D. Lee, W.B. Arthur, A.C. Kelley, G. Rogers, and T.N. Srinivasan (eds). Oxford: Clarendon Press, 1988.

'Food Consumption, Nutrient Intake and Agricultural Production in India', Nutrition Economics Group, USDA–USAID, October 1986.

'Observations on Institutions, Infrastructure, Technology and Women in Rice Farms', In *Women in Rice Farming Systems*, L. Unnevhr (ed). Los Banos, Philippines: International Rice Research Institute, 1985.

'Discussion, Price Policies in Developing Countries', In *The Role of Markets in the World Food Economy*, D. Johnson and G. Schuh (eds). Boulder: Westview Press, 1984.

'Nutrition, Work and Demographic Behavior in Rural Philippine Households' (with E. King-Quizon and B. Popkin), In *Rural Households in Asia*. Singapore: Singapore University Press, 1979.

'Time Allocation and Home Production in Philippine Rural Households' (with E. King-Quizon), In *Women in Poverty*, W. McGreemey (ed), Washington, DC: International Center for Research on Women, 1979.

About the Editors and Contributors

Editors

Lakhwinder Singh is a Professor, Department of Economics and Coordinator, Centre for Development Economics and Innovation Studies (CDEIS), at Punjabi University, Patiala. Prior to this, he was a faculty member of the University of Delhi and National Institute of Public Finance and Policy, New Delhi. He was a Ford Foundation Postdoctoral Fellow in Economics at Yale University, USA and a Visiting Research Fellow, Seoul National University, South Korea. He has been awarded Asia Fellowship by the Institute of International Education, New York, 2001. Singh is the Founding Editor of the *Millennial Asia: An International Journal of Asian Studies* jointly published by SAGE and Association of Asia Scholars (AAS). His current research interests include the national innovation system, international knowledge spillovers, pattern of development, globalization and agrarian distress in developing economies. Apart from publishing more than 50 research papers in peer-reviewed journals and chapters in the books, he has the following books to his credit: *Economic and Environmental Sustainability of the Asian Region*, Routledge, 2010; *Economic Cooperation and Infrastructural Linkages between Two Punjabs: Way Ahead*, CRRID, 2010; and *Punjab's Economic Development in the Era of Globalisation*, 2014.

K.J. Joseph is the Ministry of Commerce Chair Professor at Centre for Development Studies, Trivandrum. He is also the Vice-President of Globelics and the Editor-in-Chief of *Innovation and Development*, published by Taylor and Francis. Professor Joseph also holds the position of expert in innovation studies in the Tianjin University of Finance and Economics appointed under the Tianjin Program of Recruitment of Global Experts. Earlier positions that he held include visiting senior fellow at Research and Information System for Developing Countries (RIS), New Delhi, and visiting professor at Jawaharlal Nehru University and consultant to United Nations Economic and Social Commission for Asia and the Pacific (UNESCAP). As a Ford Foundation Postdoctoral Fellow, he undertook research on technology licensing in India under the supervision of Robert Evenson at Yale University and published jointly with him. Apart from over 80 research papers, he has the following books to his credit: *Industry under Economic Liberalization: The Case of Indian Electronics* (Sage Publications); *Information Technology, Innovation System and Trade Regime in Developing Countries: India and the ASEAN* (Palgrave Macmillan); *Export Competitiveness of Knowledge Intensive Industries*, edited with Nagesh Kumar (Oxford University Press) and the *Handbook on Innovation Systems in Developing Countries* (Edward Elgar) edited with B.A. Lundvall, Cristina Chaminade and Jan Vang.

Daniel K.N. Johnson is the Chair of the Economics and Business Department at Colorado College, and a tenured Associate Professor of Economics. He has a Ph.D. in Economics from Yale University (1998), which he completed under Bob Evenson's supervision. He previously completed an M.Sc. from the London School of Economics (1992) and a B.Soc.Sc. Honours from the University of Ottawa, Canada (1991). He is the author of over 50 refereed journal articles, commissioned pieces and book chapters. Several of those are co-authored with Bob Evenson, and he takes special pleasure in talking about those with students as models of collaboration. He frequently co-authors with his undergraduate students, inspired by Bob's methodology. Specializing in the economics of innovation and technological change, he frequently advises and lectures internationally on public policy related to intellectual property rights. In addition to his primary research on knowledge spillovers, he enjoys writing projects that apply economic models to unusual questions: predicting Olympic medal counts, explaining game show contestant behaviour, evaluating the impact of Walmart on residential property values, exploring philanthropic behaviour patterns, improving the marginal impact of microfinance lending programmes, explaining the adoption of new election equipment in the wake of vote count scandals, analysing where consumers can find the cheapest gas and evaluating pedagogical techniques and outcomes in economics. Feeling that it was not enough to simply study innovation and entrepreneurship, he founded and now runs three start-up companies, all serving the interests of higher education: Economics of Technology Consulting, Lightning Abstracts and BookCheetah.

Contributors

Vinoj Abraham is an Associate Professor at Centre for Development Studies, Trivandrum, Kerala, India.

Dinesh Abrol is currently working as a Professor at the Institute of Studies in Industrial Development (ISID) affiliated to the Indian Council of Social Science Research (ICSSR), India.

Sanjaya DeSilva is an Associate Professor of Economics at Bard College in New York, USA.

Wallace E. Huffman is currently the C.F. Curtiss Distinguished Professor of Agriculture and Life Sciences and Professor of Economics, Iowa State University, USA.

Leonardo A. Lanzona, Jr is a Professor and former Chair of Economics at the Ateneo de Manila University, Philippines.

Kristina M. Lybecker is the Gerald R. Schlessman Professor of Economics and Associate Chair of the Department of Economics and Business at Colorado College in Colorado Springs, CO, USA.

M.A. Oommen, Honorary Professor, Centre for Development Studies, formerly Chairman, 4th State Finance Commission, Government of Kerala, is an economist of repute with a rich collection of professional papers and more than 25 books to his credit.

Binay Kumar Pattnaik, Professor of Sociology at IIT Kanpur, is currently working as the Director of Institute for Social and Economic Change, Bangalore, India.

Baldev Singh Shergill is currently serving as an Assistant Professor of Economics at Punjabi University, Guru Kashi College, Damdama Sahib, Bathinda, India.

Index

abbreviated new drug applications (ANDAs), 292–93
Agricultural Handbook on Soils, 113
Agricultural Research Service (ARS) of USDA, 107, 118, 122
agriculture/agricultural
  competition among farmers, 108
  competitive structure of, 108
  development, xxviii–xxx, 6
  econometric techniques to link agricultural productivity, 108
  extension, 6–7
  model of productivity, 109–10
  public agricultural research capital (*see* Public agricultural research capital)
  technologies, 108
agro-ecological zones, 124
American Journal of Agricultural Economics, xx
Annual Survey of Industries (ASI), 243–44, 267
applied research expenditures, 113
appropriability of returns, 44
Arrovian legacy, xxiv
Asian Development Bank, 85
Asia Pacific Design Centre of SGS-Thompson, 217
ASTRA-AB Research Centre India, Bangalore, 217

barangay (village) survey, 148
basic education, 5, 11–12, 16, 28
below poverty line (BPL), 97
Beneficiary Committee (BC) system, 94

Bertrand market, 67
Bicol Multipurpose Survey, Philippines, 148, 173n2
Bicol River Basin Development Program, Philippines, xxix, 148
biotechnology, xvii, xxiv, xxxiii, 52, 190, 217–18, 233
Brazil
  local affiliates of MNCs, 228
  MNCs engaged in asset-exploiting, 228
  R&D-related FDI, 227–28
  technological developments in, 228
Brazil, Russia, India, and China (BRIC), 278
Breusch–Pagan test, 34
business process outsourcing (BPO), 210, 226

call centres, 210, 226
capital accumulation, 17
capital-intensive system, 66
capitalist development model based, xxvii
Census of Agriculture, 113
centrally sponsored schemes (CSSs), 93
Central Statistical Organisation, 243
Centre for Development Studies (CDS), 81
Chandrasekhar, K.B., 225
child development, 11
China
  academic entrepreneurs in entrepreneurial universities, 222–23
  economic liberalization, 223

internal and external perspectives changes, 220–21
location for R&D affiliates, 181
transnational capitalists class emergence, 221
chlorofluorocarbons (CFCs), 56
clinical research organization (CRO), 276
Coase theorem, 61
Cobb–Douglas production function, 140n1
collectivist messianism, 221
college, quality and specialized training for, 38
Committee on Decentralisation of Powers (1996–99), 90. *See also* Sen Committee
community-based informal arrangements, 150
community-based organizations (CBOs), 95, xxviii
community development society (CDS), 95
competitive economies, 22
competitive federalism, 71
complementarity, 19, 40n5
contract research and manufacturing services (CRAMS), 290, 292
Cooperative States Research Service (CSRS), 141n13
Copenhagen Economics study, 49–50
Copenhagen Summit on Climate Change (2009), 49
correlation coefficients between technology capability index and growth rate of employment, 266
cost efficiency estimation on rice farming, 167–72
Council of Scientific and Industrial Research (CSIR), 302–04
Cournot environment, 67
creative destruction, 112
cross-country relationships, 18
cross-licensing, 53

Current Research Information System (CRIS), USDA, 118–19, 120–22, 138, 140

Daimler Benz Research Centre India, Bangalore, 217
dalits, 101n10
data regression model, 245
decentralization process, 88
decomposition analysis, 244
decomposition of effects on real wages growth, 260–62
Department of Science and Technology (DST), 302
depreciation of knowledge, 112
Deshpande, Gururaj, 225
Dhan, Vinod, 225
diffusion, xxiii, 49, 63, 181
of knowledge, 195
of technology/new technology, xxiv, 108, 239
District Level Expert Committee (DLEC), 93
District Planning Committee (DPC), 93, 99
District Rural Development Agency (DRDA), 93
domestic capital goods intensity, 270n1
dot-com boom, 226
drug master filings (DMF), 307n6
Drug Prices Control Order (DPCO), 307n10
Drug Promotion Research Programme (DPRP), 303
Drugs and Pharmaceuticals Research Programme (DPRP), 302
dynamic complementarity, 11–12
Dynamic Integrated model of Climate and the Economy (DICE) model, 60–61, 65

eco-innovations, 43, 45, 57, 63–64
approaches for, 47–48
challenges of, 49–72

government subsidies as threat to, 70
intellectual property right (IPR) (*see* Intellectual property right [IPR])
eco-innovative products, 64
ecological forbearance, 71
e-commerce, 210, 225
econometric model of agricultural productivity, 110
economic development, xxiii–xxiv, xxv–xxviii
Economic Growth Center of Yale University, 148
economic inequalities, 83
Economic Intelligence Unit (EIU), 227
economic reforms, xxv, 84
  initiated in early 1980s, 239
  jobless growth after, 256
Economics Research Service (ERS), 117–18
economies of scale, 45, 49, 56–58, 63, 72
education/education system/ educational system, 16
  ability to transform agriculture, 7
  in developing countries, 40n4
  individual and household returns to, 5
  provisions of, 22
  quality, estimation of, 34–37
  reforms
    challenges in, 6
    conceptual framework and development, 8–22
    estimate of success, 31
efficiency, achievement of, 51
elasticity of state agricultural TFP, 136–37
Eleventh Plan, 84
emerging market economies, 50
employment, 21, 57, 64, 85–86, 149, 224
  estimated fixed effects models, 265
  growth rate of, 266–67
  in India's manufacturing sector
    changing structure of, 245–49
    elasticity across, 256–58
    in pre- and post-reform period, 249–55
    pre-and post-reform period elasticity, determinants of, 259–63
  industrial (*see* industrial employment)
  outcomes of industrial development, 263–64
energy consumption, 60
  full-cost pricing reduce, 70
energy prices, 60
environmental competition statute, 69
environmental innovations, 43, 48, 58–59. *See also* eco-innovations
environmental policy(ies)/ policymaking, 59, 62, 65, 68–69
  on renewable energy technology, 69
environmental regulations, xxvii, 61–62, 69–70
environmental sustainability, xxiv
environmental technology(ies), 47–48, 59–61
European MNCs, 217
  offshoring of R&D, 232
European Union (EU), 181, 293–94
exclusive marketing rights (EMRs), 278

farm firms, xxix, 107
federal incentives, 61
fibre-optic bubble, 226
Food and Agriculture Organization (FAO), 124
foreign capital goods intensity, 270n1
foreign direct investment (FDI), xxxiv, 55, 210, 217, 276, 280–83
foreign disembodied technology intensity, 270n1

foreign universities, interaction of IT firms with, 192–95
formal markets, 149–50
fuel-cell technology, 66

General Agreement on Tariffs and Trade (GATT), 209, 217, 272–73
General Electric (GE) Research Centre India, Bangalore, 217
genetically modified (GM) patents, 53
geo-climate sensitive, 108
Ginarte-Park index, 306n1
global competitiveness index (GCI), 22
Global Competitiveness Report of WEF, 7, 22
global delinking, 216
global innovation networks (GINs), xxxi–xxxii, 187–88, 205–6
   centrifugal forces, prevalence of, 185
   domestic capability building, 183
   emergence and spread of, challenges on, 183
   expansion of, 181
   global innovation traps or poisoned chalice, 184
   India's presence in, 183
   interactions, types of, 182
   on knowledge production and diffusion, 181–82
globalization, xxxii, xxiii–xxiv, 19, 22, 38
   as an economic phenomenon, 209
   of centre, 211
   in developing countries
      effect of, 211
      industrial firms and universities, 214
      public research organizations, 215
   of industrial R&D (*see* Industrial R&D, globalization of)
   meaning of, 209, 212–13
   of R&D activities, 213–14
   of S&T regimes
      in developing countries, 212
      features, 210–11
Global Strategy and Plan of Action (GSPOA), 304
Global Talent Acceleration Programme (GTAP), 204
good manufacturing practices (GMP), 290
gram panchayat (GP), 98
gram sabha, 88, 90–91, 96, 99, 101n8
Green Revolution technology, 147
gross domestic product (GDPPC), per capita, 24–25, 28, 30, 37–38
growth–employment dilemma, 239

Hatch-Waxman Act 1887, 61, 142n22
Hatch-Waxman Act of 1984, 279, 293
heterogeneity in policy regulations, 71
hierarchical pattern of R&D, 220
high-technology industrial employment growth rates, 255
Hind Swaraj, concept of, 83
horizontal integration, 63
household decisions model, 8–10, 12, 21
   outcomes, types of, 11
Huffman, Wallace, xxviii–xxix, 107–40
human capital formation, 20
Human Development Report (HDR), Kerala, 99
humanity, problems faced by, xxvi
hydrofluorocarbons (HCFCs), 56

imperfect credit markets, 17–18
imported technologies, 241–42
incentives, 58–64, 72
inclusive democracy, 85
inclusive development, 82–86. *See also* social inclusion

inclusive growth, 82–86, 97. *See also* social inclusion
income inequality, 25
Indian Council of Medical Research (ICMR), 303
Indian Patent Act, 1970, 278, 305, 307n10
induced innovation, 58
industrial capitalism, 211
industrial development, xxx–xxxv, 97, 212, 238–42, 263–64
  historical experience of, 240
  jobless growth phenomenon, 243
industrial employment, 242
  database and methodology for study, 243–45
  growth, quantitative assessment of, 240
  high-technology shares, 249
  low-technology and medium low-technology shares, 247–48
  medium high-technology shares, 248
industrial firms in developing countries, 214
industrial R&D, multinationalization and globalization of
  flat technological world regime, emergence of, 220–26
  globalization of local firms from developing countries, 229–30
  international division of labour in R&D, emergence of
    first phase, 215–16
    second phase, 216–19
    third phase, 219–20
  multinationalization of R&D firms from developing countries, 226–29
  offshoring by firms from developed countries, 231–34
Industrial Revolution, xxx
industrial sector, 238
  labour force slow absorption in, 242

industrial structure
  in developed countries, evolution of, 240–41
  and employment, 263–66
industrial technological capability-based approach, xxxiii, 239–40, 263–66
industries name, classified on basis of technology characteristics and industrial codes, 268–69
industry–university interaction, 182
inequality, 85, 100n3
informal institutions' role in developing agrarian economics, 149–50
information and communications technology (ICT), 182–83, 186, 212, 218, 264. *See also* global innovation networks (GINs)
firms, xxxii
industry in India, 223
international knowledge flows in India
  interaction with local *vs* foreign universities, 192–95
  low level of university–industry interaction, 197–200
  own R&D *vs* university–industry interaction, 195–97
  R&D activities in firms, 187–89
  scale and pattern of interaction with universities, 189–90
  sectoral and regional patterns, 190–92
Infosys, 225
innovation(s) system, xxiv, xxv–xxviii, xxx–xxxv, 22, 43, 64
  challenge to policymakers, 52
  eco-innovation (*see* eco-innovations)
  effective, strategies for, 71–72
  energy prices effect on, 60
  environmental (*see* environmental innovations)

and government policies,
relationship between, xxiv
government subsidies for, 59
hurdles in, 58–59
intellectual property right (IPR) (*see* intellectual property right [IPR])
kinds of, 63
in less developed nations, successful, 72
market-based firms, 58
and policy agendas, 46–47
takes place in context of rules, 45
innovative institutions as change agents, 86–87
innovators, 44–45, 58–59, 72–73, 87, 108
agricultural, xxi
environmental, 48
legal right, 49
original, 52
institutional developments contribution to economic growth, 13
institutional environment, 18, 152
institutional innovation, xxvii–xxviii, 48, 82, 86–87
intellectual property(IP)/intellectual property rights (IPRs), xxiv, xxvi–xxvii, xxxiv, xxiii, 45, 72, 272, 274, 306n3
for eco-innovations in emerging economies, 50
impact on local innovation in developing countries, 55
importance to innovation, 55
profitability of commercial research, 51
-protected technologies, 49
revisiting the making of policy on post-TRIPs evidence
analysis of activity-wise FDI, 282–83
directions of overseas R&D in India, 286–95

domestic firms' R&D directions, 295–300
foreign direct investment, 280–82
relations of public sector R&D with domestic firms, emerging, 301–5
technology transfer from MNCs to India, evidence on, 283–86
technology strategy and developing countries, 274–79
International Energy Agency (1999), 60, 70
international GDP, 24–25
internationalization, xxxii, 184, 197–200, 206–7, 212–14, 226–27, 300
international subcontracting/ outsourcing process, 216–17
intra-state public agricultural research capital, 129
Inventory of Agricultural Research, 119
ITES industry, 226
IT firms in India, survey of, 186–87
IT small-and medium-scale enterprises (SMEs), 224

Janadhikara Kalajathas, 90
Japanese MNCs offshoring R&D, 231–32
job creation in industrial sector, 242
jobless growth, 238, 243, 266–67
joint ventures, xxiii, 53, 210

Kerala, 100
development experience, 81–82
local government reform initiatives of, 89–95
People's Plan Campaign (PPC) 1996, 82, 89–91
Khosla, Vinod, 225
knowledge capital, 110

knowledge generation, xxiii, 108, 181, 204–5
Kudumbashree (KDS), xxviii, 95

labour markets, xxv, xxxiii, 5–7, 10, 12, 16–18, 30–31
  rigidity, 242–43
labour, spot markets for, 149
learning
  higher, factors to increase, 21–22
  low demand, reasons for, 5–6
  post-school, 11
Left Democratic Front (LDF) Government, Kerala, 90
less developed countries (LDCs), 15, 59, 66
lifelong education, 9, 11
livestock research, 113
local universities, interaction of IT firms with, 192–95
low-technology industrial employment growth rates, 254

Mahatma Gandhi, 83
manufacturing sector/firms, 19–20
  China
    firms operations in large sites in developed countries, 227
    low cost in, 57
  India, 235
    employment in India's manufacturing sector (*see* employment in India's manufacturing sector)
  Italian, 64
  skill gaps in, 38–39
Maoist socialism, 221
market access effect, step-by-step regression estimates of, 160–61
marketplace, valuation by, 44
markets, 72
  -based firms, 58
  for eco-innovation, 61
  for eco-innovations, 61, 63–64
  failure, 59
  need for regulations, 61–62
  transactions, 149
math education assessments, 37–38
medium low-technology industrial employment growth rates, 254–55
Microsoft R&D activities in China, 219–20
Millennium Development Goals (MDGs), 39
Mincer-Becker-Chiswick rate of return, 4
modern science, 107
Motorola R&D activities in China, 219–20
Motorola VLSI Design Centre, Hyderabad, 217
multinational corporations (MNCs), xxxi–xxxii, 181, 184, 187, 212, 216. *See also* industrial R&D, multinationalization and globalization of
  FDI and technology, 231
  global innovations networks, 218
  information and communications technology (ICT) (*see* information and communications technology [ICT])
  interaction with universities and public research institutes, 201–5
  R&D activities, effects of, 233–34
  R&D centres, 218
  R&D outsourcing from developing countries, 219
  R&D research in China, 217
multinational enterprises (MNEs), 182
multinationalization of industrial R&D. *See* Industrial R&D, multinationalization and globalization of

National Academy of Sciences/ National Research Council (NAS/NRC), 278, 306n4

National Association of Software and
  Services Companies (NASSCOM),
  186–87
National Common Minimum
  Programme, 100n6
National Industrial Classification
  (NIC), 243–44, 275
national innovation system, xxvii
National Sample Survey (NSS), 86
natural resources, 47, 142n28
  reasons for underpriced, 59, 70
new chemical entities (NCEs), 305
new drug delivery system (NDDS),
  287
new lower-cost production locations,
  241
New Millennium Indian Technology
  Leadership Initiative (NMITLI) of
  CSIR, 302–3
non-resident Indians (NRIs), 224

offshoring of R&D activities from
  developed countries, 231–34
oligopolistic nature of distribution, 63
open-source drug discovery (OSDD),
  273
ordinary least squares (OLS), 34, 120,
  162, 165
Organisation for Economic
  Co-operation and Development
  (OECD), 48, 62, 67
organizational innovations, 48
original innovation, 52
ownership–location–internalization
  (OLI) framework, 184

Panchyati Raj Institutions (PRIs),
  xxviii, 83, 88–99. *See also* social
  inclusion
  social inclusion, claim for, 88–99
panel-corrected standard errors
  (PCSE), 135
parent–child dynamics, 8
patent reward system, 54

patents, xxi, 63. *See also* intellectual
  property (IP)/intellectual property
  rights (IPRs)
  lenthen life of drugs by
    pharmaceutical firms, 307n6
  role of, 49–56
peripheral villages, 149, 153
pharmaceutical compounds, 53–54
pharmaceutical fine chemicals (PFC)
  industry, 291–92
pharmaceutical industry of India,
  xxviii–xxxiv
  post-TRIPS changes in, 278–79
policy(ies), 12–13, 16–17
  environmental policy (*see also*
    environmental policy)
  implications, 72–73
  regulation on environmental
    protection, impact of, 66–71
  role in eco-innovation, 66
Porter hypothesis, 62, 69–70
post-reform period, employment in
  manufacturing industries, 249–55
post-reform spurt in Indian economic
  growth, 238
post-Trade-Related Aspects of
  Intellectual Property Rights
  (TRIPS) Agreement, xxxiv, 209,
  272–73, 278
pragmatic scientism of Western
  science, 221
Prais–Winsten estimator, 134–35
pre-reform period, employment in
  manufacturing industries, 249–55
primary school graduates, quality and
  specialized training for, 38
primary schooling/schools, 12, 38
private returns, 4, 23, 39n1, 63
productivity, 22–23
pro-poor growth, 84, 100n1
Prowess database of Centre for
  Monitoring Indian Economy,
  270n1
public accountability, 18

public agricultural research capital, 109
    contribution to state agricultural
        productivity, 132–37
    expenditures, 140n2
    measurement of
        early measures and their
            performance, 110–17
        later measures of investments,
            117–27
        reassessment of timing and spatial
            weights, 122–32
public–private partnerships (PPPs),
    276
public research institutions (PRIs), 182
public research organizations in
    developing countries, 215
Public Works Department (PWD), 94
pure science, 107

Reaganomics, 209
regional indicators, 134
regression analysis, 31, 57, 151
regulatory landscape, 44
research and development (R&D),
    xxiii, xxxii, 7, 14, 31, 34, 37,
    184–85, 210, 270n1, 273
    activities of foreign firms active, 289
    activity in firms, 187–89
    on agricultural/farm output or
        productivity, impact of, 128
    capital, 15
    centralization of, 185
    expenditure
        domestic efforts in, 242
        structure of, 242
    expenditure impact on GDP, xxvi
    globalization of, 181
    investment, 55
    MNCs share in India and
        subsidiaries in activities of, 187
    social return to, 63
    standards and taxes yield higher
        incentives for, 67
    subsidies, 59

research institutions (RIs), 181–82,
    xxviii, xxvii
research investments, 110–11, 113,
    121–22, 133, 138
research problem areas (RPAs),
    118–20, 121
resource appraisal team, 101n8
returns to education, 4
rice farm/farming
    communities in South and
        Southeast Asia, 147
    data on, 148
    empirical analysis of markets at
        village level
        technical and cost efficiency,
            156–59, 167–72
        yields and unit costs, 151–55,
            160–67
    theoretical and empirical
        background, 148–51
rich commodity, 118
romantic scientisim of Soviet science,
    221
rural households, tax burden on, 10

schedule castes (SCs), 92, 101n10
scheduled tribes (STs), 92, 101n10
schools/schooling
    assessment by type and by GDP
        per capita quartiles and by year,
        29
    assessment of math and science
        education and training and
        vocational education, 30
    average assessment by level and
        GDP per capita quartiles, 25
    characteristics, 10–11
    in developing countries, 40n4
    disincentives within, 17
    enrolment rates and years in, 3, 28
    expenditures in, 4
    means and standard deviations of
        explanatory variables for quality
        assessments, 32–33

measures to reduce dropouts, 3
poor households wage returns, 4
quality, means and standard deviation of correlates, 26–27
Schultz hypothesis, xxx
Schumpeterian paradigm, xxiv
science and technology (S&T), 45, 209–10, 289
science education assessments, 37–38
secondary schooling, 11–12, 162
self-productivity, 11–12
Sen Committee, 90
shocks, 110
Silicon Valley NRI IT firms, 224
Silicon Valley Zhongguancun, China, 221–22
skill(s)/skilled, 13
 -based technical changes, 17
 demand for workers, 6
 formation, 8, 10–12, 16, 21 (see also household decisions model)
 labour, 15
 training, 15
Small Business Innovation Research Initiative (SBIRI) of DBT, 302–03
social arrangements for public provision, 83
social inclusion, 81–82, 88–99
 claims by PRIs for, 88–99
 meaning of, 83
 PRIs amendment for promoting, 83
social innovation, 48
social investments in education, 4–5
social mobility, 84
social returns, 4–5, 15, 19, 39n1
spatial weights reassessment on agricultural productivity, 122–32
special component plan (SCP), 92
standard setting, 53
State Agricultural Experiment Stations (SAESs), 107, 111–13, 118, 121, 138
 research investments, 121–22
state finance commission (SFC), 88, 99

subsidy(ies), 60
 government, xxiii, 59
 MNCs, xxxi
 for public education, 39n1
system complexity, 65–66
systemic reforms, 85

Taiwan, education sector of and its impact on labour, 19–20
Tata Consultancy Services (TCS), 225
technical advisory group (TAG), 93–94
technical and vocational education and training (TVET), 16, 22
technical diasporas return to India, 223
technical efficiency estimation on rice farming, 167–72
 computation using Stochastic Frontier Models, 173–74
 determinants of, 175
technical skills, 14, 38, 91
technological capability index, 240–41, 265–66, 270n1
technological determination, 14
technological innovations, xxvi, 6, 14–15, 18, 31, 37, 39, 148, 203, 211, 213, 223
technological periodicity, 14
technological progress, xxviii–xxx, 6, 14, 239
technology(ies), xxiv
 acquisition, 290
 intensity, 270n1
 transfer, xxx–xxxv
 use by LDCs/the South, 15
Technology Development Board (TDB) of DST, 302
Texas Instruments India, Bangalore, 217
Thatcherism, 209
The Indus Entrepreneurs (TiE), 225
timing weights reassessment on agricultural productivity, 122–32

Tobit estimation, 67
total factor productivity (TFP),
  109, 116, 131–32, 134, 136, 139,
  143n31
  marginal impacts of agricultural
    research and extension on, 137
tradable permits for pollution, 69
trade barriers, removal of, xxiii
Trade-Related Aspects of Intellectual
  Property Rights (TRIPS)
  Agreement, 291, 293, 304–5
Trade-Related Investment Measures
  (TRIMs), 209
trend, 110, 133–38, 140
triadization of innovation, 219
tribal sub-plan (TSP), 92
tropicalization, 228
12th Five-Year Plan (2012–17), 81

uncertainty(ies), 43–44, 64–65
  about policy regulations, 71
  stimulates innovation, 45
unit cost estimation of market access
  in rice farming, 160–67
  computation using Stochastic
    Frontier Models, 173–74
  determinants of, 175
United Nations, 39, 81
United Nations Conference on Trade
  and Development (UNCTAD),
  232–33, 264
United Nations Development
  Programme (UNDP), 99
United Progressive Alliance (UPA)
  government, 100n6
United States, 181
  agricultural geo-climatic regions
    and sub-regions, 113–15
  outsourcing to India, 225
  post-Y2K recessionary trends in, 226
  reward for first generic
    manufacturer, 294

universities in developing countries,
  214
university-industry interaction,
  195–200
unskilled workers, 17
urban-industrial centre, 148–49
urban-industrial impact hypothesis,
  xxx, 148, 172
Uruguay Round Agreement Act
  (URAA), 294
Uruguay Round TRIPS and TRIMs,
  209
US agricultural TFP growth rate, 140
US Department of Agriculture
  (USDA), 107–8, 111–13, 118, 121,
  138, 141n16, 143n30
US Food and Drug Administration
  (USFDA), 290, 293–94

venture capital (VC) funds, 222–25
vertical consolidation, 53
voluntary technical corps (VTC), 93

Wipro, 225
Women Component Plan, Kerala, 92
World Development Report (1992),
  59, 70
World Economic Forum (WEF),
  xxv–xxvi, 7, 22–24, 31, 40n9
World Health Organization (WHO),
  273, 304
world system perspective, 212
World Trade Organization (WTO),
  xxxiv, 209, 278

Y2K bug, 224
Yale University Economic Growth
  Center, xxix
yield estimation of market access in
  rice farming, 160–67

zero market value, 44–45